T0214956

Lecture Notes in Mathematics

Edited by J.-M. Morel, F. Takens and B. Teissier

Editorial Policy for Multi-Author Publications: Summer Schools / Intensive Courses

1. GENERAL. Lecture Notes aim to report new developments in all areas of mathematics and their applications – quickly, informally and at a high level. Manuscripts should be reasonably self-contained and rounded off. Thus they may, and often will, present not only results of the author but also related work by other people. They should provide sufficient motivation, examples and applications. There should also be an introduction making the text comprehensible to a wider audience. This clearly distinguishes Lecture Notes from journal articles or technical reports which normally are very concise. Articles intended for a journal but too long to be accepted by most journals, usually do not have this "lecture notes" character.

2. In general SUMMER SCHOOL and other similar INTENSIVE COURSES are held to present mathematical topics that are close to the frontiers of recent research to an audience at the beginning or intermediate graduate level, who may want to continue with this area of work, for a thesis or later. This makes demands on the didactic aspects of the presentation. Because the subjects of such schools are advanced, there often exists no textbook, and so ideally, the publication resulting from such a school could be a first approximation to such a textbook. Usually several authors are involved in the writing, so it is not always simple to obtain a unified approach to the presentation.

   For prospective publication in LNM, the resulting manuscript should not be just a collection of course notes, each of which has been developed by an individual author with little or no co-ordination with the others, and with little or no common concept. The subject matter should dictate the structure of the book, and the authorship of each part or chapter should take secondary importance. Of course the choice of authors is crucial to the quality of the material at the school and in the book, and the intention here is not to belittle their impact, but simply to say that the book should be planned to be written by these authors jointly, and not just assembled as a result of what these authors happen to submit.

   This represents considerable preparatory work (as it is imperative to ensure that the authors know these criteria before they invest work on a manuscript), and also considerable editing work afterwards, to get the book into final shape. Still it is the form that holds the most promise of a successful book that will be used by its intended audience, rather than yet another volume of proceedings for the library shelf.

3. Manuscripts should be submitted (preferably in duplicate) either to one of the series editors or to Springer-Verlag, Heidelberg. Volume editors are expected to arrange for the refereeing, to the usual scientific standards, of the individual contributions. If the resulting reports can be forwarded to us (series editors or Springer) this is very helpful. If no reports are forwarded or if other questions remain unclear in respect of homogeneity etc, the series editors may wish to consult external referees for an overall evaluation of the volume. A final decision to publish can be made only on the basis of the complete manuscript, however a preliminary decision can be based on a pre-final or incomplete manuscript. The strict minimum amount of material that will be considered should include a detailed outline describing the planned contents of each chapter.

   Volume editors and authors should be aware that incomplete or insufficiently close to final manuscripts almost always result in longer evaluation times. They should also be aware that parallel submission of their manuscript to another publisher while under consideration for LNM will in general lead to immediate rejection.

Continued on inside back-cover

Lecture Notes in Mathematics 1812

Editors:
J.–M. Morel, Cachan
F. Takens, Groningen
B. Teissier, Paris

Subseries:
Fondazione C.I.M.E., Firenze
Adviser: Pietro Zecca

Springer
Berlin
Heidelberg
New York
Hong Kong
London
Milan
Paris
Tokyo

L. Ambrosio   K. Deckelnick   G. Dziuk
M. Mimura   V.A. Solonnikov   H.M. Soner

Mathematical Aspects of Evolving Interfaces

Lectures given at the C.I.M.-C.I.M.E.
joint Euro-Summer School
held in Madeira, Funchal,
Portugal, July 3–9, 2000

Editors: P. Colli
        J.F. Rodrigues

Fondazione
C.I.M.E.

Springer

Authors

Luigi Ambrosio
Scuola Normale Superiore
Piazza dei Cavalieri 7
56126 Pisa, Italy

e-mail: luigi@ambrosio.sns.it

Klaus Deckelnick
Institute for Analysis and Numerical Analysis
Otto-von-Guericke University of Magdeburg
Universitätsplatz 2
39106 Magdeburg, Germany

e-mail:Klaus.Deckelnick@mathematik.uni-magdeburg.de

Gerhard Dziuk
Institute for Applied Mathematics
University of Freiburg
Hermann-Herderstr. 10
79104 Freiburg, Germany

e-mail: gerd@mathematik.uni-freiburg.de

Masayasu Mimura
Department of Mathematical
  and Life Sciences
Institute of Nonlinear Sciences
  and Applied Mathematics
Graduate School of Sciences
Hiroshima University
Higashi-Hiroshima 739-8526, Japan

e-mail: mimura@math.sci.hiroshima-u.ac.jp

Vsevolod A. Solonnikov
Saint-Petersburg Department of the
Steklov Mathematical Institute
Fontanka 27
191011 St. Petersburg, Russia

e-mail: solonnik@pdmi.ras.ru

Halil Mete Soner
Department of Mathematics
Koc University
Rumeli Fener Yolu Sariyer
80910 Istanbul, Turkey

e-mail: msoner@ku.edu.tr

Editors

Pierluigi Colli
Department of Mathematics 'F. Casorati'
University of Pavia
Via Ferrata 1
27100 Pavia, Italy

e-mail: pier@imati.cnr.it

José-Francisco Rodrigues
CMAF, University of Lisboa
Av. Prof. Gama Pinto 2
1649-003 Lisboa, Portugal

e-mail: rodrigue@ptmat.fc.ul.pt

Cataloging-in-Publication Data applied for
Bibliographic information published by Die Deutsche Bibliothek

Die Deutsche Bibliothek lists this publication in the Deutsche Nationalbibliografie;
detailed bibliographic data is available in the Internet at http://dnb.ddb.de

Mathematics Subject Classification (2000): 35-02, 35Bxx, 35QXX, 35K57, 53C44, 76D05, 80A22

ISSN 0075-8434
ISBN 3-540-14033-6 Springer-Verlag Berlin Heidelberg New York

Springer-Verlag Berlin Heidelberg New York a member of BertelsmannSpringer
Science + Business Media GmbH

http://www.springer.de

© Springer-Verlag Berlin Heidelberg 2003
Printed in Germany

Typesetting: Camera-ready TEX output by the authors

SPIN: 10922378   41/3142/du - 543210 - Printed on acid-free paper

Preface

The Euro-Summer School on Mathematical Aspects of Evolving Interfaces gathered senior experts and young researchers at the University of Madeira, Funchal, Portugal, in the week July 3-9, 2000. This meeting arose as a joint school of CIM (Centro Internacional de Matemática, Portugal) and CIME (Centro Internazionale Matematico Estivo, Italy).

The school was intended to present an advanced introduction and state of the art of recent analytic, modeling and numerical techniques to the mathematical representation and description of moving interfaces. Five complementary courses were delivered and this volume collects the notes of the lectures.

Interfaces are geometrical objects modeling free or moving boundaries and arise in a wide range of phase change problems in physical and biological sciences, in particular in material technology and in dynamics of patterns. Especially in the end of last century, the rigorous study of evolving interfaces in a number of applied fields becomes increasingly important, so that the possibility of describing their dynamics through suitable mathematical models became one of the most challenging and interdisciplinary problems in applied mathematics.

It was recognized that essential problems related to evolving interfaces can be modelled by means of partial differential equations and systems in domains whose boundary depends on time. In many complicated cases these boundaries are themselves unknown, and correspond, e.g., to a particular level set, or to the discontinuity set, of some physical quantity. In particular, free boundary problems are boundary value problems for differential equations set in a domain where part of the boundary is "free" and further conditions allow to exclude undeterminacy.

Although the first modern work in a free boundary problem was written by Lamé and Clapeyron in 1831, who considered a simple model for the solidification of a liquid sphere, in the last decades of the XXth century this interdisciplinary field developed tremendously with many new computational demands and new problems from industry and applied sciences, as well as with increasing contributions from Mathematics. Indeed, problems of this sort are concerned with several phenomena of high applied interest. Examples include Stefan type problems, where, tipically, the free boundary is the moving interface between liquid and solid, as well as, more general models of phase transitions. Another important example arises in filtration through porous media; here free boundaries occur as fronts between saturated and unsaturated regions. Interesting examples also come from reaction-diffusion models, fluid dynamics, contact mechanics, superconductivity and so on. Several of these problems are also of direct industrial interest, and offer an interesting opportunity of collaboration among theoretical analysts, mathematical physicists and applied scientists.

The Madeira school reported on mathematical advances in some theoretical, modeling and numerical issues concerned with dynamics of interfaces and free boundaries. Specifically, the five courses dealt with an assessment of recent results on the optimal transportation problem (L. Ambrosio), the numerical approximation of moving fronts evolving by mean curvature (G. Dziuk), the dynamics of patterns and interfaces in some reaction-diffusion systems with chemical-biological applications (M. Mimura), evolutionary free boundary problems of parabolic type or for Navier-Stokes equations (V.A. Solonnikov), and a variational approach to evolution problems for the Ginzburg-Landau functional (H.M. Soner).

We expect that these lecture notes will be useful not only to experienced readers, to find a detailed description of results and a presentation of techniques, but also to the beginners that aim to learn some of the mathematical aspects behind the different fields.

The editors

Pierluigi Colli, Pavia
José Francisco Rodrigues, Lisboa

CIME's activity is supported by:
Ministero dell'Istruzione, dell'Università e della Ricerca; Ministero degli Affari Esteri – Direzione Generale per la Promozione e la Cooperazione – Ufficio V; Istituto Nazionale di Alta Matematica "Francesco Severi"; E.U. under the Training and Mobility of Researchers Programme and UNESCO-ROSTE, Venice Office.

Table of Contents

Reaction-Diffusion Systems Arising in Biological and Chemical Systems: Application of Singular Limit Procedures

Lectures on Evolution Free Boundary Problems: Classical Solutions

Lecture Notes on Optimal Transport Problems

Luigi Ambrosio*

Scuola Normale Superiore
Piazza dei Cavalieri 7, I-56126, Pisa, Italy
luigi@ambrosio.sns.it

Contents

Introduction

These notes are devoted to the Monge–Kantorovich optimal transport problem. This problem, in the original formulation of Monge, can be stated as follows: given two distributions with equal masses of a given material $g_0(x)$, $g_1(x)$ (corresponding for instance to an embankment and an excavation), find a transport map ψ which carries the first distribution into the second and minimizes the transport cost

$$C(\psi) := \int_X |x - \psi(x)| g_0(x) \, dx.$$

The condition that the first distribution of mass is carried into the second can be written as

$$\int_{\psi^{-1}(B)} g_0(x) \, dx = \int_B g_1(y) \, dy \qquad \forall B \subset X \text{ Borel} \tag{1}$$

---

* This work has been partially supported by the GNAFA project "Problemi di Monge–Kantorovich e strutture deboli"

or, by the change of variables formula, as

$$g_1(\psi(x)) \, |\det \nabla \psi(x)| = g_0(x) \qquad \text{for } \mathcal{L}^n\text{-a.e. } x \in B$$

if ψ is one to one and sufficiently regular.

More generally one can replace the functions g_0, g_1 by positive measures f_0, f_1 with equal mass, so that (1) reads $f_1 = \psi_\# f_0$, and replace the euclidean distance by a generic cost function $c(x, y)$, studying the problem

$$\min_{\psi_\# f_0 = f_1} \int_X c(x, \psi(x)) \, df_0(x). \tag{2}$$

The infimum of the transport problem leads also to a c-dependent distance between measures with equal mass, known as Kantorovich–Wasserstein distance.

The optimal transport problem and the Kantorovich–Wasserstein distance have a very broad range of applications: Fluid Mechanics [7, 8]; Partial Differential Equations [32, 29]; Optimization [13, 14] to quote just a few examples. Moreover, the 1-Wasserstein distance (corresponding to $c(x, y) = |x - y|$ in (2)) is related to the so-called flat distance in Geometric Measure Theory, which plays an important role in its development (see [6], [24], [30], [27], [44]). However, rather than showing specific applications (for which we mainly refer to the Evans survey [21] or to the introduction of [7]), the main aim of the notes is to present the different formulations of the optimal transport problem and to compare them, focussing mainly on the linear case $c(x, y) = |x - y|$. The main sources for the preparation of the notes have been the papers by Bouchitté–Buttazzo [13, 14], Caffarelli–Feldman–McCann [15], Evans–Gangbo [22], Gangbo–McCann [26], Sudakov [42] and Evans [21].

The notes are organized as follows. In Section 1 we discuss some basic examples and in Section 2 we discuss Kantorovich's generalized solutions, i.e. the transport plans, pointing out the connection between them and the transport maps. Section 3 is entirely devoted to the one dimensional case: in this situation the order structure plays an important role and considerably simplifies the theory. Sections 4 and 5 are devoted to the ODE and PDE based formulations of the optimal transport problem (respectively due to Brenier and Evans–Gangbo); we discuss in particular the role of the so-called transport density and the equivalence of its different representations. Namely, we prove that any transport density μ can be represented as $\int_0^1 \pi_{t\#}(|y - x||\gamma) \, dt$, where γ is an optimal planning, as $\int_0^1 |E_t| \, dt$, where E_t is the "velocity field" in the ODE formulation, or as the solution of the PDE $\operatorname{div}(\nabla_\mu u \mu) = f_1 - f_0$, with no regularity assumption on f_1, f_0. Moreover, in the same generality we prove convergence of the p-laplacian approximation.

In Section 6 we discuss the existence of the optimal transport map, following essentially the original Sudakov approach and filling a gap in his original proof (see also [15, 43]). Section 7 deals with recent results, related to those obtained in [25], on the regularity and the uniqueness of the transport density.

Section 8 is devoted to the connection between the optimal transport problem and the so-called mass optimization problem. Finally, Section 9 contains a self contained list of the measure theoretic results needed in the development of the theory.

Main notation

X	a compact convex subset of an Euclidean space $\mathbb{R}^n$
$\mathcal{B}(X)$	Borel σ-algebra of X
$\mathcal{L}^n$	Lebesgue measure in $\mathbb{R}^n$
$\mathcal{H}^k$	Hausdorff k-dimensional measure in $\mathbb{R}^n$
$\mathrm{Lip}(X)$	real valued Lipschitz functions defined on X
$\mathrm{Lip}_1(X)$	functions in $\mathrm{Lip}(X)$ with Lipschitz constant not greater than 1
Σ_u	the set of points where u is not differentiable
π_t	projections $(x, y) \mapsto x + t(y - x)$, $t \in [0, 1]$
$\mathcal{S}_o(X)$	open segments $]x, y[$ with $x, y \in X$
$\mathcal{S}_c(X)$	closed segments $[x, y]$ with $x, y \in X$, $x \neq y$
$\mathcal{M}(X)$	signed Radon measures with finite total variation in X
$\mathcal{M}_+(X)$	positive and finite Radon measures in X
$\mathcal{M}_1(X)$	probability measures in X
$\|\mu\|$	total variation of $\mu \in [\mathcal{M}(X)]^n$
μ^+, μ^-	positive and negative part of $\mu \in \mathcal{M}(X)$
$f_{\#}\mu$	push forward of μ by f

1 Some elementary examples

In this section we discuss some elementary examples that illustrate the kind of phenomena (non existence, non uniqueness) which can occur. The first one shows that optimal transport maps need not exist if the first measure f_0 has atoms.

Example 1.1 (Non existence). Let $f_0 = \delta_0$ and $f_1 = (\delta_1 + \delta_{-1})/2$. In this case the optimal transport problem has no solution simply because there is no map ψ such that $\psi_{\#}f_0 = f_1$.

The following two examples deal with the case when the cost function c in $X \times X$ is $|x - y|$, i.e. the euclidean distance between x and y. In this case we will use as a test for optimality the fact that the infimum of the transport problem is always greater than

$$\sup \left\{ \int_X u \, d(f_1 - f_0) : u \in \mathrm{Lip}_1(X) \right\}. \tag{3}$$

Indeed,

$$\int_X u \, d(f_1 - f_0) = \int_X u(\psi(x)) - u(x) \, df_0(x) \leq \int_X |\psi(x) - x| \, df_0(x)$$

for any admissible transport ψ. Actually we will prove this lower bound is sharp if f_0 has no atom (see (6) and (13)).

Our second example shows that in general the solution of the optimal transport problem is not unique. In the one-dimensional case we will obtain (see Theorem 3.1) uniqueness (and existence) in the class of nondecreasing maps.

Example 1.2 (Book shifting). Let $n \geq 1$ be an integer and $f_0 = \chi_{[0,n]} \mathcal{L}^1$ and $f_1 = \chi_{[1,n+1]} \mathcal{L}^1$. Then the map $\psi(t) = t + 1$ is optimal. Indeed, the cost relative to ψ is n and, choosing the 1-Lipschitz function $u(t) = t$ in (3), we obtain that the supremum is at least n, whence the optimality of ψ follows. But since the minimal cost is n, if $n > 1$ another optimal map ψ is given by

$$\psi(t) = \begin{cases} t + n & \text{on } [0,1] \\ t & \text{on } [1,n]. \end{cases}$$

In the previous example the two transport maps coincide when $n = 1$; however in this case there is one more (and actually infinitely many) optimal transport map.

Example 1.3. Let $f_0 = \chi_{[0,1]} \mathcal{L}^1$ and $f_1 = \chi_{[1,2]} \mathcal{L}^1$ (i.e. $n = 1$ in the previous example). We have already seen that $\psi(t) = t+1$ is optimal. But in this case also the map $\psi(t) = 2 - t$ is optimal as well.

In all the previous examples the optimal transport maps ψ satisfy the condition $\psi(t) \geq t$. However it is easy to find examples where this does not happen.

Example 1.4. Let $f_0 = \chi_{[-1,1]} \mathcal{L}^1$ and $f_1 = \delta_{-1} + \delta_1$. In this case the optimal transport map ψ is unique (modulo $\mathcal{L}^1$-negligible sets); it is identically equal to -1 on $[-1,0)$ and identically equal to 1 on $(0,1]$. The verification is left to the reader as an exercise.

We conclude this section with some two dimensional examples.

Example 1.5. Assume that $2f_0$ is the sum of the unit Dirac masses at $(1,1)$ and $(0,0)$, and that $2f_1$ is the sum of the unit Dirac masses at $(1,0)$ and $(0,1)$. Then the "horizontal" transport and the "vertical" transport are both optimal. Indeed, the cost of these transports is 1 and choosing $u(x_1, x_2) = x_1$ in (3) we obtain that the infimum of the transport problem is at least 1.

Example 1.6. Assume that f_1 is the sum of two Dirac masses at $A, B \in \mathbb{R}^2$ and assume that f_0 is supported on the middle axis between them. Then

$$\int_X |x - \psi(x)| \, df_0(x) = \int_X |x - A| \, df_0(x)$$

whenever $\psi(x) \in \{A, B\}$, hence *any* admissible transport is optimal.

2 Optimal transport plans: existence and regularity

In this section we discuss Kantorovich's approach to the optimal transport problem. His idea has been to look for optimal transport "plans" , i.e. probability measures γ in the product space $X \times X$, rather than optimal transport maps. We will see that this more general viewpoint can be used in several situations to prove that actually optimal transport maps exist (this intermediate passage through a weak formulation of the problems is quite common in PDE and Calculus of Variations).

(MK) Let f_0, $f_1 \in \mathcal{M}_1(X)$. We say that a probability measure γ in $\mathcal{M}_1(X \times X)$ is *admissible* if its marginals are f_0 and f_1, i.e.

$$\pi_{0\#}\gamma = f_0, \qquad \pi_{1\#}\gamma = f_1.$$

Then, given a Borel cost function $c : X \times X \to [0, \infty]$, we minimize

$$I(\gamma) := \int_{X \times X} c(x, y) \, d\gamma(x, y)$$

among all admissible γ and we denote by $\mathcal{F}_c(f_0, f_1)$ the value of the infimum.

We also call an admissible γ a *transport plan*. Notice also that in Kantorovich's setting no restriction on f_0 or f_1 is necessary to produce admissible transport plans: the product measure $f_0 \times f_1$ is always admissible. In particular the following definition is well posed and produces a family of distances in $\mathcal{M}_1(X)$.

Definition 2.1 (Kantorovich–Wasserstein distances). *Let $p \geq 1$ and f_0, $f_1 \in \mathcal{M}_1(X)$. We define the p-Wasserstein distance between f_0 and f_1 by*

$$\mathcal{F}_p(f_0, f_1) := \left(\min_{\pi_{0\#}\gamma = f_0, \, \pi_{1\#}\gamma = f_1} \int_{X \times X} |x - y|^p \, d\gamma \right)^{1/p}. \tag{4}$$

The difference between transport maps and transport plans can be better understood with the following proposition.

Proposition 2.1 (Transport plans versus transport maps). *Any Borel transport map $\psi : X \to X$ induces a transport plan γ_ψ defined by*

$$\gamma_\psi := (Id \times \psi)_{\#} f_0. \tag{5}$$

Conversely, a transport plan γ is induced by a transport map if γ is concentrated on a γ-measurable graph Γ.

Proof. Let ψ be a transport map. Since $\pi_0 \circ (Id \times \psi) = Id$ and $\pi_1 \circ (Id \times \psi) = \psi$, we obtain immediately that $\pi_{0\#}\gamma_\psi = f_0$ and $\pi_{1\#}\gamma_\psi = \psi_\# f_0 = f_1$. Notice also that, by Lusin's theorem, the graph of ψ is γ_ψ-measurable.

Conversely, let $\Gamma \subset X \times X$ be a γ-measurable graph on which γ is concentrated and write

$$\Gamma = \{(x, \phi(x)) : x \in \pi_0(\Gamma)\},$$

for some function $\phi : \pi_0(\Gamma) \to X$. Let (K_h) be an increasing sequence of compact subsets of Γ such that $\gamma(\Gamma \setminus K_h) \to 0$ and notice that

$$f_0\left(\pi_0(K_h)\right) = \gamma\left(\pi_0^{-1}(\pi_0(K_h))\right) \geq \gamma(K_h) \to 1.$$

Hence, $\pi_0(\Gamma) \supset \cup_h \pi_0(K_h)$ is f_0-measurable and with full measure in X. Moreover, representing γ as $\gamma_x \otimes f_0$, as in (58) we get

$$0 = \lim_{h \to \infty} \gamma(X \times X \setminus K_h) = \int_X \gamma_x\left(\{y : (x, y) \notin \cup_h K_h\}\right) df_0(x)$$

$$\geq \int_{\pi_0(\Gamma)}^* \gamma_x(X \setminus \{\phi(x)\}) df_0(x)$$

(here $\int^*$ denotes the outer integral). Hence γ_x is the unit Dirac mass at $\phi(x)$ for f_0-a.e. $x \in X$. Since

$$x \mapsto \psi(x) := \int_X y \, d\gamma_x(y)$$

is a Borel map coinciding with ϕ f_0-a.e., we obtain that γ_x is the unit Dirac mass at $\psi(x)$ for f_0-a.e. x. For $A, B \in \mathcal{B}(X)$ we get

$$\gamma(A \times B) = \int_A \gamma_x(B) \, df_0(x) = f_0\left(\{x : (x, \psi(x)) \in A \times B\}\right) = \gamma_\psi(A \times B)$$

and therefore $\gamma = \gamma_\psi$.

The existence of optimal transport plans is a straightforward consequence of the w^*-compactness of probability measures and of the lower semicontinuity of I.

Theorem 2.1 (Existence of optimal plans). *Assume that c is lower semicontinuous in $X \times X$. Then there exists $\gamma \in \mathcal{M}_1(X \times X)$ solving (MK). Moreover, if c is continuous and real valued we have*

$$\min (MK) = \inf_{\psi_\# f_0 = f_1} \int_X c(x, \psi(x)) \, df_0(x) \tag{6}$$

provided f_0 has no atom.

Proof. Clearly the set of admissible γ's is closed, bounded and w^*-compact for the w^*-convergence of measures (i.e. in the duality with continuous functions in $X \times X$). Hence, it suffices to prove that

$$I(\gamma) \leq \liminf_{h \to \infty} I(\gamma_h)$$

whenever γ_h w^*-converge to γ. This lower semicontinuity property follows by the fact that c can be approximated from below by an increasing sequence of continuous and real valued functions c_h (this is a well known fact: see for instance Lemma 1.61 in [4]). The functionals I_h induced by c_h converge monotonically to I, whence the lower semicontinuity of I follows.

In order to prove the last part we need to show the existence, for any $\gamma \in \mathcal{M}_+(X \times X)$ with $\pi_{0\#}\gamma = f_0$ and $\pi_{1\#}\gamma = f_1$, of Borel maps $\psi_h : X \to X$ such that $\psi_{h\#}f_0 = f_1$ and $\delta_{\psi_h(x)} \otimes f_0$ weakly converge to γ in $\mathcal{M}(X \times X)$. The approximation Theorem 9.3 provides us, on the other hand, with a Borel map $\varphi : X \to X$ such that $\varphi_{\#}f_0$ has no atom, is arbitrarily close to f_1 and $\delta_{\varphi(x)} \otimes f_0$ is arbitrarily close (with respect to the weak topology) to γ. We will build ψ_h by an iterated application of this result.

By a standard approximation argument we can assume that $L = \mathrm{Lip}(c)$ is finite and that

$$\delta|x - y| \leq c(x, y) \qquad \forall x, y \in X,$$

for some $\delta > 0$. Possibly replacing c by c/δ, we assume $\delta = 1$.

Fix now an integer h, set $f_0^0 = f_0$ and choose $\varphi_0 : X \to X$ such that $\varphi_{0\#}f_0^0$ has no atom and

$$\mathcal{F}_1(f_1, \varphi_{0\#}f_0^0) < 2^{-h}, \qquad \int_X c(\varphi_0(x), x)\, df_0^0(x) < \mathcal{F}_c(f_1, f_0^0) + 2^{-h}.$$

Then, we set $f_0^1 = \varphi_{0\#}f_0^0$ and find $\varphi_1 : X \to X$ such that $\varphi_{1\#}f_0^1$ has no atom and

$$\mathcal{F}_1(f_1, \varphi_{1\#}f_0^1) < 2^{-1-h}, \qquad \int_X c(\varphi_1(x), x)\, df_0^1(x) < \mathcal{F}_c(f_1, f_0^1) + 2^{-1-h}.$$

Proceeding inductively and setting $f_0^k = \varphi_{(k-1)\#}f_0^{k-1}$, we find φ_k such that

$$\mathcal{F}_1(f_1, \varphi_{k\#}f_0^k) < 2^{-k-h}, \qquad \int_X c(\varphi_k(x), x)\, df_0^k(x) < \mathcal{F}_c(f_1, f_0^k) + 2^{-k-h},$$

and $\varphi_{k\#}f_0^k$ has no atom. Then, we set $\phi_0(x) = x$ and $\phi_k = \varphi_{k-1} \circ \cdots \circ \varphi_0$ for $k \geq 1$, so that $f_0^k = \phi_{k\#}f_0$. We claim that (ϕ_k) is a Cauchy sequence in $L^1(X, f_0; X)$. Indeed,

$$\sum_{k=0}^{\infty} \int_X |\phi_{k+1}(x) - \phi_k(x)|\, df_0(x) = \sum_{k=0}^{\infty} \int_X |\varphi_k(y) - y|\, df_0^k(y)$$

$$\leq 2^{1-h} + \sum_{k=0}^{\infty} \mathcal{F}_1(f_1, f_0^k) < \infty.$$

Denoting by ψ_h the limit of ϕ_k, passing to the limit as $k \to \infty$ we obtain $\psi_{h\#} f_0 = f_1$; moreover, we have

$$\int_X c(\phi_k(x), x)\, df_0(x)$$

$$\leq \int_X c(\varphi_0(x), x)\, df_0(x) + L \sum_{i=1}^{k} \int_X |\phi_i(x) - \phi_{i-1}(x))|\, df_0(x)$$

$$\leq \mathcal{F}_c(f_1, f_0) + 2^{-h} + L \sum_{i=1}^{k} \int_X |\varphi_i(y) - y|\, df_0^i(y) \leq \mathcal{F}_c(f_1, f_0) + 2^{-h}(1 + 2L).$$

Passing to the limit as $k \to \infty$ we obtain

$$\int_X c(\psi_h(x), x)\, df_0(x) \leq \mathcal{F}_c(f_1, f_0) + 2^{-h}(1 + 2L)$$

and the proof is achieved.

For instance in the case of Example 1.1 (where transport maps do not exist at all) it is easy to check that the unique optimal transport plan is given by

$$\frac{1}{2}\delta_0 \times \delta_{-1} + \frac{1}{2}\delta_0 \times \delta_1.$$

In general, however, uniqueness fails because of the linearity of I and of the convexity of the class of admissible plans γ. In Example 1.2, for instance, any measure

$$t(Id \times \psi_1)_{\#} f_0 + (1 - t)(Id \times \psi_2)_{\#} f_0$$

is optimal, with $t \in [0, 1]$ and ψ_1, ψ_2 optimal transport maps.

In order to understand the regularity properties of optimal plans γ we introduce, following [26, 36], the concept of cyclical monotonicity.

Definition 2.2 (Cyclical monotonicity). *Let $\Gamma \subset X \times X$. We say that Γ is c-cyclically monotone if*

$$\sum_{i=1}^{n} c(x_{i+1}, y_i) \geq \sum_{i=1}^{n} c(x_i, y_i) \tag{7}$$

whenever $n \geq 2$ and $(x_i, y_i) \in \Gamma$ for $1 \leq i \leq n$, with $x_{n+1} = x_1$.

The cyclical monotonicity property can also be stated in a (apparently) stronger form

$$\sum_{i=1}^{n} c(x_{\sigma(i)}, y_i) \geq \sum_{i=1}^{n} c(x_i, y_i) \tag{8}$$

for any permutation $\sigma : \{1, \ldots, n\} \to \{1, \ldots, n\}$. The equivalence can be proved either directly (reducing to the case when σ has no nontrivial invariant set) or verifying, as we will soon do, that any cyclically monotone set is contained in the c-superdifferential of a c-concave function and then checking that the superdifferential fulfils (8).

Theorem 2.2 (Regularity of optimal plans). *Assume that c is continuous and real valued. Then, for any optimal γ the set* $\mathrm{spt}\,\gamma$ *is c-cyclically monotone. Moreover, the union of* $\mathrm{spt}\,\gamma$ *as γ range among all optimal plans is c-cyclically monotone.*

Proof. Assume by contradiction that there exist an integer $n \geq 2$ and points $(x_i, y_i) \in \mathrm{spt}\,\gamma$, $i = 1, \ldots, n$, such that

$$f\left((x_i), (y_i)\right) := \sum_{i=1}^{n} c(x_{i+1}, y_i) - c(x_i, y_i) < 0,$$

with $x_{n+1} = x_1$. For $1 \leq i \leq n$, let U_i, V_i be compact neighbourhoods of x_i and y_i, respectively, such that $\gamma(U_i \times V_i) > 0$ and $f\left((u_i), (v_i)\right) < 0$ whenever $u_i \in U_i$ and $v_i \in V_i$.

Set now $\lambda = \min_i \gamma(U_i \times V_i)$ and denote by $\gamma_i \in \mathcal{M}_1(U_i \times V_i)$ the normalized restriction of γ to $U_i \times V_i$. We can find a compact space Y, a probability measure σ in Y, and Borel maps $\eta_i = u_i \times v_i : X \to U_i \times V_i$ such that $\gamma_i = \eta_{i\#}\sigma$ for $i = 1, \ldots, n$ (it suffices for instance to define Y as the product of $U_i \times V_i$, so that η_i are the projections on the i-coordinate) and define

$$\gamma' := \gamma + \frac{\lambda}{n} \sum_{i=1}^{n} (u_{i+1} \times v_i)_{\#}\sigma - (u_i \times v_i)_{\#}\sigma.$$

Since $\lambda \eta_{i\#}\sigma = \lambda\gamma_i \leq \gamma$ we obtain that $\gamma' \in \mathcal{M}_+(X \times X)$; moreover, it is easy to check that $\pi_{0\#}\gamma' = f_0$ and $\pi_{1\#}\gamma' = f_1$. This leads to a contradiction because

$$I(\gamma') - I(\gamma) = \frac{\lambda}{n} \int_Y \sum_{i=1}^{n} c(u_{i+1}, v_i) - c(u_i, v_i)\, d\sigma < 0.$$

In order to show the last part of the statement we notice that the collection of optimal transport plans is w^*-closed and compact. If $(\gamma_h)_{h \geq 1}$ is a countable dense set, then

$$\bigcup_{i=1}^{h} \mathrm{spt}\,\gamma_i = \mathrm{spt}\left(\sum_{i=1}^{h} \frac{1}{h}\gamma_i\right)$$

is c-cyclically monotone for any $h \geq 1$. Passing to the limit as $h \to \infty$ we obtain that the closure of the union of $\mathrm{spt}\,\gamma_h$ is c-cyclically monotone. By the density of (γ_h), this closure contains $\mathrm{spt}\,\gamma$ for any optimal plan γ.

Next, we relate the c-cyclical monotonicity to suitable concepts (adapted to c) of concavity and superdifferential.

Definition 2.3 (c-concavity). *We say that a function $u : X \to \mathbb{R}$ is c-concave if it can be represented as the infimum of a family (u_i) of functions given by*

$$u_i(x) := c(x, y_i) + t_i,$$

for suitable $y_i \in X$ and $t_i \in \mathbb{R}$.

Remark 2.1. **[Linear and quadratic case]** In the case when $c(x, y)$ is a symmetric function satisfying the triangle inequality, the notion of c-concavity is equivalent to 1-Lipschitz continuity with respect to the metric d_c induced by c. Indeed, given $u \in \mathrm{Lip}_1(X, d_c)$, the family of functions whose infimum is u is simply

$$\{c(x, y) + u(y) : \; y \in X\}.$$

In the quadratic case $c(x, y) = |x - y|^2/2$ a function u is c-concave if and only if $u - |x|^2/2$ is concave. Indeed, $u = \inf_i c(\cdot, y_i) + t_i$ implies

$$u(x) - \frac{1}{2}|x|^2 = \inf_i \langle x, -y_i \rangle + \frac{1}{2}|y_i|^2 + t_i$$

and therefore the concavity of $u - |x|^2/2$. Conversely, if $v = u - |x|^2/2$ is concave, from the well known formula

$$v(x) = \inf_{y, \, p \in \partial^+ v(y)} v(y) + \langle p, x - y \rangle$$

(here $\partial^+ v$ is the superdifferential of v in the sense of convex analysis) we obtain

$$u(x) = \inf_{y, \, -p \in \partial^+ v(y)} \frac{1}{2}|p - x|^2 + c(p, y).$$

Definition 2.4 (c-superdifferential). *Let $u : X \to \mathbb{R}$ be a function. The c-superdifferential $\partial_c u(x)$ of u at $x \in X$ is defined by*

$$\partial_c u(x) := \{y : \; u(z) \leq u(x) + c(z, y) - c(x, y) \; \forall z \in X\}. \tag{9}$$

The following theorem ([40], [41], [36]) shows that the graphs of superdifferentials of c-concave functions are maximal (with respect to set inclusion) c-cyclically monotone sets. It may considered as the extension of the well known result of Rockafellar to this setting.

Theorem 2.3. *Any c-cyclically monotone set Γ is contained in the graph of the c-superdifferential of a c-concave function. Conversely, the graph of the c-superdifferential of a c-concave function is c-cyclically monotone.*

Proof. This proof is taken from [36]. We fix $(x_0, y_0) \in \Gamma$ and define

$$u(x) := \inf c(x, y_n) - c(x_n, y_n) + \cdots + c(x_1, y_0) - c(x_0, y_0) \qquad \forall x \in X,$$

where the infimum runs among all collections $(x_i, y_i) \in \Gamma$ with $1 \leq i \leq n$ and $n \geq 1$. Then u is c-concave by construction and the cyclical monotonicity of Γ gives $u(x_0) = 0$ (the minimum is achieved with $n = 1$ and $(x_1, y_1) = (x_0, y_0)$).

We will prove the inequality

$$u(x) \leq u(x') + c(x, y') - c(x', y') \tag{10}$$

for any $x \in X$ and $(x', y') \in \Gamma$. In particular (choosing $x = x_0$) this implies that $u(x') > -\infty$ and $y' \in \partial_c u(x')$. In order to prove (10), we fix $\lambda > u(x')$ and find $(x_i, y_i) \in \Gamma$, $1 \leq i \leq n$, such that

$$c(x', y_n) - c(x_n, y_n) + \cdots + c(x_1, y_0) - c(x_0, y_0) < \lambda.$$

Then, setting $(x_{n+1}, y_{n+1}) = (x', y')$, we find

$$u(x) \leq c(x, y_{n+1}) - c(x_{n+1}, y_{n+1}) + c(x_{n+1}, y_n) - c(x_n, y_n)$$
$$+ \cdots + c(x_1, y_0) - c(x_0, y_0) \leq c(x, y') - c(x', y') + \lambda.$$

Since λ is arbitrary, (10) follows.

Finally, if v is c-concave, $y_i \in \partial_c v(x_i)$ for $1 \leq i \leq n$ and σ is a permutation we can add the inequalities

$$v(x_{\sigma(i)}) - v(x_i) \leq c(x_{\sigma(i)}, y_i) - c(x_i, y_i)$$

to obtain (8).

In the following corollary we assume that the cost function is symmetric, continuous and satisfies the triangle inequality, so that c-concavity reduces to 1-Lipschitz continuity with respect to the distance induced by c.

Corollary 2.1 (Linear case). *Let $\gamma \in \mathcal{M}_1(X \times X)$ with $\pi_{0\#}\gamma = f_0$ and $\pi_{1\#}\gamma = f_1$. Then γ is optimal for (MK) if and only if there exists $u : X \to \mathbb{R}$ such that*

$$|u(x) - u(y)| \leq c(x, y) \qquad \forall (x, y) \in X \times X, \tag{11}$$
$$u(x) - u(y) = c(x, y) \qquad \text{for } (x, y) \in \text{spt}\,\gamma. \tag{12}$$

In addition, there exists u satisfying (11) and such that (12) holds for any optimal planning γ. We will call any function u with these properties a maximal Kantorovich potential.

Proof. (Sufficiency) Let γ' be any admissible transport plan; by applying (11) first and then (12), we get

$$I(\gamma') \geq \int_{X \times X} (u(x) - u(y))\, d\gamma' = \int_X u\, df_0 - \int_X u\, df_1$$
$$= \int_X (u(x) - u(y))\, d\gamma = I(\gamma).$$

(Necessity) Let Γ be the closure of the union of spt γ', as γ' varies among all optimal plans for (MK). Then, we know that Γ is c-cyclically monotone, hence there exists a c-concave function u such that $\Gamma \subset \partial_c u$. Then, (11) follows by the c-concavity of u, while the inclusion $\Gamma \subset \partial_c u$ implies

$$u(y) - u(x) \leq c(y,y) - c(x,y) = -c(x,y)$$

for any $(x,y) \in$ spt $\gamma \subset \Gamma$. This, taking into account (11), proves (12).

A direct consequence of the proof of sufficiency is the identity

$$\min (\text{MK}) = \max \left\{ \int ud(f_0 - f_1) : \ u \in \text{Lip}_1(X, d_c) \right\} \tag{13}$$

(where d_c is the distance in X induced by c) and the maximum on the right is achieved precisely whenever u satisfies (12).

In the following corollary, instead, we consider the case when $c(x,y) = |x - y|^2/2$. The result below, taken from [26], was proved first by Brenier in [9, 10] under more restrictive assumptions on f_0, f_1 (see also [26] for the general case $c(x,y) = h(|x - y|)$). Before stating the result, we recall that the set Σ_v of points of nondifferentiability of a real valued concave function v (i.e. the set of points x such that $\partial^+ v(x)$ is not a singleton) is countably (CC) regular (see [45] and also [1]). This means that Σ_v can be covered with a countable family of (CC) hypersurface, i.e. graphs of differences of convex functions of $n - 1$ variables. This property is stronger than the canonical $\mathcal{H}^{n-1}$-rectifiability: it implies that $\mathcal{H}^{n-1}$-almost all of Σ_v can be covered by a sequence of C^2 hypersurfaces.

Corollary 2.2 (Quadratic case). *Assume that any (CC) hypersurface is f_0-negligible. Then the optimal planning γ is unique and is induced by an optimal transport map ψ. Moreover ψ is the gradient of a convex function.*

Proof. Let $u : X \to \mathbb{R}$ be a c-concave function such that the graph Γ of its superdifferential contains the support of any optimal planning γ. As $c(x,y) = |x - y|^2/2$, an elementary computation shows that $(x_0, y_0) \in \Gamma$ if and only if

$$-y_0 \in \partial^+ v(x_0),$$

where $v(x) = u(x) - |x|^2/2$ is the concave function already considered in Remark 2.1. Then, by the above mentioned results on differentiability of concave functions, the set of points where v is not differentiable is f_0-negligible. Hence, for f_0-a.e. $x \in X$, there is a unique $y_0 = -\nabla v(x_0) \in X$ such that $(x_0, y_0) \in \Gamma$. As spt $\gamma \subset \Gamma$, by Proposition 2.1 we infer that

$$\gamma = (Id \times \psi)_\# f_0$$

for any optimal planning γ, with $\psi = -\nabla v$.

Example 2.1 (Brenier polar factorization theorem). A remarkable consequence of Corollary 2.2 is the following result, known as polar factorization theorem. A vector field $r : X \to X$ such that

$$\mathcal{L}^n(r^{-1}(B)) = 0 \qquad \text{whenever} \qquad \mathcal{L}^n(B) = 0 \qquad (14)$$

can be written as $\nabla u \circ \eta$, with u convex and η measure preserving.

It suffices to apply Corollary 2.2 to the measures $f_0 = r_\#(\mathcal{L}^n \llcorner X)$ (absolutely continuous, due to (14)) and $f_1 = \mathcal{L}^n \llcorner X$; we have then

$$\mathcal{L}^n \llcorner X = (\nabla v)_\#(r_\#(\mathcal{L}^n \llcorner X)) = (\nabla v \circ r)_\#(\mathcal{L}^n \llcorner X),$$

for a suitable convex function v, hence $\eta = (\nabla v) \circ r$ is measure preserving. The desired representation follows with $u = v^*$, since $\nabla u = (\nabla v)^{-1}$.

Let us assume that $f_0 = \mathcal{L}^n \llcorner X$ and let $f_1 \in \mathcal{M}_+(X)$ be any other measure such that $f_1(X) = \mathcal{L}^n(X)$; in general the problem of mapping f_0 into f_1 through a Lipschitz map has no solution (it suffices to consider, for instance, the case when $X = \overline{B}_1$ and $f_1 = \frac{1}{n}\mathcal{H}^1 \llcorner \partial B_1$). However, another remarkable consequence of Corollary 2.2 is that the problem has solution if we require the transport map to be only a function of bounded variation: indeed, bounded monotone functions (in particular gradients of Lipschitz convex functions) are functions of bounded variation (see for instance Proposition 5.1 of [2]). Moreover, we can give a sharp quantitative estimate of the error made in the approximation by Lipschitz transport maps.

Theorem 2.4. *There exists a constant $C = C(n, X)$ with the following property: for any $\mu \in \mathcal{M}_+(X)$ with $\mu(X) = \mathcal{L}^n(X)$ and any $M > 0$ there exist a Lipschitz function $\phi : X \to X$ and $B \in \mathcal{B}(X)$ such that $\mathrm{Lip}(\phi) \leq M$, $\mathcal{L}^n(X \setminus B) \leq C/M$ and*

$$\mu = \phi_\#(\mathcal{L}^n \llcorner B) + \mu^s,$$

with $\mu^s \in \mathcal{M}_+(X)$ and $\mu^s(X) \leq \mathcal{L}^n(X \setminus B)$.

Proof. By Corollary 2.2 we can represent $\mu = \psi_\#(\mathcal{L}^n \llcorner X)$ with $\psi : X \to X$ equal to the gradient of a convex function. Let Ω be the interior of X; by applying Proposition 5.1 of [2] (valid, more generally, for monotone operators) we obtain that the total variation $|D\psi|(\Omega)$ can be estimated with a suitable constant C depending only on n and X. Therefore, by applying Theorem 5.34 of [4] we can find a Borel set $B \subset X$ (it is a suitable sublevel set of the maximal function of $|D\psi|$) such that $\mathcal{L}^n(X \setminus B) \leq c(n)|D\psi|(\Omega)/M$ and the restriction of ψ to B is a M-Lipschitz function, i.e. with Lipschitz constant not greater than M. By Kirszbraun theorem (see for instance [24]) we can extend $\psi|_B$ to a M-Lipschitz function $\phi : X \to X$. Setting $B^c = X \setminus B$ we have then

$$\mu = \psi_\#(\mathcal{L}^n \llcorner B) + \psi_\#(\mathcal{L}^n \llcorner B^c) = \phi_\#(\mathcal{L}^n \llcorner B) + \psi_\#(\mathcal{L}^n \llcorner B^c)$$

and, setting $\mu^s = \psi_\#(\mathcal{L}^n \llcorner B^c)$, the proof is achieved.

3 The one dimensional case

In this section we assume that $X = I$ is a closed interval of the real line; we also assume for simplicity that the transport cost is $c(x, y) = |x - y|^p$ for some $p \geq 1$.

Theorem 3.1 (Existence and uniqueness). *Assume that f_0 is a diffuse measure, i.e. $f_0(\{t\}) = 0$ for any $t \in I$. Then*

(i) there exists a unique (modulo countable sets) nondecreasing function ψ : spt $f_0 \to X$ such that $\psi_\# f_0 = f_1$;
(ii) the function ψ in (i) is an optimal transport, and, if $p > 1$, it is the unique optimal transport.

In the one-dimensional case these results are sharp: we have already seen that transport maps need not exist if f_0 has atoms (Example 1.1) and that, without the monotonicity constraint, are not necessarily unique when $p = 1$.

Proof. (i) Let $m = \min I$ and define

$$\psi(s) := \sup \{t \in I : \ f_1([m, t]) \leq f_0([m, s])\}. \tag{15}$$

It is easy to check that the following properties hold:

(a) ψ is non decreasing;
(b) $\psi(I) \supset$ spt f_1;
(c) if $\psi(s)$ is not an atom of f_1, we have

$$f_1([m, \psi(s)]) = f_0([m, s]). \tag{16}$$

Let T be the at most countable set made by the atoms of f_1 and by the points $t \in I$ such that $\psi^{-1}(t)$ contains more than one point; then ψ^{-1} is well defined on $\psi(I) \setminus T$ and $\psi^{-1}([t, t']) = [\psi^{-1}(t), \psi^{-1}(t')]$ whenever $t, t' \in \psi(I) \setminus T$ with $t < t'$. Then (c) gives

$$f_1([t, t']) = f_1([m, t']) - f_1([m, t]) = f_0([m, \psi^{-1}(t')]) - f_0([m, \psi^{-1}(t)])$$
$$= f_0([\psi^{-1}(t), \psi^{-1}(t')]) = f_0(\psi^{-1}([t, t']))$$

(notice that only here we use the fact that f_0 is diffuse). By (b) the closed intervals whose endpoints belong to $\psi(I) \setminus T$ generate the Borel σ-algebra of spt f_1, and this proves that $\psi_\# f_0 = f_1$.

Let ϕ be any nondecreasing function such that $\phi_\# f_0 = f_1$ and assume, possibly modifying ϕ on a countable set, that ϕ is right continuous. Let

$$T := \{s \in \text{spt } f_0 : (s, s') \cap \text{spt } f_0 = \emptyset \text{ for some } s' > s \}$$

and notice that T is at most countable (since we can index with T a family of pairwise disjoint open intervals). We claim that $\phi \geq \psi$ on spt $f_0 \setminus T$; indeed, for $s \in$ spt $f_0 \setminus T$ and $s' > s$ we have the inequalities

$$f_1([m, \phi(s')]) = f_0(\phi^{-1}([m, \phi(s')])) \geq f_0([m, s']) > f_0([m, s])$$

and, by the definition of ψ, the inequality follows letting $s' \downarrow s$. In particular

$$\int_I \phi - \psi \, df_0 = \int_I \phi \, df_0 - \int_I \psi \, df_0 = \int_I 1 \, df_1 - \int_I 1 \, df_1 = 0,$$

whence $\phi = \psi$ f_0-a.e. in I. It follows that $\phi(s) = \psi(s)$ at any continuity point $s \in$ spt f_0 of ϕ and ψ.

(ii) By a continuity argument it suffices to prove that ψ is the unique solution of the transport problem for any $p > 1$ (see also [26, 15]). Let γ be an optimal planning and notice that the cyclical monotonicity proved in Theorem 2.2 gives

$$|x - y'|^p + |x' - y|^p \geq |x - y|^p + |x' - y'|^p$$

whenever (x, y), $(x', y') \in$ spt γ. If $x < x'$, this condition implies that $y \leq y'$ (this is a simple analytic calculation that we omit, and here the fact that $p > 1$ plays a role). This means that the set

$$T := \{x \in \text{spt } f_0 : \text{card}(\{y : (x, y) \in \text{spt } \gamma\}) > 1\}$$

is at most countable (since we can index with T a family of pairwise disjoint open intervals) hence f_0-negligible. Therefore, for f_0-a.e. $x \in I$, there exists a unique $y = \tilde{\psi}(x) \in I$ such that $(x, y) \in$ spt γ (the existence of at least one y follows by the fact that the projection of spt γ on the first factor is spt f_0). Notice also that $\tilde{\psi}$ is nondecreasing in its domain.

Arguing as in Proposition 2.1, we obtain that

$$\gamma = (Id \times \tilde{\psi})_\# f_0.$$

In particular,

$$f_1 = \pi_{1\#} \gamma = \pi_{1\#} \left((Id \times \tilde{\psi})_\# f_0 \right) = \tilde{\psi}_\# f_0$$

and, since $\tilde{\psi}$ is non decreasing, it follows that $\tilde{\psi} = \psi$ (up to countable sets) on spt f_0.

4 The ODE version of the optimal transport problem

In this and in the next section we rephrase the optimal transport problem in differential terms. In the following we consider a fixed auxiliary open set Ω containing X; we assume that Ω is sufficiently large, namely that the open r-neighbourhood of X is contained in Ω, with $r > \text{diam}(X)$ (the necessity of this condition will be discussed later on).

The first idea, due to Brenier ([8, 7] and also [11]) is to look for all the paths f_t in $\mathcal{M}_+(X)$ connecting f_0 to f_1. In the simplest case when $f_t = \delta_{x(t)}$, it turns out that the velocity field $E_t = \dot{x}(t)\delta_{x(t)}$ is related to f_t by the equation

$$\dot{f}_t + \nabla \cdot E_t = 0 \qquad \text{in } (0,1) \times \Omega \tag{17}$$

in the distribution sense. Indeed, given $\varphi \in C_c^\infty(0,1)$ and $\phi \in C_c^\infty(\Omega)$, it suffices to take $\varphi(t)\phi(x)$ as test function in (17) and to use the definitions of f_t and E_t to obtain

$$(\dot{f}_t + \nabla \cdot E_t)(\varphi\phi) = -\int_0^1 \dot{\varphi}(t)\phi(x(t)) + \varphi(t)\langle \nabla\phi(x(t)), \dot{x}(t)\rangle \, dt$$

$$= -\int_0^1 \frac{d}{dt}\left[\varphi(t)\phi(x(t))\right] \, dt = 0.$$

More generally, regardless of any assumption on $(f_t, E_t) \in \mathcal{M}_+(X) \times [\mathcal{M}(\Omega)]^n$, it is easy to check that (17) holds in the distribution sense if and only if

$$\dot{f}_t(\phi) = \nabla\phi \cdot E_t \quad \text{in } (0,1) \qquad \forall \phi \in C_c^\infty(\Omega) \tag{18}$$

in the distribution sense. We will use both interpretations in the following.

One more interpretation of (17) is given in the following proposition. Recall that a map f defined in $(0,1)$ with values in a metric space (E, d) is said to be absolutely continuous if for any $\varepsilon > 0$ there exists $\delta > 0$ such that

$$\sum_i (y_i - x_i) < \delta \qquad \Longrightarrow \qquad \sum_i d(f(y_i), f(x_i)) < \varepsilon$$

for any family of pairwise disjoint intervals $(x_i, y_i) \subset (0,1)$.

Proposition 4.1. *If some family $(f_t) \subset \mathcal{M}_1(X)$ fulfils (17) for suitable measures $E_t \in [\mathcal{M}(\Omega)]^n$ satisfying $\int_0^1 |E_t|(\Omega)\, dt < \infty$, then f is an absolutely continuous map between $(0,1)$ and $\mathcal{M}_1(X)$, endowed with the 1-Wasserstein distance (4) and*

$$\lim_{h \to 0} \frac{\mathcal{F}_1(f_{t+h}, f_t)}{|h|} \leq |E_t|(\Omega) \qquad \text{for } \mathcal{L}^1\text{-a.e. } t \in (0,1). \tag{19}$$

Conversely, if f_t is an absolutely continuous map, we can choose E_t so that equality holds in (19).

Proof. In (13) we can obviosuly restrict to test functions u such that $|u| \leq r = \text{diam}(X)$ on X; by our assumption on Ω any of these function can be extended to $\mathbb{R}^n$ in such a way that the Lipschitz constant is still less than 1 and $u \equiv 0$ in a neighbourhood of $\mathbb{R}^n \setminus \Omega$. In particular, choosing an optimal u and setting $u_\varepsilon = u * \rho_\varepsilon$, we get

$$\mathcal{F}_1(f_s, f_t) = \lim_{\varepsilon \to 0+} \int_X u_\varepsilon \, d(f_t - f_s) = \lim_{\varepsilon \to 0+} \int_s^t \nabla \cdot E_\tau(u_\varepsilon) \, d\tau$$

$$= -\lim_{\varepsilon \to 0+} \int_s^t E_\tau \cdot \nabla u_\varepsilon \, d\tau \leq \int_s^t |E_\tau|(\Omega) \, d\tau \qquad (20)$$

whenever $0 \leq s \leq t \leq 1$ and this easily leads to (19).

In the proof of the converse implication we can assume with no loss of generality (up to a reparameterization by arclength) that f is a Lipschitz map. We can consider $\mathcal{M}_1(X)$ as a subset of the dual $Y = G^*$, where

$$G := \{\phi \in C^1(\mathbb{R}^n) \cap \text{Lip}(\mathbb{R}^n) : \phi \equiv 0 \text{ on } \mathbb{R}^n \setminus \Omega\}$$

endowed with the norm $\|\phi\| = \text{Lip}(\phi)$. By using convolutions and (13), it is easy to check that $\mathcal{F}_1(\mu, \nu) = \|\mu - \nu\|_Y$, so that $\mathcal{M}_1(X)$ is isometrically embedded in Y. Moreover, using Hahn–Banach theorem it is easy to check that any $y \in Y$ is representable as the divergence of a measure $E \in [\mathcal{M}(\Omega)]^n$ with $\|y\| = |E|(\Omega)$ (E is not unique, of course).

By a general result proved in [5], valid also for more than one independent variable, any Lipschitz map f from $(0, 1)$ into the dual Y of a separable Banach space is weakly*-differentiable for $\mathcal{L}^1$-almost every t, i.e.

$$\exists w^* - \lim_{h \to 0} \frac{f_{t+h} - f_t}{h} =: \dot{f}(t)$$

and $f_t - f_s = \int_s^t \dot{f}(\tau) \, d\tau$ for $s, t \in (0, 1)$. In addition, the map is also metrically differentiable for $\mathcal{L}^1$-almost every t, i.e.

$$\exists \lim_{h \to 0} \frac{\|f_{t+h} - f_t\|}{|h|} =: m\dot{f}(t).$$

Although $\dot{f}$ is only a w^*-limit of the difference quotients, it turns out (see [5]) that the metric derivative $m\dot{f}$ is $\mathcal{L}^1$-a.e. *equal* to $\|\dot{f}\|$.

Putting together these informations the conclusion follows.

According to Brenier, we can formulate the optimal transport problem as follows.

(ODE) Let $f_0, f_1 \in \mathcal{M}_1(X)$ be given probability measures. Minimize

$$J(E) := \int_0^1 |E_t|(\Omega) \, dt \qquad (21)$$

among all Borel maps $f_t : [0,1] \to \mathcal{M}_+(X)$ and $E_t : [0,1] \to [\mathcal{M}(\Omega)]^n$ such that (17) holds.

Example 4.1. In the case considered in Example 1.3 the measures

$$f_t = \chi_{[t,t+1]}\mathcal{L}^1, \qquad E_t = \chi_{[t,t+1]}\mathcal{L}^1$$

provide an admissible and optimal flow, obviously related to the optimal transport map $x \mapsto x + 1$. But, quite surprisingly, we can also define an optimal flow by

$$f_t = \begin{cases} \frac{1}{1-2t}\chi_{[2t,1]}\mathcal{L}^1 & \text{if } 0 \leq t < 1/2 \\ \delta_1 & \text{if } t = 1/2 \\ \frac{1}{2t-1}\chi_{[1,2t]}\mathcal{L}^1 & \text{if } 1/2 < t \leq 1 \end{cases},$$

whose "velocity field" is

$$E_t = \begin{cases} \frac{2(1-x)}{(1-2t)^2}\chi_{[2t,1]}\mathcal{L}^1 & \text{if } 0 \leq t < 1/2 \\ \frac{2(x-1)}{(2t-1)^2}\chi_{[1,2t]}\mathcal{L}^1 & \text{if } 1/2 < t \leq 1 \end{cases}.$$

It is easy to check that $\dot{f}_t = \nabla \cdot E_t = 0$ and that $|E_t|$ are probability measures for any $t \neq 1/2$, hence $J(E) = 1$. The relation of this new flow with the optimal transport map $x \mapsto 2 - x$ will be seen in the following (see Remark 4.1(3)).

In order to relate solutions of (ODE) to solutions of (MK), we will need an uniqueness theorem, under regularity assumptions in the space variable, for the ODE $\dot{f}_t + \nabla \cdot (g_t f_t) = 0$. If f_t, g_t are smooth (say C^2) with respect to both the space and time variables, uniqueness is a consequence of the classical method of characteristics (see for instance §3.2 of [20]), which provides the representation

$$f_t(x_t) = f_0(x) \exp\left(-\int_0^t c_s(x_s)\, ds\right),$$

where $c_t = \nabla \cdot g_t$ and x_t solves the ODE

$$\dot{x}_t = g_t(x_t), \qquad x_0 = x, \qquad t \in (0,1).$$

See also [33] for more general uniqueness and representation results in a weak setting.

Theorem 4.1. *Assume that*

$$\dot{f}_t + \nabla \cdot (g_t f_t) = 0 \qquad in\ (0,1) \times \mathbb{R}^n, \tag{22}$$

where $\int_0^1 |f_t|(\mathbb{R}^n)\, dt < \infty$ and $|g_t| + \mathrm{Lip}(g_t) \leq C$, with C independent of t and $f_0 = 0$. Then $f_t = 0$ for any $t \in (0,1)$.

Proof. Let g_t^ε be obtained from g_t by a mollification with respect to the space and time variables and define $X^\varepsilon(s, t, x)$ as the solution of the ODE $\dot x = g_s^\varepsilon(x)$ (with s as independent variable) such that $X^\varepsilon(t, t, x) = x$. Define, for $\psi \in C_c^\infty((0, 1) \times \mathbb{R}^n)$ fixed,

$$\varphi^\varepsilon(t, x) := -\int_t^1 \psi(s, X^\varepsilon(s, t, x))\, ds.$$

Since $X^\varepsilon(s, t, X^\varepsilon(t, 0, x)) = X^\varepsilon(s, 0, x)$, we have

$$\varphi^\varepsilon(t, X^\varepsilon(t, 0, x)) = -\int_t^1 \psi(s, X^\varepsilon(s, 0, x))\, ds$$

and, differentiating both sides, we infer

$$\left[\frac{\partial \varphi^\varepsilon}{\partial t} + g_t^\varepsilon \cdot \nabla \varphi^\varepsilon\right](t, X^\varepsilon(t, 0, x)) = \psi(t, X^\varepsilon(t, 0, x))$$

whence $\psi = (\partial \varphi^\varepsilon/\partial t + g_t^\varepsilon \cdot \nabla \varphi^\varepsilon)$ in $(0, 1) \times \mathbb{R}^n$.

Insert now the test function φ^ε in (22) and take into account that $f_0 = 0$ to obtain

$$0 = \int_0^1 \int_{\mathbb{R}^n} f_t\left(\frac{\partial \varphi^\varepsilon}{\partial t} + g_t \cdot \nabla \varphi^\varepsilon\right) dx\, dt = \int_0^1 \int_{\mathbb{R}^n} f_t\psi + f_t(g_t - g_t^\varepsilon) \cdot \nabla \varphi^\varepsilon\, dx\, dt.$$

The proof is finished letting $\varepsilon \to 0^+$ and noticing that $|g_t^\varepsilon| + |\nabla \varphi^\varepsilon| \le C$, with C independent of ε.

In the following theorem we show that (MK) and (ODE) are basically equivalent. Here the assumption that Ω is large enough plays a role: indeed, if for instance $X = [x_0, x_1]$, $f_0 = \delta_{x_0}$, and $f_1 = \delta_{x_1}$, the infimum of (ODE) is easily seen to be less than

$$\text{dist}(x_0, \partial\Omega) + \text{dist}(x_1, \partial\Omega).$$

Therefore, in the general case when f_0, f_1 are arbitrary measures in X, we require that $\text{dist}(\partial\Omega, X) > \text{diam}(X)$.

Theorem 4.2 ((MK) versus (ODE)). *The problem (ODE) has at least one solution and* min *(ODE)* = min *(MK). Moreover, for any optimal planning* $\gamma \in \mathcal{M}_1(X \times X)$ *for (MK) the measures*

$$f_t := \pi_{t\#}\gamma, \qquad E_t := \pi_{t\#}((y - x)\gamma) \qquad t \in [0, 1], \qquad (23)$$

with $\pi_t(x, t) = x + t(y - x)$, *solve (ODE).*

Proof. Let (f_t, E_t) be defined by (23). For any $\phi \in C^\infty(\mathbb{R}^n)$ we compute

$$\frac{d}{dt} \int_X \phi \, df_t = \frac{d}{dt} \int_{X \times X} \phi \, (x + t(y - x)) \, d\gamma$$

$$= \int_{X \times X} \nabla \phi \, (x + t(y - x)) \cdot (y - x) \, d\gamma$$

$$= \sum_{i=1}^n \int_X \nabla_i \phi \, dE_{t,i} = -\nabla \cdot E_t(\phi);$$

hence the ODE (17) is satisfied. Then, we simply evaluate the energy $J(E)$ in (21) by

$$J(E) = \int_0^1 |\pi_{t\#}((y - x)\gamma)|(X) \, dt \le \int_0^1 \pi_{t\#}(|y - x|\gamma)(X) \, dt \qquad (24)$$

$$= \int_0^1 \int_{X \times X} |x - y| \, d\gamma dt = I(\gamma).$$

This shows that $\inf (ODE) \le \min (MK)$.

In order to prove the opposite inequality we first use Proposition 4.1 and then we present a different strategy, which provides more geometric informations.

By Proposition 4.1 we have

$$\mathcal{F}_1(f_0, f_1) \le \int_0^1 \frac{d}{dt} \mathcal{F}_1(f_t, f_0) \, dt \le \int_0^1 |E_t|(\Omega) \, dt$$

for any admissible flow (f_t, E_t). This proves that $\min (MK) \le \inf (ODE)$.

For given (and admissible) (f_t, E_t), assuming that

$$\mathrm{spt} \int_0^1 |E_t| \, dt \subset\subset \Omega,$$

we exhibit an optimal planning γ with $I(\gamma) \le J(E)$.

To this aim we fix a cut-off function $\theta \in C_c^\infty(\Omega)$ with $0 \le \theta \le 1$ and $\theta \equiv 1$ on $X \cup \mathrm{spt} \int_0^1 |E_t| \, dt$; then, we define

$$f_t^\varepsilon := f_t * \rho_\varepsilon + \varepsilon\theta \quad \text{and} \quad E_t^\varepsilon := E_t * \rho_\varepsilon,$$

where ρ is any convolution kernel with compact support and $\rho_\varepsilon(x) = \varepsilon^{-n}\rho(x/\varepsilon)$. Notice that f_t^ε are strictly positive on $\mathrm{spt} \, E_t$; moreover (17) still holds, $\mathrm{spt} \, E_t^\varepsilon \subset \Omega$ for ε small enough and $\mathcal{L}^1$-a.e. t, and by (53) we have

$$\int_{\mathbb{R}^n} |E_t^\varepsilon|(y) \, dy \le |E_t|(\Omega) \qquad \forall t \in [0, 1]. \qquad (25)$$

We define $g_t^\varepsilon = E_t^\varepsilon / f_t^\varepsilon$ and denote by $\psi_t^\varepsilon(x)$ the semigroup in $[0, 1]$ associated to the ODE

$$\dot{\psi}_t^\varepsilon(x) = g_t^\varepsilon\left(\psi_t^\varepsilon(x)\right), \qquad \psi_0^\varepsilon(x) = x. \tag{26}$$

Notice that this flow for ε small enough maps Ω into itself and leaves $\mathbb{R}^n \setminus \Omega$ fixed. Now, we claim that $f_t^\varepsilon = \psi_{t\#}^\varepsilon(f_0^\varepsilon \mathcal{L}^n)$ for any $t \in [0,1]$. Indeed, since by definition

$$\dot{f}_t^\varepsilon + \nabla \cdot (g_t^\varepsilon f_t^\varepsilon) = \dot{f}_t^\varepsilon + \nabla \cdot E_t^\varepsilon = 0,$$

by Theorem 4.1 and the linearity of the equation we need only to check that also $\nu_t^\varepsilon = \psi_{t\#}^\varepsilon(f_0^\varepsilon \mathcal{L}^n)$ satisfies the ODE $\dot{\nu}_t^\varepsilon + \nabla \cdot (g^\varepsilon \nu_t^\varepsilon) = 0$. This is a straightforward computation based on (26).

Using this representation, we can view ψ_1^ε as approximate solutions of the optimal transport problem and define

$$\gamma^\varepsilon := (Id \times \psi_1^\varepsilon)_\#(f_0^\varepsilon \mathcal{L}^n),$$

i.e.

$$\int \varphi(x,y)\, d\gamma^\varepsilon(x) = \int_{\mathbb{R}^n} \varphi\left(x, \psi_1^\varepsilon(x)\right) f_0^\varepsilon(x)\, dx$$

for any bounded Borel function φ in $\mathbb{R}^n \times \mathbb{R}^n$.

In order to evaluate $I(\gamma^\varepsilon)$, we notice that

$$\text{Length}(\psi_t^\varepsilon(x)) = \int_0^1 |g_t^\varepsilon|(\psi_t^\varepsilon(x))\, dt;$$

hence, by multiplying by f_0^ε and integrating we get

$$\int_{\mathbb{R}^n} \text{Length}(\psi_t^\varepsilon(x)) f_0^\varepsilon(x)\, dx = \int_0^1 \int_{\mathbb{R}^n} |g_t^\varepsilon(y)| f_t^\varepsilon(y)\, dy\, dt \tag{27}$$

$$= \int_0^1 \int_{\mathbb{R}^n} |E_t^\varepsilon|(y)\, dy\, dt \le \int_0^1 |E_t|(\Omega)\, dt.$$

As

$$I(\gamma^\varepsilon) = \int_{\mathbb{R}^n} |\psi_1^\varepsilon(x) - x| f_0^\varepsilon(x)\, dx \le \int_{\mathbb{R}^n} \text{Length}(\psi_t^\varepsilon(x))\, df_0(x),$$

this proves that $I(\gamma^\varepsilon) \le J(E)$. Passing to the limit as $\varepsilon \downarrow 0$ and noticing that

$$\pi_{0\#}\gamma^\varepsilon = f_0^\varepsilon \mathcal{L}^n, \qquad \pi_{1\#}\gamma^\varepsilon = f_1^\varepsilon \mathcal{L}^n,$$

we obtain, possibly passing to a subsequence, an admissible planning $\gamma \in \mathcal{M}_1(\mathbb{R}^n \times \mathbb{R}^n)$ for f_0 and f_1 such that $I(\gamma) \le J(E)$. We can turn γ in a planning supported in $X \times X$ simply replacing γ by $(\pi \times \pi)_\#\gamma$, where $\pi : \mathbb{R}^n \to X$ is the orthogonal projection.

Remark 4.1. (1) Starting from (f_t, E_t), the (possibly multivalued) operation leading to γ and then to the new flow

$$\tilde{f}_t := \pi_{t\#}\gamma, \qquad \tilde{E}_t := \pi_{t\#}((y-x)\gamma)$$

can be understood as a sort of "arclength reparameterization" of (f_t, E_t). However, since this operation is local in space (consider for instance the case of two line paths, one parameterized by arclength, one not), in general there is no function $\varphi(t)$ such that

$$(\tilde{f}_t, \tilde{E}_t) = (f_{\varphi(t)}, E_{\varphi(t)}).$$

(2) The solutions (f_t, E_t) in Example 4.1 are built as $\pi_{t\#}\gamma$ and $\pi_{t\#}((y-x)\gamma)$ where $\gamma = (Id \times \psi)_{\#} f_0$ and $\psi(x) = x+1$, $\psi(x) = 2-x$. In particular, in the second example, $f_{1/2} = \delta_1$ because 1 is the midpoint of *any* transport ray.

(3) By making a regularization first in space and then in time of (f_t, E_t) one can avoid the use of Theorem 4.1, using only the classical representation of solutions of (22) with characteristics.

Remark 4.2. [**Optimality conditions**] (1) If (f_t, E_t) is optimal, then

$$\mathrm{spt} \int_0^1 |E_t|\, dt \subset\subset \Omega.$$

Indeed, we can find $\Omega' \subset\subset \Omega$ which still has the property that the open r-neighbourhood of X is contained in Ω', with $r = \mathrm{diam}(X)$, hence

$$\int_0^1 |E_t|(\Omega)\, dt = \min(\mathrm{MK}) = \int_0^1 |E_t|(\Omega')\, dt.$$

(2) Since the two sides in the chain of inequalities (24) are equal when γ is optimal, we infer

$$|\pi_{t\#}((y-x)\gamma)| = \pi_{t\#}(|y-x|\gamma) \qquad \text{for } \mathcal{L}^1\text{-a.e. } t \in (0,1). \tag{28}$$

Analogously, we have

$$\left| \int_0^1 E_t\, dt \right| = \int_0^1 |E_t|\, dt \tag{29}$$

whenever (f_t, E_t) is optimal. If the strict inequality $<$ holds, then

$$\hat{f}_t = f_0 + t(f_1 - f_0) \quad \text{and} \quad \hat{E}_t = \int_0^1 E_\tau\, d\tau$$

provide an admissible pair for (ODE) with strictly less energy.

Remark 4.3. [**Nonlinear cost**] When the cost function $c(x,y)$ is $|x-y|^p$ for some $p > 1$ the corresponding problem (MK) is still equivalent to (ODE), provided we minimize, instead of J, the energy

$$J_p(f, E) := \int_0^1 \Phi_p(f_t, E_t)\, dt,$$

where

$$\Phi_p(\sigma,\nu) = \begin{cases} \int_X \left|\dfrac{\nu}{\sigma}\right|^p d\sigma & \text{if } |\nu| << \sigma; \\[2ex] +\infty & \text{otherwise.} \end{cases}$$

The proof of the equivalence is quite similar to the one given in Theorem 4.2. Again the essential ingredient is the inequality

$$\Phi_p(\nu * \rho_\varepsilon, \sigma * \rho_\varepsilon) \le \Phi_p(\nu,\sigma).$$

The latter follows by Jensen's inequality and the convexity of the map $(z,t) \mapsto |z|^p t^{1-p}$ in $\mathbb{R}^n \times (0,\infty)$, which provide the pointwise estimate

$$\left|\frac{\nu * \rho_\varepsilon}{\sigma * \rho_\varepsilon}\right|^p \sigma * \rho_\varepsilon \le \left(\left|\frac{\nu}{\sigma}\right|^p \sigma\right) * \rho_\varepsilon.$$

In this case, under the same assumptions on f_0 made in Corollary 2.2, due to the uniqueness of the optimal planning $\gamma = (Id \times \psi_\#) f_0$, we obtain that, for optimal (f_t, E_t), $\gamma^\varepsilon = (Id \times \psi_1^\varepsilon)_\# f_0^\varepsilon$ converge to γ as $\varepsilon \to 0^+$. As a byproduct, the maps ψ_1^ε converge to ψ as Young measures. If $f_0 << \mathcal{L}^n$ it follows that ψ_h converge to ψ in $[L^r(f_0)]^n$ for any $r \in [1,\infty)$ (see Lemma 9.1 and the remark following it).

See §2.6 of [4] for a systematic analysis of the continuity and semicontinuity properties of the functional $(\sigma,\nu) \mapsto \Phi_p(\sigma,\nu)$ with respect to weak convergence of measures.

I do not know whether in the previous proof it is actually possible to show full convergence of γ^ε as $\varepsilon \to 0^+$. However, using a more refined estimate and a geometric lemma (see Lemma 4.1 below) we will prove that any limit measure γ satisfies the condition

$$\int_0^1 \pi_{t\#}(|y-x|\gamma) \, dt = \int_0^1 |E_t| \, dt. \tag{30}$$

We call *transport density* any of the measures in (30). This double representation will be relevant in Section 7, where under suitable assumptions we will obtain the uniqueness of the transport density.

Corollary 4.1. *Let (f_t, E_t) be optimal for (ODE). Then there exists an optimal planning γ such that (30) holds. In particular spt $\int_0^1 |E_t| \, dt \subset X$.*

Proof. Let $m = \min(MK) = \min(ODE)$ and recall that, by Remark 4.2(1), the measure $\int_0^1 |E_t| \, dt$ has compact support in Ω. It suffices to build an optimal planning γ such that

$$\int_0^1 \pi_{t\#}((y-x)\gamma) \, dt = \int_0^1 E_t \, dt. \tag{31}$$

Indeed, by (29) we get

$$\int_0^1 \pi_{t\#}(|y - x||\gamma) \geq \int_0^1 |E_t| \, dt$$

and the two measures coincide, having both mass equal to m.

Keeping the same notation used in the final part of the proof of Theorem 4.2, we define γ_t^ε as $(Id \times \psi_t^\varepsilon)_\# f_0^\varepsilon$ and we compute

$$I(\gamma^\varepsilon) = \int_0^1 \frac{d}{dt} I(\gamma_t^\varepsilon) \, dt = \int_0^1 \int_{B_R} \frac{\psi_t^\varepsilon(x) - x}{|\psi_t^\varepsilon(x) - x|} \cdot \dot{\psi}_t^\varepsilon(x) f_0^\varepsilon(x) \, dx dt.$$

Since $I(\gamma^\varepsilon) \to m$ as $\varepsilon \to 0^+$ and since (by (25))

$$\int_0^1 \int_{B_R} |\dot{\psi}_t^\varepsilon(x)| f_0^\varepsilon(x) \, dx dt = \int_0^1 \int_{B_R} |E_t^\varepsilon| \, dx dt \leq \int_0^1 |E_t|(\Omega) \, dt = m,$$

we infer

$$\lim_{\varepsilon \to 0^+} \int_{B_R} \int_0^1 \left| \frac{\dot{\psi}_t^\varepsilon(x)}{|\dot{\psi}_t^\varepsilon(x)|} - \frac{\psi_t^\varepsilon(x) - x}{|\psi_t^\varepsilon(x) - x|} \right|^2 |\dot{\psi}_t^\varepsilon(x)| f_0^\varepsilon(x) \, dt dx = 0. \qquad (32)$$

Now, using Lemma 4.1 and the Young inequality $2ab \leq \delta a^2 + b^2/\delta$, for any $\phi \in C_c^\infty(\mathbb{R}^n)$ we obtain

$$\left| \int_0^1 [\pi_{t\#}((y - x)\gamma^\varepsilon) - E_t^\varepsilon](\phi) \, dt \right|$$

$$= \left| \int_X \int_0^1 [(\psi_1^\varepsilon(x) - x)\phi(t(\psi_1^\varepsilon(x) - x)) - \dot{\psi}_t^\varepsilon(x)\phi(\psi_t^\varepsilon(x))] f_0^\varepsilon(x) \, dt dx \right|$$

$$\leq R \operatorname{Lip}(\phi)\delta \int_X \operatorname{Length}(\psi_t^\varepsilon) f_0^\varepsilon(x) \, dx$$

$$+ \frac{R \operatorname{Lip}(\phi)}{\delta} \int_X \int_0^1 \left| \frac{\dot{\psi}_t^\varepsilon(x)}{|\dot{\psi}_t^\varepsilon(x)|} - \frac{\psi_t^\varepsilon(x) - x}{|\psi_t^\varepsilon(x) - x|} \right|^2 |\dot{\psi}_t^\varepsilon(x)| f_0^\varepsilon(x) \, dt dx.$$

with $R = \operatorname{diam}(\Omega)$. Passing to the limit first as $\varepsilon \to 0^+$, taking (32) into account, and then passing to the limit as $\delta \to 0^+$, we obtain

$$\int_0^1 \pi_{t\#}((y - x)\gamma)(\phi) \, dt = \int_0^1 E_t(\phi) \, dt.$$

Remark 4.4. [**Decomposition of vectorfields**] Let E be a vectorfield with $\nabla \cdot E = f$ in Ω and assume that $|E|(\Omega) = \min(\text{MK})$ with $f_0 = f^+$ and $f_1 = f^-$. Then, choosing $E_t = E$ and $f_t = f_0 + t(f_1 - f_0)$ in (31), we obtain that E can be represented as the superposition of elementary vectorfields

$$E = \int_{\Omega \times \Omega} E_{xy} \, d\gamma(x,y) \qquad \text{with} \qquad E_{xy} := (y - x) \, \mathcal{H}^1 \llcorner [x,y]$$

without cancellations, i.e. in such a way that

$$|E| = \int_{\Omega \times \Omega} |E_{xy}| \, d\gamma(x,y), \qquad |\nabla \cdot E| = \int_{\Omega \times \Omega} |\nabla \cdot E_{x,y}| \, d\gamma(x,y).$$

The possibility to represent a generic vectorfield whose divergence is a measure as a superposition without cancellations of elementary vectorfields associated to rectifiable curves is discussed in [39]. It turns out that this property is not true in general, at least if $n \geq 3$; hence the optimality condition $|E|(\Omega) = \min (\text{MK})$ is essential.

Lemma 4.1. *Let $\psi \in \text{Lip}([0,1], \mathbb{R}^n)$ with $\psi(0) = 0$ and $\phi \in \text{Lip}(\mathbb{R}^n)$. Then*

$$\left| \int_0^1 \psi(1)\phi(t\psi(1)) - \dot{\psi}(t)\phi(\psi(t)) \, dt \right| \leq LR \int_0^1 \left| \frac{\dot{\psi}(t)}{|\dot{\psi}(t)|} - \frac{\psi(t)}{|\psi(t)|} \right| |\dot{\psi}(t)| \, dt,$$

with $L = \text{Lip}(\phi)$, $R = \sup |\psi|$.

Proof. We start from the elementary identity

$$\frac{d}{dt} [\phi(s\psi(t))\psi_i(t)] - \frac{d}{ds} [s\phi(s\psi(t))\dot{\psi}_i(t)] \tag{33}$$

$$= s \sum_{j=1}^n \frac{\partial \phi}{\partial x_j}(s\psi(t)) \left(\psi_i(t)\dot{\psi}_j(t) - \psi_j(t)\dot{\psi}_i(t) \right)$$

and integrate both sides in $[0,1] \times [0,1]$. Then, the left side becomes exactly the integral that we need to estimate from above. The right side, up to the multiplicative constant $\text{Lip}(\phi)$, can be estimated with

$$\int_0^1 |\psi(t) \wedge \dot{\psi}(t)| \, dt = \int_0^1 |\psi(t)| \left| \dot{\psi}(t) \wedge \left(\frac{\dot{\psi}(t)}{|\dot{\psi}(t)|} - \frac{\psi(t)}{|\psi(t)|} \right) \right| dt$$

$$\leq \sup |\psi| \int_0^1 \left| \frac{\dot{\psi}(t)}{|\dot{\psi}(t)|} - \frac{\psi(t)}{|\psi(t)|} \right| |\dot{\psi}(t)| \, dt.$$

Remark 4.5. The geometric meaning of the proof above is the following: the integral to be estimated is the action on the 1-form ϕdx_i of the closed and rectifiable current T associated to the closed path starting at 0, arriving at $\psi(1)$ following the curve $\psi(t)$, and then going back to 0 through the segment $[0, \psi(1)]$; for any 2-dimensional current G such that $\partial G = T$, as

$$T(\phi dx_i) = G(d\phi \wedge dx_i), \tag{34}$$

this action can be estimated by the mass of G times $\text{Lip}(\phi)$. The cone construction provides a current G with $\partial G = T$, whose mass can be estimated by

$$\int_0^1 \int_0^1 s|\psi(t) \wedge \dot\psi(t)| \, ds dt.$$

With this choice of G the identity (34) corresponds to (33).

In order to relate also (f_t, E_t) to the Kantorovich potential, we define the transport rays and the transport set and we prove the differentiability of the potential on the transport set.

Definition 4.1 (Transport rays and transport set). *Let $u \in \text{Lip}_1(X)$. We say that a segment $]x, y[\subset X$ is a transport ray if it is a maximal open oriented segment whose endpoints x, y satisfy the condition*

$$u(x) - u(y) = |x - y|. \tag{35}$$

The transport set $\mathcal{T}_u$ is defined as the union of all transport rays. We also define $\mathcal{T}_u^e$ as the union of the closures of all transport rays.

Denoting by $\mathcal{F}$ the compact collection of pairs (x, y) with $x \neq y$ such that (35) holds, the transport set is also given by

$$\mathcal{T}_u = \bigcup_{t \in (0,1)} \bigcup_{(x,y) \in \mathcal{F}} \{x + t(y - x)\}$$

and therefore is a Borel set (precisely a countable union of closed sets).

We can now easily prove that the u is differentiable at any point in $\mathcal{T}_u$.

Proposition 4.2 (Differentiability of the potential). *Let $u \in \text{Lip}_1(X)$. Then u is differentiable at any point $z \in \mathcal{T}_u$. Moreover $-\nabla u(z)$ is the unit vector parallel to the transport ray containing z.*

Proof. Let x, $y \in X$ be such that (35) holds. By the triangle inequality and the 1-Lipschitz continuity of u we get

$$u(x) - u(x + t(y - x)) = t|y - x| \qquad \forall t \in [0, 1].$$

This implies that, setting $\nu = (y - x)/|y - x|$, the partial derivative of u along ν is equal to -1 for any internal point z of the segment. For any unit vector ξ perpendicular to ν we have

$$u(z + h\xi) - u(z) = u(z + h\xi) - u(z + \sqrt{|h|}\nu) + u(z + \sqrt{|h|}\nu) - u(z)$$
$$\leq \sqrt{|h|^2 + |h|} - \sqrt{|h|} = O(|h|^{3/2}) = o(|h|).$$

A similar argument also proves that $u(z + h\xi) - u(z) \geq o(|h|)$. This proves the differentiability of u at z and the identity $\nabla u(z) = -\nu$.

In Section 6 we will need a mild Lipschitz property of the potential ∇u on $\mathcal{T}_u^e$ (see also [15, 43] for the case of general strictly convex norms). We will use this property to prove in Corollary 6.1 that $\mathcal{T}_u^e \setminus \mathcal{T}_u$ is Lebesgue negligible. This property can also be used to prove that ∇u is approximately differentiable $\mathcal{L}^n$-a.e. on $\mathcal{T}_u$ (see for instance Theorem 3.1.9 of [24]).

Theorem 4.3 (Countable Lipschitz property). *Let $u \in \mathrm{Lip}_1(X)$. There exists a sequence of Borel sets T_h covering $\mathcal{L}^n$-almost all of $\mathcal{T}_u^e$ such that ∇u restricted to T_h is a Lipschitz function.*

Proof. Given a direction $\nu \in \mathbf{S}^{n-1}$ and $a \in \mathbb{R}$, let R be the union of the half closed transport rays $[x, y[$ with $\langle y - x, \nu \rangle \geq 0$ and $\langle y, \nu \rangle \geq a$. It suffices to prove that the restriction of ∇u to

$$T := R \cap \{x : x \cdot \nu < a\} \setminus \Sigma_u$$

has the countable Lipschitz property stated in the theorem. To this aim, since BV_{loc} funtions have this property (see for instance Theorem 5.34 of [4] or [24]), it suffices to prove that ∇u coincides $\mathcal{L}^n$-a.e. in T with a suitable function $w \in [BV_{\mathrm{loc}}(S_a)]^n$, where $S_a = \{x : x \cdot \nu < a\}$. To this aim we define

$$\tilde{u}(x) := \min\{u(y) + |x - y| : y \in Y_a\},$$

where Y_a is the collection of all right endpoints of transport rays with $y \cdot \nu \geq a$. By construction $\tilde{u} \geq u$ and equality holds on $R \supset T$.

We claim that, for $b < a$, $\tilde{u} - C|x|^2$ is concave in S_b for $C = C(b)$ large enough. Indeed, since $|x - y| \geq a - b > 0$ for any $y \in Y_a$ and any $x \in S_b$, the functions

$$u(y) + |x - y| - C|x|^2, \qquad y \in Y_a$$

are all concave in H for C large enough depending on $a - b$. In particular, as gradients of real valued concave functions are BV_{loc} (see for instance [2]), we obtain that

$$w := \nabla \tilde{u} = \nabla(u - C|x|^2) + 2Cx$$

is a BV_{loc} function in S_a. Since $\nabla u = w$ $\mathcal{L}^n$-a.e. in T the proof is achieved.

By a similar argument, using semi-convexity in place of semi-concavity, one can take into account the right extreme points of the transport rays.

As a byproduct of the equivalence between (MK) and (ODE) we can prove that (ODE) has "regular" solutions, related to the transport set and to the gradient of any maximal Kantorovich potential u.

Theorem 4.4 (Regularity of (f_t, E_t)). *For any solution (f_t, E_t) of (ODE) representable as in Theorem 4.2 for a suitable optimal planning γ, there exists a 1-Lipschitz function u such that*

(i) *f_t is concentrated on the transport set $\mathcal{T}_u$ and $|E_t| \leq Cf_t$ for $\mathcal{L}^1$-a.e. $t \in (0, 1)$, with $C = \mathrm{diam}(X)$;*

(ii) $E_t = -\nabla u |E_t|$ *for $\mathcal{L}^1$-a.e. $t \in (0,1)$;*

(iii) *for any convolution kernel ρ we have*

$$\lim_{\varepsilon \to 0^+} \int_0^1 \int_X |\nabla u - \nabla u * \rho_\varepsilon|^2 \, d|E_t| \, dt = 0; \qquad (36)$$

(iv) $|E_t|(X) = \int_0^1 |E_\tau|(X) \, d\tau$ *for $\mathcal{L}^1$-a.e. $t \in (0,1)$.*

Proof. (i) Let u be any maximal Kantorovich potential and let $\mathcal{T} = \mathcal{T}_u$ be the associated transport set. For any $t \in (0,1)$ we have

$$f_t(X \setminus \mathcal{T}) = \int_{X \times X} \chi_{\{(x,y): \, x+t(y-x) \notin \mathcal{T}\}} \, d\gamma = 0$$

because, by (12), any segment $]x, y[$ with $(x,y) \in \operatorname{spt} \gamma$ is contained in a transport ray, hence contained in $\mathcal{T}$. The inequality $|E_t| \leq C f_t$ simply follows by the fact that $|x - y| \leq C$ on $\operatorname{spt} \gamma$.

(ii) Choosing t satisfying (28) and taking into account Proposition 4.2, for any $\phi \in C(X)$ we obtain

$$\int_X \phi \nabla u \, d|E_t| = \int_{X \times X} \phi(\pi_t) \nabla u(\pi_t) |y - x| \, d\gamma$$

$$= - \int_{X \times X} \phi(\pi_t)(y - x) \, d\gamma = -E_t(\phi).$$

(iii) Let $m = J(E) = \min (\text{MK})$. By (13) we get

$$m = \int_X u \, d(f_0 - f_1) = \lim_{\varepsilon \to 0^+} \int_X u * \rho_\varepsilon \, d(f_1 - f_0)$$

$$= \lim_{\varepsilon \to 0^+} \int_X \int_0^1 u * \rho_\varepsilon \, d\dot{f}_t \, dt = \lim_{\varepsilon \to 0^+} \int_0^1 \nabla u * \rho_\varepsilon \cdot E_t \, dt$$

$$\leq \int_0^1 |E_t|(X) \, dt = m.$$

This proves that

$$\lim_{\varepsilon \to 0^+} \int_0^1 \left[|E_t|(X) - \int_X \langle \nabla u * \rho_\varepsilon, \nabla u \rangle \, d|E_t| \right] dt = 0,$$

whence, taking into account that $|\nabla u| = 1$ and $|\nabla u * \rho_\varepsilon| \leq 1$, (36) follows.

(iv) The inequality $|E_t|(X) \leq I(\gamma) = J(E)$ has been established during the proof of Theorem 4.2. By minimality equality, it holds for $\mathcal{L}^1$-a.e. $t \in (0,1)$.

Remark 4.6. (1) It is easy to produce examples of optimal flows (f_t, E_t) satisfying (i), (ii), (iii), (iv) which are not representable as in Theorem 4.2 (again it suffices to consider the sum of two paths, with constant sum of velocities). It is not clear which conditions must be added in order to obtain this representation property.

(2) Notice that condition (i) actually implies that f is a Lipschitz map from $[0,1]$ into $\mathcal{M}_1(X)$ endowed with the 1-Wasserstein metric.

5 The PDE version of the optimal transport problem and the p-laplacian approximation

In this section we see how, in the case when the cost function is linear, the optimal transport problem can be rephrased using a PDE. As in the previous section we consider an auxiliary bounded open set Ω such that the open r-neighbourhood of X is contained in Ω, with $r > \text{diam}(X)$; we assume also that Ω has a Lipschitz boundary.

Specifically, we are going to consider the following problem and its connections with (MK) and (ODE).

(PDE) Let $f \in \mathcal{M}(X)$ be a measure with $f(X) = 0$. Find $\mu \in \mathcal{M}_+(\Omega)$ and $u \in \text{Lip}_1(\Omega)$ such that

(i) there exist smooth functions u_h uniformly converging to u on X, equal to 0 on $\partial\Omega$ and such that the functions ∇u_h converge in $[L^2(\mu)]^n$ to a function $\nabla_\mu u$ satisfying

$$|\nabla_\mu u| = 1 \qquad \mu\text{-a.e. in } \Omega; \tag{37}$$

(ii) the following PDE is satisfied in the sense of distributions

$$-\nabla \cdot (\nabla_\mu u \, \mu) = f \qquad \text{in } \Omega. \tag{38}$$

Given (μ, u) admissible for (PDE), we call μ the *transport density* (the reason for that soon will be clear) and $\nabla_\mu u$ the *tangential gradient* of u.

Remark 5.1. (1) Choosing u_h as test function in (38) gives

$$\mu(\Omega) = \int_\Omega |\nabla_\mu u|^2 \, d\mu = \lim_{h \to \infty} \int_\Omega \langle \nabla_\mu u, \nabla u_h \rangle \, d\mu \tag{39}$$
$$= \lim_{h \to \infty} f(u_h) = f(u),$$

so that the total mass $\mu(\Omega)$ depends only on f and u. We will prove in Theorem 5.1 that actually $\mu(\Omega)$ depends only on f and that μ is concentrated on X.

(2) It is easy to prove that, given (μ, u), there is at most one tangential gradient $\nabla_\mu u$: indeed, if $\nabla u_h \to \nabla_\mu u$ and $\nabla \tilde{u}_h \to \tilde{\nabla}_\mu u$, we have

$$\int_\Omega \nabla_\mu u \cdot \tilde{\nabla}_\mu u \, d\mu = \lim_{h \to \infty} \int_\Omega \nabla_\mu u \cdot \nabla \tilde{u}_h \, d\mu = \lim_{h \to \infty} f(\tilde{u}_h) = f(u) = \mu(\Omega),$$

whence $\nabla_\mu u = \tilde{\nabla}_\mu u$ μ-a.e. On the other hand, Example 5.1 below shows that to some u there could correspond more than one measures μ solving (PDE). We will obtain uniqueness in Section 7 under absolute continuity assumptions on f^+ or f^-.

We will need the following lemma, in which the assumption that Ω is "large enough" plays a role.

Lemma 5.1. *If (μ, u) solves (PDE) then* spt $\mu \subset\subset \Omega$.

Proof. Let $\Omega' \subset\subset \Omega'' \subset\subset \Omega$ be such that $|x - y| > r = \text{diam}(X)$ for any $x \in X$ and any $y \in \mathbb{R}^n \setminus \Omega'$. We choose a minimum point x_0 for the restriction of u to X and define

$$w(x) := \min\left\{[u(x) - u(x_0)]^+, \text{dist}(x, \mathbb{R}^n \setminus \Omega')\right\}.$$

Then, it is easy to check that $w(x) = u(x) - u(x_0)$ on X and that $w \equiv 0$ on $\mathbb{R}^n \setminus \Omega'$. Since

$$\mu(\Omega) = f(u) = f(u - u(x_0)) = \lim_{\epsilon \to 0^+} f(w * \rho_\epsilon)$$

$$= \lim_{\epsilon \to 0^+} \int_\Omega \nabla_\mu u \cdot \nabla w * \rho_\epsilon \, dx \leq \mu(\Omega''),$$

we conclude that $\mu(\Omega \setminus \Omega'') = 0$.

In the following theorem we build a solution of (PDE) choosing μ as a transport density of (MK) with $f_0 = f^+$ and $f_1 = f^-$. As a byproduct, by Corollary 4.1 we obtain that the support of μ is contained in X and that μ is representable as the right side in (30).

Theorem 5.1 ((ODE) versus (PDE)). *Assume that (f_t, E_t) with*

$$\text{spt} \int_0^1 |E_t| \, dt \subset\subset \Omega$$

solves (ODE) and let u be a maximal Kantorovich potential relative to f_0, f_1. Then, setting $f = f_0 - f_1$, the pair (μ, u), with $\mu = \int_0^1 |E_t| \, dt$, solves (PDE) with $\nabla_\mu u = \nabla u$ and, in particular, spt $\mu \subset X$.

Conversely, if (μ, u) solves (PDE), setting $f_0 = f^+$ and $f_1 = f^-$, the measures (f_t, E_t) defined by

$$f_t := f_0 + t(f_1 - f_0), \qquad E_t := -\nabla_\mu u \, \mu$$

solve (ODE) and spt $\mu \subset X$. *In particular*

$$\mu(X) = \min (ODE) = \min (MK).$$

Proof. By Corollary 4.1 we can assume that μ is representable as in (30) for a suitable optimal planning γ. Then, condition (i) in (PDE) with $\nabla_\mu u = \nabla u$ follows by (36). By (17) and condition (ii) of Theorem 4.4 we get

$$\nabla \cdot (\nabla_\mu u |F_t|) = \nabla \cdot (\nabla u |F_t|) = \dot{f}_t,$$

with $F_t = \pi_{t\#}((y - x)\gamma)$. Integrating in time and taking into account the definition of μ, we obtain (38). Notice also that $\mu(X) = J(E)$.

If (f_t, E_t) are defined as above, we get

$$\dot{f}_t + \nabla \cdot E_t = f_1 - f_0 - \nabla \cdot (\nabla_\mu u \mu) = f_1 - f_0 + f = 0.$$

Also in this case $J(E) = \mu(X)$.

Example 5.1 (Non uniqueness of μ). In general, given f, the measure which solves (PDE) for a given $u \in \mathrm{Lip}_1(X)$ is not unique. As an example one can consider the situation in Example 1.5, where μ can be concentrated either on the horizontal sides of the square or on the vertical sides of the square.

As shown in [22], an alternative construction can be obtained by solving the problem "$-\Delta_\infty u = f$", i.e. studying the following problems

$$\begin{cases} -\nabla \cdot (|\nabla u_p|^{p-2} \nabla u_p) = f & \text{in } \Omega \\ u_p = 0 & \text{on } \partial\Omega \end{cases}, \qquad (40)$$

as $p \to \infty$. In this case μ is the limit, up to subsequences, of $|\nabla u_p|^{p-1} \mathcal{L}^n$. Under suitable regularity assumptions, Evans and Gangbo prove that $\mu = a\mathcal{L}^n$ for some $a \in L^1(\Omega)$ and use (a, u) to build an optimal transport ψ; their construction is based on a careful regularization corresponding to the one used in Theorem 4.2 in the special case $E_t = a\mathcal{L}^n$ and $f_t = f_0 + t(f_1 - f_0)$. However, since the goal in Theorem 4.2 is to build only an optimal planning and not an optimal transport, our proof is much simpler and works in a much greater generality (i.e., with no special assumption on f).

Now we also prove that the limiting procedure of Evans and Gangbo leads to solutions of (PDE), regardless of any assumption on the data f_0, f_1.

Theorem 5.2. *Let u_p be solutions of (40) with $p \geq n + 1$. Then*

(i) the measures $H_p = |\nabla u_p|^{p-2} \nabla u_p \mathcal{L}^n \llcorner \Omega$ are equi-bounded in Ω;
(ii) (u_p) are equibounded and equicontinuous in Ω;
(iii) If $(H, u) = \lim_j (H_{p_j}, u_{p_j})$ with $p_j \to \infty$, then (μ, u) solves (PDE) with $\mu = |H|$.

Proof. (i) Using u_p as test function and the Sobolev embedding theorem, we get

$$\int_\Omega |\nabla u_p|^p \, dx = f(u_p) \leq |f|(X)\|u_p\|_\infty \leq C \left(\int_\Omega |\nabla u_{p_0}|^{p_0} \, dx \right)^{1/p_0},$$

with $p_0 = n + 1$. Using Hölder inequality we infer

$$\int_\Omega |\nabla u_p|^p \, dx \leq C^{\frac{p}{p-1}} \mathcal{L}^n(\Omega)^{\frac{p-p_0}{p_0(p-1)}}.$$

(ii) Follows by (i) and the Sobolev embedding theorem.

(iii) Clearly $-\nabla \cdot H = f$ in Ω. By (i) and Hölder inequality we infer

$$\sup_{q \geq n+1} \int_\Omega |\nabla u|^q \, dx < \infty,$$

whence $|\nabla u| \leq 1$ $\mathcal{L}^n$-a.e. in Ω.

We notice first that the functional

$$\nu \mapsto \int_\Omega \left| \frac{\nu}{|\nu|} - w \right|^2 \phi \, d|\nu|$$

is lower semicontinuous with respect to the weak convergence of measures for any nonnegative $\phi \in C_c(\Omega)$, $w \in [C(\Omega)]^n$. The verification of this fact is straightforward: it suffices to expand the squares. Second, we notice that

$$\lim_{\epsilon \to 0^+} \limsup_{p \to \infty} \int_\Omega \left| \frac{H_p}{|H_p|} - \nabla u_\epsilon \right|^2 d|H_p| = 0$$

whenever $u_\epsilon \in \mathrm{Lip}_1(\mathbb{R}^n)$ are smooth functions uniformly converging to u in X and equal to 0 on $\partial\Omega$. Indeed, as $|\nabla u_\epsilon| \leq 1$, we have

$$\int_\Omega \left| \frac{H_p}{|H_p|} - \nabla u_\epsilon \right|^2 d|H_p| \leq 2 \int_\Omega |\nabla u_p|^{p-1} \left(1 - \frac{\nabla u_\epsilon \cdot \nabla u_p}{|\nabla u_p|} \right) dx$$

$$\leq 2 \int_\Omega |\nabla u_p|^{p-2} (|\nabla u_p|^2 - \nabla u_\epsilon \cdot \nabla u_p) \, dx + \omega_p$$

$$= 2f(u_p) - 2f(u_\epsilon) + \omega_p,$$

where $\omega_p = \sup_{t \geq 0} t^{p-1} - t^p$ tends to 0 as $p \to \infty$.

Taking into account these two remarks, setting $p = p_j$ and passing to the limit as $j \to \infty$, we obtain

$$\lim_{\epsilon \to 0^+} \int_\Omega \left| \frac{H}{|H|} - \nabla u_\epsilon \right|^2 \phi \, d|H| = 0.$$

for any nonnegative $\phi \in C_c(\Omega)$. This implies that ∇u_ϵ converge in $L^2_{\mathrm{loc}}(|H|)$ to the Radon–Nikodým derivative $H/|H|$. Since $|\nabla u_\epsilon| \leq 1$ we have also convergence in $[L^2(|H|)]^n$.

We now set $\mu = |H|$ and $\nabla_\mu u = \lim_\epsilon \nabla u_\epsilon$, so that $H = \nabla_\mu u \mu$ and (38) is satisfied.

6 Existence of optimal transport maps

In this section we prove the existence of optimal transport maps in the case when $c(x, y) = |x - y|$ and f_0 is absolutely continuous with respect to $\mathcal{L}^n$, following essentially the original Sudakov approach and filling a gap in his original proof (see the comments after Theorem 6.1). Using a maximal Kantorovich potential we decompose almost all of X in transport rays and we build an optimal transport maps by gluing the 1-dimensional transport maps obtained in each ray. The assumption $f_0 << \mathcal{L}^n$ is used to prove that the conditional measures f_{0C} within any transport ray C are non atomic (and even absolutely continuous with respect to $\mathcal{H}^1 \llcorner C$), so that Theorem 3.1 is applicable. Therefore the proof depends on the following two results whose proof is based on the countable Lipschitz property of ∇u stated in Theorem 4.3.

Theorem 6.1. *Let $B \in \mathcal{B}(X)$ and let $\pi : B \to \mathcal{S}_c(X)$ be a Borel map satisfying the conditions*

(i) $\pi(x) \cap \pi(x') = \emptyset$ whenever $\pi(x) \neq \pi(x')$;
(ii) $x \in \pi(x)$ for any $x \in B$;
(iii) the direction $\nu(x)$ of $\pi(x)$ is a $\mathbf{S}^{n-1}$-valued countably Lipschitz map on B.

Then, for any measure $\lambda \in \mathcal{M}_+(X)$ absolutely continuous with respect to $\mathcal{L}^n \llcorner B$, setting $\mu = \pi_\# \lambda \in \mathcal{M}_+(\mathcal{S}_c(X))$, the measures λ_C of Theorem 9.1 are absolutely continuous with respect to $\mathcal{H}^1 \llcorner C$ for μ-a.e. $C \in \mathcal{S}_c(X)$.

Proof. Being the property stated stable under countable disjoint unions we may assume that

(a) there exists a unit vector ν such that $\nu(x) \cdot \nu \geq \frac{1}{2}$ for any $x \in B$;
(b) $\nu(x)$ is a Lipschitz map on B;
(c) B is contained in a strip

$$\{x : \ a - b \leq x \cdot \nu \leq a\}$$

with $b > 0$ sufficiently small (depending only on the Lipschitz constant of ν) and $\pi(x)$ intersects the hyperplane $\{x : \ x \cdot \nu = a\}$.

Assuming with no loss of generality $\nu = e_n$ and $a = 0$, we write $x = (y, z)$ with $y \in \mathbb{R}^{n-1}$ and $z < 0$. Under assumption (a), the map $T : \pi(B) \to \mathbb{R}^{n-1}$ which associates to any segment $\pi(x)$ the vector $y \in \mathbb{R}^{n-1}$ such that $(y, 0) \in \pi(x)$ is well defined. Moreover, by condition (i), T is one to one. Hence, setting $f = T \circ \pi : B \to \mathbb{R}^{n-1}$,

$$\nu = T_\# \mu = f_\# \lambda, \qquad C(y) = T^{-1}(y) \supset f^{-1}(y)$$

and representing $\lambda = \eta_y \otimes \nu$ with $\eta_y = \lambda_{C(y)} \in \mathcal{M}_1(f^{-1}(y))$, we need only to prove that $\eta_y << \mathcal{H}^1 \llcorner C(y)$ for ν-a.e. y.

To this aim we examine the Jacobian, in the y variables, of the map $f(y, t)$. Writing $\nu = (\nu_y, \nu_t)$, we have

$$f(y, t) = y + \tau(y, t)\nu_y(y, t) \qquad \text{with} \qquad \tau(y, t) = -\frac{t}{\nu_t(y, t)}.$$

Since $\nu_t \geq 1/2$ and $\tau \leq 2b$ on B, we have

$$\det\left(\nabla_y f(y,t)\right) = \det\left(Id + \tau\nabla_y\nu_y + \frac{t}{\nu_t^2}\nabla_y\nu_t \otimes \nu_y\right) > 0$$

if b is small enough, depending only on $\mathrm{Lip}(\nu)$.

Therefore, the coarea factor

$$\mathbf{C}f := \sqrt{\sum_A \det^2 A}$$

(where the sum runs on all $(n-1) \times (n-1)$ minors A of ∇f) of f is strictly positive on B and, writing $\lambda = g\mathcal{L}^n$ with $g = 0$ out of B, Federer's coarea formula (see for instance [4], [38]) gives

$$\lambda = \frac{g}{\mathbf{C}f}\mathbf{C}f\mathcal{L}^n = \frac{g}{\mathbf{C}f}\mathcal{H}^1 \llcorner f^{-1}(y) \otimes \mathcal{L}^{n-1} = \eta_y' \otimes \nu',$$

with $L = \{y \in \mathbb{R}^{n-1} : \mathcal{H}^1(f^{-1}(y)) > 0\}$ and

$$\eta_y' := \frac{\frac{g}{\mathbf{C}f}\mathcal{H}^1 \llcorner f^{-1}(y)}{\int_{f^{-1}(y)} g/\mathbf{C}f \, d\mathcal{H}^1}, \qquad \nu' = \left(\int_{f^{-1}(y)} \frac{g}{\mathbf{C}f} \, d\mathcal{H}^1\right)\mathcal{L}^{n-1}\llcorner L.$$

By Theorem 9.2 we obtain $\nu = \nu'$ and $\eta_y = \eta_y'$ for ν-a.e. y, and this concludes the proof.

Remark 6.1. (1) In [42] V.N.Sudakov stated the theorem above (see Proposition 78 therein) for maps $\pi : B \to S_o(X)$ without the countable Lipschitz assumption (iii) and also in a greater generality, i.e. for a generic "Borel decomposition" of the space in open affine regions, even of different dimensions. However, it turns out that the assumption (iii) is *essential* even if we restrict to 1-dimensional decompositions. Indeed, G.Alberti, B.Kirchheim and D.Preiss [3] have recently found an example of a compact family of open and pairwise disjoint segments in $\mathbb{R}^3$ such that the collection B of the midpoints of the segments has strictly positive Lebesgue measure. In this case, of course, if $\lambda = \mathcal{L}^n\llcorner B$ the conditional measures λ_C are unit Dirac masses concentrated at the midpoint of C, so that the conclusion of Theorem 6.1 fails.

If $n = 2$ it is not hard to prove that actually condition (iii) follows by (ii), so that the counterexample mentioned above is optimal.

(2) Since any open segment can be approximated from inside by closed segments, by a simple approximation argument one can prove that Theorem 6.1 still holds as well for maps $\pi : B \to S_o(X)$ or for maps with values into half open $[x,y[$ segments.

Also the assumption (i) (that the segments do not intersect) can be relaxed. For instance we can assume that the segments can intersect only at their extreme points, provided the collection of these extreme points is Lebesgue negligible. This is precisely the content of the next corollary; again, a counterexample in [28] shows that this property need not be true for a generic compact family of disjoint line segments in $\mathbb{R}^n$, $n \geq 3$.

Corollary 6.1 (Negligible extreme points). *Let $u \in \mathrm{Lip}_1(X)$. Then the collection of the extreme points of the transport rays is Lebesgue negligible.*

Proof. We prove that the collection L of all left extreme points is negligible, the proof for the right ones being similar. Let $B = L \setminus \Sigma_u$ and set

$$\pi(x) := [x, x - \frac{r(x)}{2} \nabla u(x)],$$

where $r(x)$ is the length of the transport ray emanating from x (this ray is unique due to the differentiability of u at x). By Theorem 4.3 the map π has the countable Lipschitz property on B and, by construction, $\pi(x) \cap \pi(x') = \emptyset$ whenever $x \neq x'$. By Theorem 6.1 we obtain

$$\lambda = \int_{\mathcal{S}_c(X)} \lambda_C \, d\nu,$$

with $\lambda = \mathcal{L}^n \llcorner B$, $\nu = \pi_\# \lambda$ and $\lambda_C << \mathcal{H}^1 \llcorner C$ probability measures concentrated on $\pi^{-1}(C)$ for ν-a.e. C. Since $\pi^{-1}(C)$ contains only one point for any $C \in \pi(B)$, we obtain $\lambda_C = 0$ for ν-a.e. C, whence $\lambda(B) = \nu(\mathcal{S}_c(X)) = 0$. $\qquad \square$

Theorem 6.2 (Sudakov). *Let $f_0, f_1 \in \mathcal{M}_1(X)$ and assume that $f_0 << \mathcal{L}^n$. Then there exists an optimal transport ψ mapping f_0 to f_1. Moreover, if $f_1 << \mathcal{L}^n$ we can choose ψ so that ψ^{-1} is well defined f_1-a.e. and $\psi_\#^{-1} f_1 = f_0$.*

Proof. Let γ be an optimal planning. In the first two steps we assume that

$$\text{for } f_0\text{-a.e. } x \text{ there exists } y \neq x \text{ such that } (x, y) \in \mathrm{spt}\,\gamma. \qquad (41)$$

This condition holds for instance if $f_0 \wedge f_1 = 0$ because in this case any optimal planning γ does not charge the diagonal Δ of $X \times X$, i.e. $\gamma(\Delta) = 0$ (otherwise $h = \pi_{0\#}(\chi_\Delta \gamma) = \pi_{1\#}(\chi_\Delta \gamma)$ would be a nonzero measure less than f_0 and f_1).

Let $u \in \mathrm{Lip}_1(X)$ be a maximal Kantorovich potential given by Corollary 2.1, i.e. a function satisfying

$$u(x) - u(y) = |x - y| \qquad \gamma\text{-a.e. in } X \times X \qquad (42)$$

for any optimal planning γ, and let $\mathcal{T}_u$ be the transport set relative to u (see Definition 4.1). In the following we set

$$\tilde{X} = \{x \in X \setminus \Sigma_u : \exists y \neq x \text{ s.t. } (x, y) \in \mathrm{spt}\,\gamma\}.$$

By (41) and the absolute continuity assumption we know that f_0 is concentrated on $\widetilde{X}$. Notice also that for any $x \in \widetilde{X}$ there exists a unique closed transport ray containing x: this follows by the fact that any $x \in \widetilde{X}$ is a differentiability point of u and by Proposition 4.2.

Step 1. We define $r : X \times X \to \mathcal{S}_c(X)$ as the map which associates to any pair (x, y) the closed transport ray containing $[x, y]$. By (42) the map r is well defined γ-a.e. out of the diagonal Δ; moreover, being f_0 concentrated on $\widetilde{X}$, r is also well defined γ-a.e. on Δ. Hence, according to Theorem 9.1 we can represent

$$\gamma = \gamma_C \otimes \nu \qquad \text{with} \qquad \nu := r_\# \gamma$$

and (42) gives

$$u(x) - u(y) = |x - y| \qquad \gamma_C\text{-a.e. in } X \times X, \tag{43}$$

for ν-a.e. $C \in \mathcal{S}_c(X)$. By the sufficiency part in Corollary 2.1, we infer that γ_C is an optimal planning relative to the probability measures

$$f_{0C} := \pi_{0\#}\gamma_C, \qquad f_{1C} := \pi_{1\#}\gamma_C$$

for ν-a.e. $C \in \mathcal{S}_c(X)$. By (56) we infer

$$\pi_{t\#}\gamma(B) = \int_{\mathcal{S}_c(X)} \pi_{t\#}\gamma_C(B)\, d\nu(C) \qquad \forall t \in [0, 1],\ B \in \mathcal{B}(X). \tag{44}$$

Notice also that

$$\mathcal{F}_1(f_0, f_1) = \int_{\mathcal{S}_c(X)} \mathcal{F}_1(f_{0C}, f_{1C})\, d\nu(C). \tag{45}$$

Indeed,

$$I(\gamma) = \int_{X \times X} |x - y|\, d\gamma = \int_{\mathcal{S}_c(X)} \int_{X \times X} |x - y|\, d\gamma_C\, d\nu(C) = \int_{\mathcal{S}_c(X)} I(\gamma_C)\, d\nu(C)$$

and we know by (43) that γ_C is optimal for ν-a.e. $C \in \mathcal{S}_c(X)$.

Step 2. We denote by $\pi : \widetilde{X} \to \mathcal{S}_c(X)$ the natural map, so that $\pi(x)$ is the closed transport ray containing x. Since $r = \pi \circ \pi_0$ on $\widetilde{X} \times \widetilde{X}$ we obtain

$$\nu = r_\# \gamma = \pi_\#(\pi_{0\#}\gamma) = \pi_\# f_0.$$

Moreover (44) gives

$$f_0(B) = \int_{\mathcal{S}_c(X)} f_{0C}(B)\, d\nu(C) \qquad \forall B \in \mathcal{B}(X). \tag{46}$$

Notice that the segments $\pi(x)$, $x \in \widetilde{X}$, can intersect only at their right extreme point and that, by Corollary 6.1, the collection of these extreme points is Lebesgue negligible. As a consequence of (46) with $\nu = \pi_\# f_0$ and Remark 6.1(2), the measures f_{0C} are absolutely continuous with respect to $\mathcal{H}^1 \llcorner C$ for μ-a.e. $C \in \mathcal{S}_c(X)$. Hence, by Theorem 3.1, for μ-a.e. $C \in \mathcal{S}_c(X)$ we can find a nondecreasing map $\psi_C : C \to C$ (this notion makes sense, since C is oriented) such that $\psi_{C\#} f_{0C} = f_{1C}$. Notice also that ψ^{-1} is well defined f_{1C}-a.e., if also $f_{1C} << \mathcal{H}^1 \llcorner C$.

Taking into account that the closed transport rays in $\pi(\widetilde{X})$ are pairwise disjoint in $\widetilde{X}$, we can glue all the maps ψ_C to produce a single Borel map $\psi : \widetilde{X} \to X$. The map ψ is Borel because we have been able to exhibit the one dimensional transport map constructively and because of the Borel property of the maps $C \mapsto f_{iC}$ (see (15) and (54)); the simple but boring details are left to the reader.

Since $\psi_\# f_{0C} = \psi_{C\#} f_{0C} = f_{1C}$ for μ-a.e. $C \in \mathcal{S}_c(X)$, taking (44) into account we infer

$$\psi_\# f_0 = \int_{\mathcal{S}_c(X)} \psi_\# f_{0C} \, d\nu(C) = \int_{\mathcal{S}_c(X)} f_{1C} \, d\nu(C) = f_1.$$

Finally ψ is an optimal transport because (45) holds and any ψ_C is an optimal transport.

Step 3. In this step we show how the assumption (41) can be removed. We define

$$X' := \{x \in X : (x, x) \in \operatorname{spt} \gamma \text{ and } (x, y) \notin \operatorname{spt} \gamma \ \forall y \neq x\}$$

and $L = \{(x, x) : x \in X'\}$. Then, we set $f_0' = f_0 \llcorner X'$ and $f_0'' = f_0 \llcorner X''$, with $X'' = X \setminus X'$, and

$$f_1' := \pi_{1\#}(\gamma \llcorner L), \qquad f_1'' := f_1 - f_1'.$$

Since $L \subset \Delta$, we have $f_1' = \pi_{0\#}(\gamma \llcorner L) = f_0'$, hence we can choose on X' the map $\psi_1 = Id$ as transport map to obtain $(\psi_1)_\# f_0' = f_1'$. Since f_0'' is concentrated on X'', the condition (41) is satisfied with f_0'' in place of f_0 and we can find an optimal transport map $\psi_2 : X'' \to X$ such that $(\psi_2)_\# f_0'' = f_1''$. Gluing these two transport maps we obtain a transport map ψ such that $\psi_\# f_0 = f_1$; since, by construction,

$$u(x) - u(\psi(x)) = |x - \psi(x)| \qquad f_0\text{-a.e. in } X,$$

we infer that ψ is optimal.

This proof strongly depends on the strict convexity of the euclidean distance, which provides the first and second order differentiability properties of the potential u on the transport set $\mathcal{T}_u$. Notice also that if $c(x, y) = \|x - y\|$ and the norm $\| \cdot \|$ is not strictly convex, then the "transport rays" need not be one-dimensional and, to our knowledge, the existence of an optimal transport map is an open problem in this situation. Indeed, this existence result is stated by Sudakov in [42] but his proof is faulty, for the reasons outlined in Remark 6.1(1).

Under special assumptions on the data f_0, f_1 (absolute continuity, separated supports, Lipschitz densities), Evans and Gangbo provided in [22] a proof based on differential methods of the existence of optimal transport maps. For strictly convex norms, the first fully rigorous proofs of the existence of an optimal transport map under the only assumption that $f_0 << \mathcal{L}^n$ have been given in [15] and [43]. As in Theorem 6.2, the proof is strongly based on the differentiability of the directions of transport rays.

7 Regularity and uniqueness of the transport density

In this section we investigate the regularity and the uniqueness properties of the transport density μ arising in (PDE). Recall that, as Theorem 5.1 shows, any such measure μ can also be represented as $\int_0^1 |E_t| \, dt$ for a suitable optimal pair (f_t, E_t) for (ODE) (even with E_t independent of t). In turn, by Corollary 4.1, any optimal measure $\mu = \int_0^1 |E_t| \, dt$ can be represented as

$$\mu = \int_0^1 \pi_{t\#}(|y - x|\gamma) \, dt \qquad (47)$$

for a suitable optimal planning γ. For this reason, in the following we restrict our attention to the representation (47), valid for any optimal measure μ for (PDE).

A more manageable formula for μ, first considered by G.Bouchitté and G.Buttazzo in [14], is given in the following elementary lemma.

Lemma 7.1. *Let* μ *be as in (47). Then*

$$\mu(B) = \int_{X \times X} \text{Length} \, (]x, y[\cap B) \, d\gamma(x, y) \qquad \forall B \in \mathcal{B}(X). \qquad (48)$$

Proof. For any Borel set B we have

$$\mu(B) = \int_0^1 \int_{\pi_t^{-1}(B)} |y - x| \, d\gamma(x, y) \, dt = \int_{X \times X} |y - x| \int_0^1 \chi_{\pi_t^{-1}(B)} \, dt \, d\gamma(x, y)$$

$$= \int_{X \times X} \text{Length} \, (]x, y[\cap B) \, d\gamma(x, y).$$

A direct consequence of (48) is that μ is concentrated on the transport set (since γ-a.e. segment $]x, y[$ is contained in a transport ray); another consequence is the density estimate

$$\frac{\mu(B_r(x))}{2r} \le f_0(X)f_1(X), \qquad (49)$$

because Length($]x,y[\cap B_r(x)$) $\leq 2r$ for any ball $B_r(x)$ and any segment $]x,y[$. The first results on μ that we state, proved by A.Pratelli in [34] (see also [19]), relate the dimensions of f_0 and f_1 to the dimension of μ and show that necessarily μ is absolutely continuous with respect to the Lebesgue measure if f_0 (or, by symmetry, f_1) has this property.

Theorem 7.1. *Assume that for some $k > 0$ we have*

$$\sup_{r \in (0,1)} \frac{f_0(B_r(x))}{r^k} < \infty \quad f_0\text{-a.e. in } X.$$

Then μ has the same property. In particular $\mu << \mathcal{L}^n$ if $f_0 << \mathcal{L}^n$.

The proof of Theorem 7.1 is based on (48); the density estimate on μ is achieved by a careful analysis of the transport rays crossing a generic ball $B_r(x)$. A similar analysis proved the following summability estimate (see [19]).

Theorem 7.2. *Assume that $f_0 = g_0 \mathcal{L}^n$ and $f_1 = g_1 \mathcal{L}^n$ with $g_0, g_1 \in L^p(X)$, $p > 1$. Then $\mu = h\mathcal{L}^n$, with $h \in L^\infty(X)$ if $p = \infty$ and $h \in L^q(X)$ for any $q < p$ if $p < \infty$.*

It is not known whether $g_0, g_1 \in L^p$ implies $h \in L^p$ for $p < \infty$.

Definition 7.1 (Hausdorff dimension of a measure). *Let $\mu \in \mathcal{M}_+(X)$. The Hausdorff dimension $\mathcal{H}\text{-dim}(\mu)$ is the supremum of all $k \geq 0$ such that $\mu << \mathcal{H}^k$.*

In other words $\mu(B) = 0$ whenever $\mathcal{H}^k(B) = 0$ for some $k < \mathcal{H}\text{-dim}(\mu)$ and for any $k > \mathcal{H}\text{-dim}(\mu)$ there exists a Borel set B with $\mu(B) > 0$ and $\mathcal{H}^k(B) = 0$. Notice that if μ is made of pieces of different dimensions, then $\mathcal{H}\text{-dim}(\mu)$ is the smallest of these dimensions.

Using the density estimates and the implications (see for instance Theorem 2.56 of [4] or [38]; here $t > 0$ and $k > 0$)

$$\limsup_{r \to 0+} \frac{\mu(B_r(x))}{\omega_k r^k} \geq t \quad \forall x \in B \quad \Longrightarrow \quad \mu(B) \geq t\,\mathcal{H}^k(B), \qquad (50)$$

$$\limsup_{r \to 0+} \frac{\mu(B_r(x))}{\omega_k r^k} \leq t \quad \forall x \in B \quad \Longrightarrow \quad \mu(B) \leq 2^k t\,\mathcal{H}^k(B), \qquad (51)$$

we can prove a natural lower bound on the Hausdorff dimension of μ.

Corollary 7.1. *Let μ be a transport density. Then*

$$\mathcal{H}\text{-dim}(\mu) \geq \max\{1, \mathcal{H}\text{-dim}(f_0), \mathcal{H}\text{-dim}(f_1)\}.$$

Proof. By (49) we infer that μ has finite (and even bounded) 1-dimensional density at any point. In particular (51) gives $\mu(B) = 0$ whenever $\mathcal{H}^1(B) = 0$, so that $\mathcal{H}\text{-dim}(\mu) \geq 1$.

Let $k = \mathcal{H}\text{-dim}(f_0)$ and $k' < k$; then f_0 has finite k'-dimensional density f_0-a.e., otherwise by (50) the set where the density is not finite would be $\mathcal{H}^{k'}$-negligible and with strictly positive f_0-measure. By Theorem 7.1 we infer that μ has finite k'-dimensional density μ-a.e. By the same argument used before with $k' = 1$ we obtain that $\mathcal{H}\text{-dim}(\mu) \geq k'$ and therefore, since $k' < k$ is arbitrary, $\mathcal{H}\text{-dim}(\mu) \geq k$. A symmetric argument proves that $\mathcal{H}\text{-dim}(\mu) \geq \mathcal{H}\text{-dim}(f_1)$.

We conclude this section proving the uniqueness of the transport density, under the assumption that either $f_0 << \mathcal{L}^n$ or $f_1 << \mathcal{L}^n$ (see Example 5.1 for a nonuniqueness example if neither f_0 nor f_1 are absolutely continuous). Similar results have been first announced by Feldman and McCann (see [25]).

We first deal with the one dimensional case, where this absolute continuity assumption is not needed.

Lemma 7.2. *Let μ be a transport density in $X \subset \mathbb{R}$ and assume that the interior (a, b) of X is a transport ray. Then $\mu = h\mathcal{L}^1$ in (a, b) with*

$$h(t) = \sigma f((a, t)) + c \quad \mathcal{L}^1\text{-a.e. in } (a, b),$$

for some constant c, where $f = f_1 - f_0$ and $\sigma = 1$ if $u' = 1$ in (a, b), $\sigma = -1$ if $u' = -1$ in (a, b). Moreover

$$c = \frac{\mathcal{F}_1(f_0, f_1) - \sigma \int_{(a,b)} (b - t)\, df(t)}{b - a}.$$

Proof. The proof of the first statement is based on the equation $\mu' = \sigma f$ and on a smoothing argument. By integrating both sides we get

$$c(b - a) = \mu(X) - \sigma \int_a^b f((a, t))\, dt = \mu(X) - \sigma \int_a^b \int_{(a,t)} 1\, df(\tau)\, dt$$

$$= \mu(X) - \sigma \int_{(a,b)} \int_\tau^b 1\, dt\, df(\tau) = \mu(X) - \sigma \int_{(a,b)} (b - \tau)\, df(\tau).$$

The conclusion is achieved taking into account that $\mu(X) = \min(\text{MK}) = \mathcal{F}_1(f_0, f_1)$.

Theorem 7.3 (Uniqueness). *Assume that either f_0 or f_1 are absolutely continuous with respect to $\mathcal{L}^n$. Then the transport density is absolutely continuous and unique.*

Proof. We already know from Theorem 7.1 that any transport density is absolutely continuous and we can assume that $f_0 << \mathcal{L}^n$. By Corollary 4.1

we know that the class of transport densities relative to (f_0, f_1) is equal to the class of transport densities relative to $(f_0 - h, f_1 - h)$ where h is any measure in $\mathcal{M}_+(X)$ such that $h \leq f_0 \wedge f_1$ (indeed, this subtraction does not change the velocity field E_t). Hence, it is not restrictive to assume that $f_0 \wedge f_1 = 0$.

We can assume that μ is representable as in (30) for a suitable optimal planning γ. Adopting the same notation of the proof of Theorem 6.1, we have

$$\gamma = \gamma_C \otimes \nu,$$

where $\nu = \pi_\# f_0$ does not depend on γ (and here the absolute continuity assumption on f_0 plays a crucial role) and γ_C is an optimal planning relative to the measures $f_{0C} = \pi_{0\#}\gamma_C$, $f_{1C} = \pi_{1\#}\gamma_C$ for ν-a.e. C. In particular

$$\mu = \int_0^1 \pi_{t\#} \left(|y - x|\gamma_C \otimes \nu \right) dt = \int_0^1 \pi_{t\#} (|y - x|\gamma_C) \, dt \otimes \nu.$$

Hence, it suffices to show that the measures

$$\mu_C := \int_0^1 \pi_{t\#} (|y - x|\gamma_C) \, dt$$

do not depend on γ (up to ν-negligible sets of course). To this aim, taking into account Lemma 7.2, it suffices to show that f_{0C} and f_{1C} do not depend on γ.

Indeed, we already know from (46) and Theorem 9.2 that f_{0C} do not depend on γ.

The argument for f_{1C} is more involved. First, since $\gamma(\Delta) = 0$ (due to the assumption $f_0 \wedge f_1 = 0$), we have $|x - y| > 0$ γ-a.e., and therefore $|x - y| > 0$ γ_C-a.e. for ν-a.e. $C \in S_c(X)$. As a consequence, setting $C = [x_C, y_C]$, we obtain that $f_{1C}(x_C) = 0$ for ν-a.e. C. Second, we examine the restriction f'_{1C} of f_{1C} to the relative interior of C noticing that (44) gives

$$f_1 \llcorner \mathcal{T}_u = \int_{S_c(X)} f_{1C} \llcorner \mathcal{T}_u \, d\nu(C) = \int_{S_c(X)} f'_{1C} \, d\nu(C),$$

because $\mathcal{T}_u \cap C$ is the relative interior of C for any $C \in \pi(\tilde{X})$. By Theorem 9.2 we obtain that f'_{1C} depend only on f_1, $\mathcal{T}_u$ and ν.

In conclusion, since

$$f_{1C} = f'_{1C} + (1 - f'_{1C}(X)) \, \delta_{y_C},$$

we obtain that f_{1C} do not depend on γ as well.

8 The Bouchitté–Buttazzo mass optimization problem

In [13, 14] Bouchitté and Buttazzo consider the following problem. Given $f \in \mathcal{M}(X)$ with $f(X) = 0$, they define

$$\mathcal{E}(\mu) := \inf \left\{ \int_X \frac{1}{2} |\nabla v|^2 \, d\mu - f(v) : \ v \in C^\infty(X) \right\}$$

for any $\mu \in \mathcal{M}_+(X)$. Then, they raised the following *mass optimization problem.*

(BB) Given $m > 0$, maximize $\mathcal{E}(\mu)$ among all measures $\mu \in \mathcal{M}_+(X)$ with $\mu(X) = m$.

A possible physical interpretation of this problem is the following: we may imagine that μ represents the conductivity of some material, thinking that the conductivity (i.e. the inverse resistivity) is zero out of spt μ; accordingly, we may imagine that $f = f^+ - f^-$ is a balanced density of positive and negative charges. Then $-\mathcal{E}(\mu)$ represents the heating corresponding to the given conductivity, so that there is an obvious interest in maximizing $\mathcal{E}(\mu)$ and, for a minimizer u, the (formal) first variation of the energy

$$-\nabla \cdot (\nabla u \, \mu) = f^+ - f^-$$

corresponds to Ohm's law.

More generally, problems of this sort appear in Shape Optimization and Linear Elasticity. In these cases u is no longer real valued and general energies of the form

$$\mathcal{F}(u) = \int_X f(x, \nabla u(x)) \, d\mu(x) \tag{52}$$

must be considered in order to take into account *light structures*, corresponding to mass distributions not absolutely continuous with respect to $\mathcal{L}^n$. This was the main motivation for Bouchitté, Buttazzo and Seppecher [12] in their development of a general differential calculus with measures. This calculus, based on a suitable concept of tangent space to a measure, enables the study of the relaxation of functionals (52) and provides an explicit formula for their lower semicontinuous envelope.

Coming back to the scalar problem (BB), a remarkable fact discovered in [13, 14] is its connection with the Monge–Kantorovich optimal transport problem, in the case when $c(x, y) = |x - y|$. It turns out that the solutions of (BB) are in one to one correspondence with constant multiples of transport densities for (PDE) (or, equivalently, for (MK) with $f_0 = f^+$ and $f_1 = f^-$).

As a byproduct we have that μ is unique and absolutely continuous with respect to $\mathcal{L}^n$ if either f_0 or f_1 are absolutely continuous with respect to $\mathcal{L}^n$.

Theorem 8.1 ((PDE) versus (BB)). *Let (μ, u) be a solution of (PDE) and set $\tilde{m} = \mu(X)$. Then $\tilde{\mu} = \frac{m}{\tilde{m}} \mu$ solves (BB) and any solution of (BB) is representable in this way. Moreover*

$$\max(BB) = -\frac{1}{2} \frac{\tilde{m}^2}{m}.$$

Proof. Setting $v = \lambda u_h$ (with u_h as in the definition of (PDE)), for any $\nu \in \mathcal{M}_+(X)$ with $\nu(X) = m$ we estimate

$$\mathcal{E}(\nu) \le \frac{m}{2}\lambda^2 - \lambda f(u_h),$$

so that, letting $h \to \infty$ and taking into account (39), we obtain

$$\mathcal{E}(\nu) \le \frac{m}{2}\lambda^2 - \tilde{m}\lambda.$$

By minimizing with respect to λ we obtain $\mathcal{E}(\nu) \le -\tilde{m}^2/(2m)$.

On the other hand, using Young inequality and choosing λ so that $\lambda\tilde{m} = m$, we can estimate

$$\int_X \frac{1}{2}|\nabla v|^2 \, d\tilde{\mu} - f(v) \ge \int_X \langle \nabla v, \lambda \nabla_\mu u \rangle \, d\tilde{\mu} - \lambda^2 \frac{m}{2} - f(v)$$

$$= -\nabla \cdot (\nabla_\mu u \, \mu)(v) - f(v) - \frac{1}{2}\frac{\tilde{m}^2}{m}$$

$$= -\frac{1}{2}\frac{\tilde{m}^2}{m},$$

for any $v \in C^\infty(X)$. This proves that $\tilde{\mu}$ is optimal for (BB).

Conversely, it has been proved in [13, 14] that for any solution σ of (BB) there exists $v \in \mathrm{Lip}(X)$ such that $|\nabla_\sigma v| = m/\tilde{m}$ σ-a.e. and

$$-\nabla \cdot (\nabla_\sigma v \sigma) = f,$$

where $\nabla_\sigma v$ is understood in the Bouchitté–Buttazzo sense. But, since (see [14] again)

$$\int_X |\nabla_\sigma v|^2 \, d\sigma$$

$$= \inf\left\{ \liminf_{h \to \infty} \int_X |\nabla v_h|^2 \, d\sigma : v_h \to v \text{ uniformly, } v_h \in C^\infty(X) \right\},$$

we obtain a sequence of smooth functions v_h uniformly converging to v such that $\nabla v_h \to \nabla_\sigma v$ in $[L^2(\sigma)]^n$. Setting $u = \frac{\tilde{m}}{m}v$ and $\mu = \frac{m}{\tilde{m}}\sigma$, this proves that (σ, u) solves (PDE).

9 Appendix: some measure theoretic results

In this section we list all the measure theoretic results used in the previous section; reference books for the content of this section are [37], [18] and [4].

Let us begin with some terminology.

• **(Measures)** Let X be a locally compact and separable metric space. We denote by $[\mathcal{M}(X)]^m$ the space of Radon measures with values in $\mathbb{R}^m$ and with finite total variation in X. We recall that the total variation measure of $\mu = (\mu_1, \ldots, \mu_m) \in [\mathcal{M}(X)]^m$ is defined by

$$|\mu|(B) := \sup \left\{ \sum_{i=1}^{\infty} |\mu(B_i)| : B = \bigsqcup_{i=1}^{\infty} B_i, \ B_i \in \mathcal{B}(X) \right\}$$

and belongs to $\mathcal{M}_+(X)$. By Riesz theorem the space $[\mathcal{M}(X)]^m$, endowed with the norm $\|\mu\| = |\mu|(X)$, is isometric to the dual of $[C(X)]^m$. The duality is given by the integral, i.e.

$$\langle \mu, u \rangle := \sum_{i=1}^{m} \int_X u_i \, d\mu_i.$$

Recall also that, for $\mu \in \mathcal{M}_+(X)$ and $f \in [L^1(X, \mu)]^m$, the measure $f\mu \in [\mathcal{M}(X)]^m$ is defined by

$$f\mu(B) := \int_B f \, d\mu \qquad \forall B \in \mathcal{B}(X)$$

and $|f\mu| = |f|\mu$.

• **(Push forward of measures)** Let $\mu \in [\mathcal{M}(X)]^m$, let Y be another metric space and let $f : X \to Y$ be a Borel map. Then the push forward measure $f_\#\mu \in [\mathcal{M}(Y)]^m$ is defined by

$$f_\#\mu(B) := \mu\left(f^{-1}(B)\right) \qquad \forall B \in \mathcal{B}(Y)$$

and satisfies the more general property

$$\int_Y u \, df_\#\mu = \int_X u \circ f \, d\mu \qquad \text{for any bounded Borel function } u : Y \to \mathbb{R}.$$

It is easy to check that $|f_\#\mu| \leq f_\#|\mu|$.

• **(Support)** We say that $\mu \in [\mathcal{M}(X)]^m$ is concentrated on a Borel set B if $|\mu|(X \setminus B) = 0$ and we denote by $\operatorname{spt} \mu$ the smallest closed set on which μ is concentrated (the existence of a smallest set follows by the separability of X, precisely by the Lindelöf property). The support is also given by the formula (sometimes taken as the definition)

$$\operatorname{spt} \mu = \{x \in X : |\mu|(B_\varrho(x)) > 0 \ \forall \varrho > 0\}.$$

• **(Convolution)** If $X \subset \mathbb{R}^n$, $\mu \in [\mathcal{M}(X)]^m$ and $\rho \in C_c^\infty(\mathbb{R}^n)$ (for instance a convolution kernel); we define

$$\mu * \rho(x) := \int_X \rho(x - y) \, d\mu(y) \qquad x \in \mathbb{R}^n.$$

Notice that $\mu * \rho \in [C^\infty(\mathbb{R}^n)]^m$ because $D^\alpha(\mu * \rho) = (D^\alpha \rho) * \mu$ for any multiindex α and

$$\sup |D^\alpha(\mu * \rho)| \leq \sup |D^\alpha \rho| \|\mu\|.$$

Moreover, using Jensen's inequality it is easy to check (see for instance Theorem 2.2(ii) of [4]) that

$$\int_C |\mu * \rho| \, dx \leq |\mu|(C_\rho), \quad \text{where} \quad C_\rho := \{x : \, \text{dist}(x, C) \leq \text{diam}(\text{spt}\,\rho)\},$$

(53)

for any closed set $C \subset \mathbb{R}^n$.

• **(Weak convergence)** Assume that X is a compact metric space. We say that a family of measures $(\mu_h) \subset \mathcal{M}(X)$ weakly converges to $\mu \in \mathcal{M}(X)$ if

$$\lim_{h \to \infty} \int_X u \, d\mu_h = \int_X u \, d\mu \quad \forall u \in C(X),$$

i.e., if μ_h weakly* converge to μ as elements of the dual of $C(X)$. In the case X is locally compact and separable, the concept is analogous, simply replacing $C(X)$ the the subspace $C_c(X)$ of functions with compact support. It is easy to check that $\mu * \rho_\varepsilon \mathcal{L}^n$ weakly converge to μ in $\mathbb{R}^n$ whenever ρ is a convolution kernel.

We mention also the following criterion for the weak convergence of positive measures (see for instance Proposition 1.80 of [4]): if $\mu_h, \mu \in \mathcal{M}_+(X)$ satisfy

$$\liminf_{h \to \infty} \mu_h(A) \geq \mu(A) \quad \forall A \subset X \text{ open}$$

and

$$\limsup_{h \to \infty} \mu_h(X) \leq \mu(X),$$

then $\int_X \phi \, d\mu_h \to \int_X \phi \, d\mu$ for any bounded continuous function $\phi : X \to \mathbb{R}$.

• **(Measure valued maps)** Let X, Y be locally compact and separable metric spaces. Let $y \mapsto \lambda_y$ be a map which assigns to any $y \in Y$ a measure $\lambda_y \in [\mathcal{M}(X)]^m$. We say that λ_y is a Borel map if $y \mapsto \lambda_y(A)$ is a real valued Borel map for any open set $A \subset X$. By a monotone class argument it can be proved that $y \mapsto \lambda_y$ is a Borel map if and only if

$$y \mapsto \lambda_y(\{x : \, (y, x) \in B\}) \quad \text{is a Borel map for any } B \in \mathcal{B}(Y \times X). \quad (54)$$

Moreover $y \mapsto |\lambda_y|$ is a Borel map whenever $y \mapsto \lambda_y$ is Borel (detailed proofs are in §2.5 of [4]). If $\mu \in \mathcal{M}_+(Y)$, analogous statements hold for $\mathcal{B}(Y)_\mu$-measurable measure valued maps, where $\mathcal{B}(Y)_\mu$ is the σ-algebra of μ-measurable sets (in this case one has to replace $\mathcal{B}(Y \times X)$ by $\mathcal{B}(Y)_\mu \otimes \mathcal{B}(X)$ in (54)).

• **(Decomposition of measures)** The following result plays a fundamental role in these notes; it is also known as disintegration theorem.

Theorem 9.1 (Decomposition of measures). *Let X, Y be locally compact and separable metric spaces and let $\pi : X \to Y$ be a Borel map. Let $\lambda \in [\mathcal{M}(X)]^m$ and set $\mu = \pi_\#|\lambda| \in \mathcal{M}_+(Y)$. Then there exist measures $\lambda_y \in [\mathcal{M}(X)]^m$ such that*

(i) $y \mapsto \lambda_y$ is a Borel map and $|\lambda_y|$ is a probability measure in X for μ-a.e. $y \in Y$;

(ii) $\lambda = \lambda_y \otimes \mu$, i.e.

$$\lambda(A) = \int_Y \lambda_y(A)\, d\mu(y) \qquad \forall A \in \mathcal{B}(X); \tag{55}$$

(iii) $|\lambda_y|\left(X \setminus \pi^{-1}(y)\right) = 0$ for μ-a.e. $y \in Y$.

The representation provided by Theorem 9.1 of λ can be used sometimes to compute the push forward of λ. Indeed,

$$f_\#(\lambda_y \otimes \mu) = f_\# \lambda_y \otimes \mu \tag{56}$$

for any Borel map $f : X \to Z$, where Z is any other compact metric space. Notice also that

$$|\lambda| = |\lambda_y| \otimes \mu. \tag{57}$$

Indeed, the inequality "$\leq$" is trivial and the opposite one follows by evaluating both measures at $B = X$, using the fact that $|\lambda_y|(X) = 1$ for μ-a.e. y.

In the case when $m = 1$ and $\lambda \in \mathcal{M}_+(X)$, the proof of Theorem 9.1 is available in many textbooks of measure theory or probability (in this case λ_y are the the so-called conditional probabilities induced by the random variable π, see for instance [18]); in the vector valued case one can argue component by component, but the fact that $|\lambda_y|$ are probability measures is not straightforward.

In Theorem 2.28 of [4] the decomposition theorem is proved in the case when $X = Y \times Z$ is a product space and $\pi(y, z) = y$ is the projection on the first variable; in this situation, since λ_y are concentrated on $\pi^{-1}(y) = \{y\} \times Z$, it is sometimes convenient to consider them as measures on Z, rather than measures on X, writing (55) in the form

$$\lambda(B) = \int_Y \lambda_y\left(\{z : (y, z) \in B\}\right) d\mu(y) \qquad \forall B \in \mathcal{B}(X). \tag{58}$$

Once the decomposition theorem is known in the special case $X = Y \times Z$ and $\pi(y, z) = z$ the general case can be easily recovered: it suffices to embed X into the product $Y \times X$ through the map $f(x) = (\pi(x), x)$ and to apply the decomposition theorem to $\tilde{\lambda} = f_\# \lambda$.

Now we discuss the uniqueness of λ_x and μ in the representation $\lambda = \lambda_x \otimes \mu$. For simplicity we discuss only the case of positive measures.

Theorem 9.2. *Let X, Y and π be as in Theorem 9.1; let $\lambda \in \mathcal{M}_+(X)$, $\mu \in \mathcal{M}_+(Y)$ and let $y \mapsto \eta_y$ be a Borel $\mathcal{M}_+(X)$-valued map defined on Y such that*

(i) $\lambda = \eta_y \otimes \mu$, i.e. $\lambda(A) = \int_Y \eta_y(A) \, d\mu(y)$ for any $A \in \mathcal{B}(X)$;
(ii) $\eta_y(X \setminus \pi^{-1}(y)) = 0$ for μ-a.e. $y \in Y$.

Then η_y are uniquely determined μ-a.e. in Y by (i) and (ii), and, setting $B = \{y : \eta_y(X) > 0\}$, the measure $\mu \llcorner B$ is absolutely continuous with respect to $\pi_\# \lambda$. In particular

$$\frac{\mu \llcorner B}{\pi_\# \lambda} \eta_y = \lambda_y \qquad \text{for } \pi_\# \lambda\text{-a.e. } y \in Y, \tag{59}$$

where λ_y are as in Theorem 9.1.

Proof. Let η_y, η_y' be satisfying (i) and (ii). We have to show that $\eta_y = \eta_y'$ for μ-a.e. y. Let (A_n) be a sequence of open sets stable by finite intersection which generates the Borel σ-algebra of X. Choosing $A = A_n \cap \pi^{-1}(B)$, with $B \in \mathcal{B}(Y)$, in (i) gives

$$\int_B \eta_y(A_n) \, d\mu(y) = \int_B \eta_y'(A_n) \, d\mu(y).$$

Being B arbitrary, we infer that $\eta_y(A_n) = \eta_y'(A_n)$ for μ-a.e. y, and therefore there exists a μ-negligible set N such that $\eta_y(A_n) = \eta_y'(A_n)$ for any $n \in \mathbb{N}$ and any $y \in Y \setminus N$. By Proposition 1.8 of [4] we obtain that $\eta_y = \eta_y'$ for any $y \in Y \setminus N$.

Let $B' \subset B$ be any $\pi_\# \lambda$-negligible set; then $\pi^{-1}(B')$ is λ-negligible and therefore (ii) gives

$$0 = \int_Y \eta_y \left(\pi^{-1}(B') \right) d\mu(y) = \int_{B'} \eta_y(X) \, d\mu(y).$$

As $\eta_y(X) > 0$ on $B \supset B'$, this implies that $\mu(B') = 0$. Writing $\mu \llcorner B = h\pi_\# \lambda$, we obtain $\lambda = h\eta_y \otimes \pi_\# \lambda$ and $\lambda = \lambda_y \otimes \pi_\# \lambda$. As a consequence, (59) holds.

• **(Young measures)** These measures, introduced by L.C.Young, arise in a natural way in the study of oscillatory phenomena and in the analysis of weak limits of nonlinear quantities (see [31] for a comprehensive introduction to this wide topic).

Specifically, assume that we are given compact metric spaces X, Y and a sequence of Borel maps $\psi_h : X \to Y$; in order to understand the limit behaviour of ψ_h, we associate to them the measures

$$\gamma_{\psi_h} = (Id \times \psi_h)_\# \mu = \int \delta_{\psi_h(x)} \, d\mu(x)$$

and we study their limit in $\mathcal{M}(X \times Y)$. Assuming, possibly for a subsequence, that $\gamma_{\psi_h} \to \gamma$, due to the fact that $\pi_{0\#}(\gamma_{\psi_h}) = \mu$ for any h, we obtain that $\pi_{0\#}\gamma = \mu$; hence, according to Theorem 9.1, we can represent γ as

$$\gamma = \gamma_x \otimes \mu,$$

for suitable probability measures γ_x in Y, with $x \mapsto \gamma_x$ Borel. The family of measures γ_x is called *Young limit* of the sequence (ψ_h); once the Young limit is known, we can compute the w^*-limit of $\gamma(\psi_h)$ in $L^\infty(X, \mu)$ for *any* $\gamma \in C(Y)$: indeed, using test function of the form $\phi(x)\gamma(y)$, we easily obtain that the limit (in the dual of $C(X)$ and therefore in $L^\infty(X, \mu)$) is given by

$$L_\gamma(x) := \int_Y \gamma(y) \, d\gamma_x(y).$$

We will use the following two well known results of the theory of Young measures.

Theorem 9.3 (Approximation theorem). *Let $\gamma \in \mathcal{M}_+(X \times Y)$ and write $\gamma = \gamma_x \otimes \mu$ with $\mu = \pi_{0\#}\gamma$. Then, if μ has no atom, there exists a sequence of Borel maps $\psi_h : X \to Y$ such that*

$$\gamma_x \otimes \mu = \lim_{h \to \infty} \delta_{\psi_h(x)} \otimes \mu.$$

Moreover, we can choose ψ_h in such a way that the measures $\psi_{h\#}\mu$ have no atom as well.

Proof. (Sketch) We assume first that $\gamma_x = \gamma$ is independent of x. By approximation we can also assume that

$$\gamma = \sum_{i=1}^p p_i \delta_{y_i},$$

for suitable $y_i \in Y$ and $p_i \in [0, 1]$. Let Q_h, $h \in \mathbb{Z}^n$, be a partition of $\mathbb{R}^n$ in cubes with side length $1/h$; since μ has no atom, by Lyapunov theorem we can find a partition $X_h^1, \dots, X_h^p$ of $X \cap Q_h$ such that $\mu(X_h^i) = p_i \mu(X \cap Q_h)$. Then we define

$$\psi_h = p_i \quad \text{on } X_h^i.$$

For any $\phi \in C(X)$ and $\varphi \in C(Y)$ we have

$$\int_{X \times Y} \phi\varphi \, d(\delta_{\psi_h} \otimes \mu) = \sum_{h \in \mathbb{Z}^n} \sum_{i=1}^p \varphi(y_i) \int_{X_h^i} \phi \, d\mu$$

$$\sim \sum_{h \in \mathbb{Z}^n} \sum_{i=1}^p p_i \varphi(y_i) \int_{X \cap Q_h} \phi \, d\mu = \int_{X \times Y} \phi\varphi \, d\mu \times \gamma$$

and this proves that

$$\mu \times \gamma = \lim_{h \to \infty} \delta_{\psi_h} \otimes \mu.$$

If we want $\psi_{h\#}\mu$ to be non atomic, the above construction needs to be modified only slightly: it suffices to take small balls B_i centered at y_i and to define ψ_h equal to ϕ_i on X_h^i, where $\phi_i : \mathbb{R}^n \to B_i$ is any Borel and one to one map.

If γ_x is piecewise constant (say in a canonical subdivision of X induced by a partition in cubes), then we can repeat the local construction above in each region where γ_x is constant. Moreover we can approximate Lipschitz functions γ_x (with respect to the 1-Wasserstein distance) with piecewise constant ones. Finally, any Borel map γ_x can be approximated by Lipschitz ones through a convolution.

We will also need the following result.

Lemma 9.1. *Let $\psi_h, \psi : X \to Y$ be Borel maps and $\mu \in \mathcal{M}_1(X)$. Then $\psi_h \to \psi$ μ-a.e. if and only if*

$$\gamma_h := \delta_{\psi_h(x)} \otimes \mu(x) \to \gamma := \delta_{\psi(x)} \otimes \mu(x)$$

weakly in $\mathcal{M}(X \times Y)$.

Proof. Assume that $\psi_h \to \psi$ μ-a.e. Since

$$\int_{X \times Y} \varphi \, d\gamma_\varphi = \int_X \varphi(x, \phi(x)) \, d\mu(x) \qquad \forall \varphi \in C(X \times Y),$$

the dominated convergence theorem gives that $\gamma_h \to \gamma$. To prove the opposite implication, fix $\varepsilon > 0$ and a compact set K such that $\psi|_K$ is continuous and $\mu(X \setminus K) < \varepsilon$. We use as test function

$$\varphi(x, y) = \chi_K(x)\gamma(y - \psi(x)),$$

with $\gamma(t) = 1 \wedge |t|/\varepsilon$ (by approximation, although not continuous in $X \times Y$, this is an admissible test function) to obtain

$$\lim_{h \to \infty} \mu(\{x \in K : |\psi_h(x) - \psi(x)| \geq \varepsilon\}) = 0.$$

The conclusion follows letting $\varepsilon \to 0^+$.

With a similar proof one can obtain a slightly more general result, namely

$$\gamma_h := \delta_{\psi_h(x)} \otimes \mu_h \to \gamma := \delta_{\psi(x)} \otimes \mu$$

implies $\psi_h \to \psi$ μ-a.e., provided $|\mu_h - \mu|(X) \to 0$.

References

[1]   G.ALBERTI: *On the structure of the singular set of convex functions*. Calc.
      Var. Partial Differential Equations, **2** (1994), 17–27.
[2]   G.ALBERTI & L.AMBROSIO: *A geometric approach to monotone functions in*
      $\mathbb{R}^n$. Math. Z., **230** (1999), 259–316.
[3]   G.ALBERTI, B.KIRCHHEIM & D.PREISS: Personal communication.
[4]   L.AMBROSIO, N.FUSCO & D.PALLARA: *Functions of Bounded Variation and*
      Free Discontinuity Problems. Oxford University Press, 2000.
[5]   L.AMBROSIO & B.KIRCHHEIM: *Rectifiable sets in metric and Banach spaces*.
      Math. Ann., **318** (2000), 527–555.
[6]   L.AMBROSIO & B.KIRCHHEIM: *Currents in metric spaces*. Acta Math., **185**
      (2000), 1–80.
[7]   J.BENAMOU & Y.BRENIER: *A computational fluid mechanics solution to the*
      Monge–Kantorovich mass transfer problem. Numerische Math., **84** (2000),
      375–393.
[8]   Y.BRENIER: *A homogenized model for vortex sheets*. Arch. Rational Mech.
      Anal., **138** (1977), 319–353.
[9]   Y. BRENIER: *Décomposition polaire et réarrangement monotone des champs*
      de vecteurs. Comm. Pure Appl. Math., **44** (1991), 375–417.
[10] Y.BRENIER: *Polar factorization and monotone rearrangement of vector-valued*
      functions. Arch. Rational Mech. Anal., **122** (1993), 323–351.
[11] Y.BRENIER: *A Monge–Kantorovich approach to the Maxwell equations*. In
      preparation.
[12] G.BOUCHITTÉ, G.BUTTAZZO & P.SEPPECHER: *Energies with respect to a*
      measure and applications to low dimensional structures. Calc. Var. Partial
      Differential Equations, **5** (1997), 37–54.
[13] G.BOUCHITTÉ, G.BUTTAZZO & P.SEPPECHER: *Shape optimization via Mon-*
      ge–Kantorovich equation. C. R. Acad. Sci. Paris Sér. I Math., **324** (1997),
      1185–1191.
[14] G.BOUCHITTÉ & G.BUTTAZZO: *Characterization of optimal shapes and*
      masses through Monge–Kantorovich equation. J. Eur. Math. Soc. (JEMS),
      3 (2001), 139–168.
[15] L.CAFFARELLI, M.FELDMAN & R.J.MCCANN: *Constructing optimal maps for*
      Monge's transport problem as a limit of strictly convex costs. J. Amer. Math.
      Soc., **15** (2002), 1–26.
[16] C.CASTAING & M.VALADIER: *Convex analysis and measurable multifunctions*.
      Lecture Notes in Mathematics **580**, Springer, 1977.
[17] A.CELLINA & S.PERROTTA: *On the validity of the maximum principle and of*
      the Euler–Lagrange equation for a minimum problem depending on the gradi-
      ent. SIAM J. Control Optim., **36** (1998), 1987–1998.
[18] C.DELLACHERIE & P.MEYER: *Probabilities and potential*. Mathematical Stud-
      ies, **29**, North Holland, 1978.
[19] L.DE PASCALE & A.PRATELLI: *Regularity properties for Monge transport den-*
      sity and for solutions of some shape optimizazion problem. Calc. Var. Partial
      Differential Equations, **14** (2002), 249–274.
[20] L.C.EVANS: *Partial Differential Equations*. Graduate Studies in Mathematics,
      19, AMS, 1998.

[21] L.C.EVANS: *Partial Differential Equations and Monge-Kantorovich Mass Transfer.* Current developments in mathematics, 65–126, Int. Press. Boston, MA, 1999.

[22] L.C.EVANS & W.GANGBO: *Differential Equation Methods for the Monge-Kantorovich Mass Transfer Problem.* Mem. Amer. Math. Soc., **137** (1999), no. 653.

[23] K.J.FALCONER: *The geometry of fractal sets.* Cambridge University Press, 1985.

[24] H.FEDERER: *Geometric measure theory.* Springer, 1969.

[25] M.FELDMAN & R.J.MCCANN: *Uniqueness and transport density in Monge's mass transport problem.* Forthcoming.

[26] W.GANGBO & R.J.MCCANN: *The geometry of optimal transportation.* Acta Math., **177** (1996), 113–161.

[27] R.JERRARD & H.M.SONER: *Functions of bounded higher variation.* Indiana Univ. Math. J., **51** (2002), 645–677.

[28] D.G.LARMAN: *A compact set of disjoint line segments in* $\mathbf{R}^3$ *whose end set has positive measure.* Mathematika, **18** (1971), 112–125.

[29] R.JORDAN, D.KINDERLEHRER & F.OTTO: *The variational formulation of the Fokker–Planck equation.* SIAM J. Math. Anal., **29** (1998), 1–17.

[30] F.MORGAN: *Geometric measure theory – A beginner's guide.* Second edition, Academic Press, 1995.

[31] S.MÜLLER: *Variational methods for microstructure and phase transitions.* Proc. CIME summer school "Calculus of variations and geometric evolution problems", Cetraro (Italy), 1996, S.Hildebrandt and M.Struwe eds, 85–210, Lecture Notes in Math., 1713, Springer, Berlin, 1999.

[32] F.OTTO: *The geometry of dissipative evolution equations: the porous medium equation.*, Comm. Partial Differential Equations, **26** (2001), 101–174.

[33] F.POUPAUD & M.RASCLE: *Measure solutions to the linear multi-dimensional transport equation with nonsmooth coefficients.* Comm. Partial Differential Equations, **22** (1997), 301–324.

[34] A.PRATELLI: *Regolarità delle misure ottimali nel problema di trasporto di Monge–Kantorovich.* Thesis, Pisa University, 2000.

[35] S.T.RACHEV & L.RÜSCHENDORF: *Mass transportation problems.* Probability and its applications, Springer, 1998.

[36] L.RÜSCHENDORF: *On c-optimal random variables.* Statistics & Probability Letters, **27** (1996), 267–270.

[37] W.RUDIN: *Real and Complex Analysis.* McGraw–Hill, New York, 1974.

[38] L.SIMON: *Lectures on geometric measure theory.* Proc. Centre for Math. Anal., Australian Nat. Univ., **3**, 1983.

[39] S.K.SMIRNOV: *Decomposition of solenoidal vector charges into elementary solenoids and the structure of normal one-dimensional currents.* St. Petersburg Math. J., **5** (1994), 841–867.

[40] C.SMITH & M.KNOTT: *On the optimal transportation of distributions.* J. Optim. Theory Appl., **52** (1987), 323–329.

[41] C.SMITH & M.KNOTT: *On Hoeffding–Fréchet bounds and cyclic monotone relations.* J. Multivariate Anal., **40** (1992), 328–334.

[42] V.N.SUDAKOV: *Geometric problems in the theory of infinite dimensional distributions.* Proc. Steklov Inst. Math., **141** (1979), 1–178.

[43] N.S.TRUDINGER & X.J.WANG: *On the Monge mass transfer problem.* Calc. Var. Partial Differential Equations, **13** (2001), 19–31.

[44] B.WHITE: *Rectifiability of flat chains*. Ann. of Math. (2), **150** (1999), 165–184.
[45] L.ZAJÍČEK: *On the differentiation of convex functions in finite and infinite dimensional spaces*. Czechoslovak Math. J., **29 (104)** (1979), 340–348.

Numerical Approximation of Mean Curvature Flow of Graphs and Level Sets

Klaus Deckelnick[1] and Gerhard Dziuk[2]

[1] Institut für Analysis und Numerik, Otto-von-Guericke-Universität Magdeburg,
Universitätsplatz 2, Postfach 4120, D-39106 Magdeburg, Germany
Klaus.Deckelnick@mathematik.uni-magdeburg.de
[2] Institut für Angewandte Mathematik, Universität Freiburg,
Hermann-Herder-Straße 10, D-79104 Freiburg, Germany
gerd@mathematik.uni-freiburg.de

Summary. The mean curvature flow problem for graphs is closely related to the mean curvature of level sets. We design algorithms for both problems in the context of finite elements together with a semi implicit time discretization. The time discretization linearises the problem and is to the effect that in every time step a linear system of equations has to be solved numerically. We prove convergence estimates of optimal order in space and time for the geometric quantities. The results are true in any space dimension. We also provide numerical tests.

1 Introduction

The subject of this paper is the numerical solution of the mean curvature flow problem for graphs and for level sets. Mean curvature flow means the motion of a surface or an interface with a velocity, which is its mean curvature. In short, this problem can be written as

$$V = -H.$$

V is the normal velocity and H is the mean curvature of a surface. The normalisation is chosen in such a way that the normal velocity of a sphere is positive and spheres shrink under this law of motion. For graphs the problem can be written as a differential equation for the height function. The resulting differential equation

$$\frac{1}{\sqrt{1 + |\nabla u|^2}} \frac{\partial u}{\partial t} - \sum_{i=1}^{n} \frac{\partial}{\partial x_i} \left(\frac{1}{\sqrt{1 + |\nabla u|^2}} \frac{\partial u}{\partial x_i} \right) = 0$$

has some unpleasant properties, especially when one is interested in computational methods for its solution. It is highly nonlinear, not uniformly parabolic and it is not in divergence form. Nevertheless there is a quite complete analysis for this problem, some of which will be presented in Section 2.2. The analysis uses the maximum principle and is therefore based on the L^∞-norm.

We are interested in deriving numerical schemes for this differential equation and in proving convergence of the discrete solutions to the continuous solution under realistic assumptions. As a numerical method we choose finite elements because of their efficiency and accessibility to modern adaptive methods. The finite element method by its nature is a method for reflexive spaces, in particular for spaces which are based on L^2. Although the equation is not in divergence form we shall derive a finite element method for the discretization with respect to space. The time discretization is such that it linearizes the equations. At the end, in every time step a linear system of equations has to be solved. This can be done efficiently by some conjugate gradient algorithm. The numerical schemes will be such that they can be incorporated easily into existing finite element codes for the solution of linear parabolic equations.

The error estimates which we shall prove are of a form, which is adapted to the geometric origin of the equation. We shall prove that normal velocity and normal vector converge of optimal order

$$\|V - V_{h\tau}\|_{L^2(0,T;L^2(\Gamma_{h\tau}))} + \|\nu - \nu_{h\tau}\|_{L^\infty(0,T;L^2(\Gamma_{h\tau}))} \leq c(\tau + h).$$

Here $V_{h\tau}$ is the normal velocity of the discrete surface $\Gamma_{h\tau}$ and $\nu_{h\tau}$ is its normal. The index h stands for the spatial grid size and τ stands for the time step size. A quite astonishing result is that we do not have any coupling of τ and h although our time discretization is not implicit, but semi-implicit only.

The level set form of mean curvature flow,

$$\frac{1}{|\nabla u|} \frac{\partial u}{\partial t} - \sum_{i=1}^{n} \frac{\partial}{\partial x_i}\left(\frac{1}{|\nabla u|} \frac{\partial u}{\partial x_i}\right) = 0,$$

for the scalar level function $u = u(x,t)$ is closely related to the the mean curvature flow problem for graphs via a regularization. By a simple scaling trick the algorithms and convergence results for graphs can be transformed to the level set problem. We prove convergence of the discrete solution of the regularised problem to the viscosity solution of the level set flow. The optimal coupling of discretization parameters τ, h and regularization parameter is an open question. Besides the general convergence result for viscosity solutions we can give numerical examples which suggest that there might be a nontrivial coupling condition.

The results for mean curvature flow can be generalized to anisotropic mean curvature flow, see [6].

Although we shall only treat the case of piecewise linear finite elements a generalization to higher order C^0-elements is straightforward. It creates technical difficulties at the boundary and higher regularity assumptions on the continuous solution.

Fried [14] used the level set finite element method for the computation of dendritic growth. Further information about numerical level set methods are contained in [18], [19] and [21].

Notations $H^{m,p}(\Omega)$ denotes the usual Sobolev space. The corresponding norm is given by

$$\|u\|_{H^{m,p}(\Omega)} = \left(\sum_{k=0}^{m} \|D^k u\|_{L^p(\Omega)}^p \right)^{\frac{1}{p}},$$

with the usual modification for $p = \infty$. For $p = 2$ we simply write $H^m(\Omega) = H^{m,2}(\Omega)$ with norm $\|\cdot\|_{H^m(\Omega)}$; furthermore we use $\|\cdot\|$ to denote the L^2-norm. We shall often use, if not otherwise indicated,

$$v^m(x) = v(x, m\tau)$$

as an abbreviation for the evaluation of a function $v = v(x, t)$ on the m-th time level $t = m\tau$ for given time step size $\tau > 0$.

2 Mean Curvature Flow of Graphs

2.1 The Differential Equation

We study the numerical computation of the isotropic evolution of n-dimensional surfaces which are graphs in $\mathbb{R}^{n+1}$. The law of motion is given by

$$V = -H. \tag{1}$$

Here V represents the normal velocity and H is the mean curvature of the graph

$$\Gamma(t) = \{(x, u(x, t)) \mid x \in \Omega\}$$

over a bounded domain $\Omega \subset \mathbb{R}^n$. The normal to Γ is chosen to be

$$\nu(u) = \frac{(\nabla u, -1)}{Q(u)}, \quad Q(u) = \sqrt{1 + |\nabla u|^2}.$$

The normal velocity of Γ then can be written as

$$V(u) = -\frac{u_t}{Q(u)}. \tag{2}$$

The mean curvature of Γ is given by

$$H(u) = \nabla \cdot \frac{\nabla u}{\sqrt{1 + |\nabla u|^2}}.$$

Equation (1) then leads to the differential equation

$$\frac{u_t}{\sqrt{1 + |\nabla u|^2}} - \nabla \cdot \frac{\nabla u}{\sqrt{1 + |\nabla u|^2}} = 0. \tag{3}$$

We shall solve the initial boundary value problem

$$u_t - \sqrt{1 + |\nabla u|^2} \, \nabla \cdot \frac{\nabla u}{\sqrt{1 + |\nabla u|^2}} = 0 \text{ in } \Omega \times (0, T)$$

$$u = u_0 \text{ on } \partial\Omega \times (0, T) \cup \Omega \times \{0\}. \tag{4}$$

The boundary data are chosen independently of time. A generalization to time dependent boundary values is possible and adds some technical problems only. In more explicit form the differential equation reads

$$u_t - \sum_{i,j=1}^{n} \left(\delta_{ij} - \frac{u_{x_i} u_{x_j}}{1 + |\nabla u|^2} \right) u_{x_i x_j} = 0. \tag{5}$$

The equation is parabolic but uniformly parabolic only if $|\nabla u|$ is uniformly bounded. With

$$a_{ij} = \delta_{ij} - \frac{u_{x_i} u_{x_j}}{1 + |\nabla u|^2}$$

one has, for arbitrary $\xi \in \mathbb{R}^n$,

$$\sum_{i,j=1}^{n} a_{ij} \xi_i \xi_j = |\xi|^2 - \frac{(\nabla u \cdot \xi)^2}{1 + |\nabla u|^2} \geq |\xi|^2 \left(1 - \frac{|\nabla u|^2}{1 + |\nabla u|^2} \right) \geq |\xi|^2 \frac{1}{1 + |\nabla u|^2}.$$

We also observe that the equation is *not* in divergence form.

2.2 Some analysis for the problem

Analytical results for the problem 4 are due to Huisken [15] and Lieberman [16].

Theorem 2.1. *Let u_0 be from $C^{2,\alpha}(\overline{\Omega})$ and let the domain Ω be mean convex with $\partial\Omega \in C^{2,\alpha}$. Assume that the function u_0 satisfies the compatibility condition $H(u_0) = 0$ on $\partial\Omega$. Then there exists a unique solution $u \in C^{2+\alpha, 1+\frac{\alpha}{2}}(\overline{\Omega} \times [0, T])$ of problem (4) for all $T < \infty$.*

The assumption that the boundary of the domain has nonnegative mean curvature is a necessary condition. If it is not satisfied the continuous solution breaks down after finite time, although, as we shall see later, the discrete solution exists for all times. The main tools in the proof of the previous theorem are a boundary gradient estimate and an evolution equation for the area element $Q = Q(u)$. The important fact in (6) below is, that the terms on the right hand side of the differential equation are negative and the maximum principle is available for an estimate of Q.

Lemma 2.1. *Let u be a smooth solution of (4). Then with $Q = \sqrt{1 + |\nabla u|^2}$ and $\nu_i = u_{x_i}/Q$ $(i = 1, \ldots, n)$ the differential equation*

$$Q_t - \sum_{i,j=1}^{n} (\delta_{ij} - \nu_i\nu_j)Q_{x_ix_j}$$

$$= -\frac{1}{Q} \sum_{i,j=1}^{n} (\delta_{ij} - \nu_i\nu_j)Q_{x_i}Q_{x_j} - \frac{1}{Q} \sum_{i,j,m=1}^{n} (\delta_{ij} - \nu_i\nu_j)u_{x_mx_i}u_{x_mx_j} \qquad (6)$$

holds in $\Omega \times (0,T)$.

Later our numerical algorithms will be based on a variational formulation of the problem, i.e. on an $L^2(\Omega)$ formulation. For that purpose integral estimates for the solution will be important. Also we will not have the maximum principle as a tool for our discrete equations and therefore will have to estimate $L^2(\Omega)$-norms.

Lemma 2.2. *For the solution of problem (4) one has the energy equation*

$$\int_\Omega \frac{u_t^2}{Q(u)} + \frac{d}{dt} \int_\Omega Q(u) = 0. \qquad (7)$$

If $u_0 = 0$ on $\partial\Omega$, then

$$\frac{d}{dt} \int_\Omega u^2 Q(u) + 2 \int_\Omega \frac{|\nabla u|^2}{Q(u)} + \int_\Omega \frac{u_t^2 u^2}{Q(u)} = 0. \qquad (8)$$

For later use it is necessary to identify the important geometric quantities in the energy equations. The equations (7) and (8) can be written as

$$\int_\Gamma V(u)^2 + \frac{d}{dt}|\Gamma| = 0,$$

$$\frac{d}{dt} \int_\Gamma u^2 + 2 \int_\Gamma |\underline{\nabla} u|^2 + \int_\Gamma V(u)^2 u^2 = 0,$$

where $\underline{\nabla}$ denotes the tangential gradient on Γ.

Proof (of Lemma 2.2). For the first equation (7) we observe that, since $u_t = 0$ on $\partial\Omega$,

$$\int_\Omega \frac{u_t^2}{Q(u)} = \int_\Omega u_t \nabla \cdot \frac{\nabla u}{Q(u)} = -\int_\Omega \nabla u_t \cdot \frac{\nabla u}{Q(u)} = -\frac{d}{dt} \int_\Omega Q(u).$$

The second equation uses the differential equation twice.

$$\frac{1}{2}\frac{d}{dt}\int_\Omega u^2 Q(u) = \int_\Omega u u_t Q(u) + \frac{1}{2}\int_\Omega u^2 \frac{\nabla u}{Q(u)}\cdot\nabla u_t$$

$$= \int_\Omega u u_t Q(u) + \frac{1}{2}\int_\Omega \frac{\nabla u}{Q(u)}\cdot\nabla(u^2 u_t) - \int_\Omega u u_t \frac{|\nabla u|^2}{Q(u)}$$

$$= \int_\Omega \frac{u u_t}{Q(u)} + \frac{1}{2}\int_\Omega \frac{\nabla u}{Q(u)}\cdot\nabla(u^2 u_t)$$

$$= \int_\Omega u\nabla\cdot\frac{\nabla u}{Q(u)} - \frac{1}{2}\int_\Omega \nabla\cdot\frac{\nabla u}{Q(u)}u^2 u_t$$

$$= -\int_\Omega \frac{|\nabla u|^2}{Q(u)} - \frac{1}{2}\int_\Omega \frac{u^2 u_t^2}{Q(u)}.$$

This is the second energy equation (8).

Our error analysis for the numerical approximation of (4) will be carried out under the assumption that (4) has a solution u which satisfies

$$N := \sup_{(0,T)} \|u\|_{H^{2,\infty}(\Omega)} + \sup_{(0,T)} \|u_t\|_{H^{1,\infty}(\Omega)}$$

$$+ \int_0^T \|u_t\|_{H^2(\Omega)}^2 + \int_0^T \|u_{tt}\|^2 < \infty. \tag{9}$$

An existence theorem in Sobolev spaces is provided in [3].

2.3 Discretization

Discretization in Space Let $\mathcal{T}_h$ be a family of triangulations of Ω with maximum mesh size $h := \max_{S\in\mathcal{T}_h} \operatorname{diam}(S)$. We denote by Ω_h the corresponding discrete domain, i.e.

$$\bar{\Omega}_h = \bigcup_{S\in\mathcal{T}_h} S$$

and assume that all vertices on $\partial\Omega_h$ also lie on $\partial\Omega$. Furthermore we suppose that the triangulation is nondegenerate in the sense that

$$\max_{S\in\mathcal{T}_h} \frac{\operatorname{diam}(S)}{\rho_S} \leq \kappa,$$

where the constant $\kappa > 0$ is independent of h and ρ_S denotes the radius of the largest ball which is contained in the simplex $\bar{S}$. The discrete space is defined by

$$X_h := \{v_h \in C^0(\bar{\Omega}_h)\,|\,v_h \text{ is a linear polynomial on each } S \in \mathcal{T}_h\}$$

and $X_{h0} := X_h \cap H_0^1(\Omega_h)$.

There exists an interpolation operator $I_h : H^2(\Omega_h) \to X_h$, mapping $H^2(\Omega_h) \cap H_0^1(\Omega)$ into X_{h0}, such that

$$\|v - I_h v\| + h\|\nabla(v - I_h v)\| \le ch^2\|v\|_{H^2(\Omega_h)} \tag{10}$$

for all $v \in H^2(\Omega_h)$.

Although the differential equation is not in divergence form, one can easily derive a variational form of (4), namely

$$\int_\Omega \frac{u_t \phi}{Q(u)} + \int_\Omega \frac{\nabla u \cdot \nabla \phi}{Q(u)} = 0, \qquad \phi \in H_0^1(\Omega), 0 < t < T. \tag{11}$$

The semi–discrete approximation of (11) then is given by the following Algorithm.

Algorithm 1 Let $u_h^0 = I_h u_0$. For $t \in [0,T)$ find $u_h(\cdot,t) \in X_h$, with $u_h(\cdot,t) - I_h u_0 \in X_{h0}$ and $u_h(\cdot,0) = u_{h0} = I_h u_0$, such that

$$\int_{\Omega_h} \frac{u_{ht}\phi_h}{Q(u_h)} + \int_{\Omega_h} \frac{\nabla u_h \cdot \nabla \phi_h}{Q(u_h)} = 0 \tag{12}$$

holds for all $t \in (0,T)$ and all discrete test functions $\phi_h \in X_{h0}$.

Fully discrete schemes and stability The derivation of fully discrete schemes for the mean curvature flow problem is an important task for applications. We have to prove stability and convergence and have to show the efficiency of the algorithm. In the following we list three different discretizations of (12) with respect to time. We use the notation

$$v^m(x) = v(x, m\tau)$$

for $m = 0, \ldots, M$, with time step $\tau > 0$ and $M \le [T/\tau]$.

A *fully implicit* scheme for (4) is given by

$$\frac{1}{\tau} \int_{\Omega_h} \frac{(u_h^{m+1} - u_h^m)\phi_h}{Q(u_h^{m+1})} + \int_{\Omega_h} \frac{\nabla u_h^{m+1} \cdot \nabla \phi_h}{Q(u_h^{m+1})} = 0$$

$(m = 0, \ldots, M-1)$ for all $\phi_h \in X_{h0}$. So, in every time step one has to solve a highly nonlinear elliptic problem.

The numerically simplest time discretization is the *explicit* one,

$$\frac{1}{\tau} \int_{\Omega_h} \frac{(u_h^{m+1} - u_h^m)\phi_h}{Q(u_h^m)} + \int_{\Omega_h} \frac{\nabla u_h^m \cdot \nabla \phi_h}{Q(u_h^m)} = 0$$

$(m = 0, \ldots, M-1)$, for all $\phi_h \in X_{h0}$, for which numerically one only has to invert the mass matrix (weighted by $Q(u_h^m)$). If one uses mass lumping, every time step only requires one matrix multiplication. However, just as for the linear heat equation, a severe restriction on the smallness of the time step size with respect to the grid size h will be necessary. It then certainly is worth to try some *semi–implicit* scheme for time discretization. We shall discuss the scheme in full detail in Section 2.4.

Algorithm 2 *Let $u_h^0 = I_h u_0$. For $m = 0, \ldots M - 1$ compute $u_h^{m+1} \in X_h$ such that $u_h - u_h^0 \in X_{h0}$ and for every $\phi_h \in X_{h0}$*

$$\frac{1}{\tau} \int_{\Omega_h} \frac{u_h^{m+1} \phi_h}{Q(u_h^m)} + \int_{\Omega_h} \frac{\nabla u_h^{m+1} \cdot \nabla \phi_h}{Q(u_h^m)} = \frac{1}{\tau} \int_{\Omega_h} \frac{u_h^m \phi_h}{Q(u_h^m)}. \tag{13}$$

Here in every time step one has to solve a discrete linear elliptic differential equation with stiffness matrix weighted by $Q(u_h^m)$.

As a stability criterion we use the basic energy norm introduced in (7). The fact that we have unconditional stability, although the scheme is not fully implicit but semi-implicit in time, is a first hint that we might expect unconditional convergence. See Section 2.4 for the details.

Theorem 2.2. *The solution $u_h^m, 0 \le m \le M$ of (2) satisfies, for every $m \in \{1, \ldots, M\}$,*

$$\tau \sum_{k=0}^{m-1} \int_{\Omega_h} |V_h^k|^2 Q(u_h^k) + \sum_{k=0}^{m-1} \int_{\Omega_h} (Q(u_h^{k+1}) - Q(u_h^k))^2 \frac{1}{Q(u_h^k)}$$

$$+ \frac{1}{2} \sum_{k=0}^{m-1} \int_{\Omega_h} |\nu(u_h^{k+1}) - \nu(u_h^k)|^2 Q(u_h^{k+1}) + \int_{\Omega_h} Q(u_h^m) = \int_{\Omega_h} Q(u_h^0), \tag{14}$$

where

$$V_h^k = -\frac{u_h^{k+1} - u_h^k}{\tau Q(u_h^k)}$$

is the discrete normal velocity.

Proof. We choose $\phi_h = u_h^{k+1} - u_h^k$ as a test function in (2) for $m = k$ and get

$$\frac{1}{\tau} \int_{\Omega_h} \frac{(u_h^{k+1} - u_h^k)^2}{Q(u_h^k)} + \int_{\Omega_h} \frac{\nabla u_h^{k+1} \cdot \nabla(u_h^{k+1} - u_h^k)}{Q(u_h^k)} = 0. \tag{15}$$

Let us use the notation $\underline{\nu}(v_h) = \frac{\nabla v_h}{Q(v_h)}$, so that $\nu(v_h) = (\underline{\nu}(v_h), \frac{-1}{Q(v_h)})$. Then

$$\frac{\nabla u_h^{k+1} \cdot \nabla(u_h^{k+1} - u_h^k)}{Q(u_h^k)} = \frac{Q(u_h^{k+1})^2 - 1}{Q(u_h^k)} - \underline{\nu}(u_h^{k+1}) \cdot \underline{\nu}(u_h^k) Q(u_h^{k+1})$$

$$= \frac{Q(u_h^{k+1})^2}{Q(u_h^k)} + \frac{1}{2} |\nu(u_h^{k+1}) - \nu(u_h^k)|^2 Q(u_h^{k+1}) - Q(u_h^{k+1})$$

$$= \frac{1}{2} |\nu(u_h^{k+1}) - \nu(u_h^k)|^2 Q(u_h^{k+1}) + Q(u_h^{k+1}) - Q(u_h^k) + \frac{(Q(u_h^{k+1}) - Q(u_h^k))^2}{Q(u_h^k)}.$$

We use this result in (15) and sum over $k = 0, \ldots, m - 1$.

The continuous solution of (4) exists globally in time only if the domain Ω is mean convex. In contrast to this, the discrete solution u_h^m exists for $m \in \mathbb{N}$, i.e. for $T = \infty$ without any geometric condition on the domain Ω_h.

Lemma 2.3. *The semi-discrete solution u_h from Algorithm 1 and the discrete solution u_h^m from Algorithm 2 exist for all times.*

Proof. This is a simple consequence of the fact that area is bounded uniformly. From Theorem 2.2 we have that

$$\sup_{m \in \mathbb{N}} \int_{\Omega_h} Q(u_h^m) \le \int_{\Omega_h} Q(u_h^0)$$

and this implies that

$$\sup_{m \in \mathbb{N}} \sup_{\Omega_h} Q(u_h^m) \le ch^{-n} \int_{\Omega_h} Q(u_h^0).$$

Thus the linear system (13) is uniquely solvable for every $m \in \mathbb{N}$. A similar argument holds for the semi-discrete algorithm.

In Figure 1 we show the discrete solution of (4) on a non convex smooth domain in the plane. Local existence of a smooth solution is known. Under certain parameters the gradient blows up near the non convex part of the boundary and forms a "free boundary" there [17]. Since the discrete solution exists for all times, the gradient at the boundary in the numerical results only becomes "nearly vertical".

Convergence of the semi-discrete scheme In [4] we proved convergence for the semi-discrete Finite Element scheme from Algorithm 1, see also [3]. Here we only give the result. The estimates depend on the quantity

$$\sup_{(0,T)} \left(\|u(\cdot, t)\|_{H^{2,\infty}(\Omega)}^2 + \|u_t(\cdot, t)\|_{H^{1,\infty}(\Omega)}^2 \right) + \int_0^T \|\nabla u_t\|_{H^2(\Omega)}^2, \qquad (16)$$

which is finite for the continuous solution under suitable assumptions on the data, see [4]. We assume that we have a solution of this quality up to time T. The error estimate in the next theorem then is valid as long as this solution exists.

Theorem 2.3. *Let u be a solution of the continuous problem (4) with finite norms (16). Then there exists a unique solution $u_h \in C^1((0,T), X_h)$ of the semi-discrete problem (12) and*

$$\int_0^T \int_{\Omega \cap \Omega_h} (V(u) - V(u_h))^2 Q(u_h) + \sup_{(0,T)} \int_{\Omega \cap \Omega_h} |\nu(u) - \nu(u_h)|^2 Q(u_h) \le ch^2.$$

The constant c depends on the norms (16).

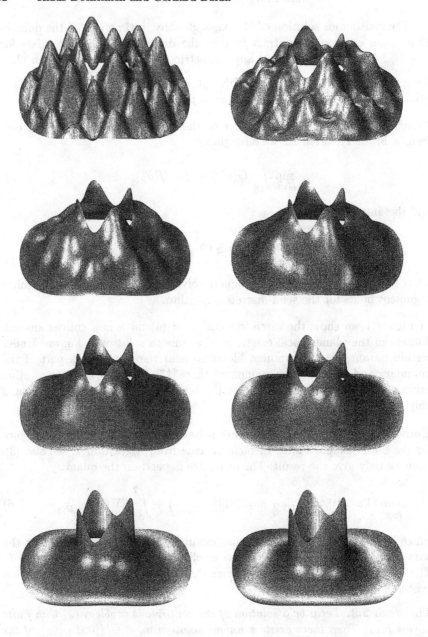

Fig. 1. Evolution of a graph over a non convex domain by mean curvature flow

2.4 Convergence of the fully discrete semi implicit scheme

For domains in the plane and under the condition $\tau \leq \delta_0 h$, for some constant $\delta_0 > 0$, the convergence of this algorithms is proved and optimal error estimates are given in [5]. In this section we are going to give a proof of the unconditional convergence together with optimal error estimates for the geometric quantities V and ν of the fully discrete semi-implicit scheme (12). Thus no coupling of time step size and spatial grid size is necessary although the algorithm is not implicit. We emphasize that the results are true for any space dimension.

Choose on open set $\Omega' \subset \mathbb{R}^n$ which contains $\bar{\Omega} \cup \Omega_h$ for all $h \leq 1$. In view of the regularity (9) of u and since $\partial\Omega$ is smooth, there exists an extension $\bar{u} : \Omega' \times [0, T] \to \mathbb{R}$ such that $\bar{u}_{|\Omega \times [0,T]} = u$ and

$$\sup_{(0,T)} \|\bar{u}\|_{H^{2,\infty}(\Omega')} + \sup_{(0,T)} \|\bar{u}_t\|_{H^{1,\infty}(\Omega')} + \int_0^T \|\bar{u}_t\|_{H^2(\Omega')}^2 + \int_0^T \|\bar{u}_{tt}\|^2 \leq cN,$$
(17)

where N appeared in (9).

Theorem 2.4. *Assume that there exists a solution of (4) on $(0, T) \times \Omega$ and that the norms (9) are finite and let u_h^m, $(m = 1, \ldots, M)$ be the solution of Algorithm 2. Then there exists a $\tau_0 > 0$ such that for all $0 < \tau \leq \tau_0$*

$$\tau \sum_{m=0}^{M} \int_{\Omega \cap \Omega_h} (V(u^m) - V_h^m)^2 Q(u_h^m) \leq c(\tau^2 + h^2),$$
(18)

$$\sup_{m=0,\ldots,M} \int_{\Omega \cap \Omega_h} |\nu(u^m) - \nu(u_h^m)|^2 Q(u_h^m) \leq c(\tau^2 + h^2),$$
(19)

where $M = [\frac{T}{\tau}]$ and

$$V_h^m = -\frac{u_h^{m+1} - u_h^m}{\tau} \frac{1}{Q(u_h^m)}$$

is the discrete normal velocity.

The *proof* of the Theorem will be based on some Lemmas, which we are going to prove now. In the first Lemma we provide the error relation by subtracting suitable forms of the discrete and the continuous equation.

Lemma 2.4. *For every $\phi_h \in X_{h0}$ we have the error relation*

$$\frac{1}{\tau} \int_{\Omega_h} \left(\frac{\bar{u}^{m+1} - \bar{u}^m}{Q(\bar{u}^m)} - \frac{u_h^{m+1} - u_h^m}{Q(u_h^m)} \right) \phi_h + \int_{\Omega_h} \left(\frac{\nabla \bar{u}^{m+1}}{Q(\bar{u}^m)} - \frac{\nabla u_h^{m+1}}{Q(u_h^m)} \right) \cdot \nabla \phi_h$$

$$= \int_{\Omega_h} \frac{\nabla(\bar{u}^{m+1} - \bar{u}^m)}{Q(\bar{u}^m)} \cdot \nabla \phi_h + \int_{\Omega_h} \left(\frac{\bar{u}^{m+1} - \bar{u}^m}{\tau} - \bar{u}_t^m \right) \frac{1}{Q(\bar{u}^m)} \phi_h$$

$$- \int_{\Omega_h \backslash \Omega} (H(\bar{u}^m) + V(\bar{u}^m)) \phi_h.$$
(20)

Proof. For every $\phi \in H_0^1(\Omega_h)$ we have that

$$
\int_{\Omega_h} \frac{\nabla \overline{u}^m}{Q(\overline{u}^m)} \cdot \nabla \phi = - \int_{\Omega_h} \nabla \cdot \frac{\nabla \overline{u}^m}{Q(\overline{u}^m)} \phi
$$

$$
= - \int_{\Omega \cap \Omega_h} \nabla \cdot \frac{\nabla u^m}{Q(u^m)} \phi - \int_{\Omega_h \setminus \Omega} \nabla \cdot \frac{\nabla \overline{u}^m}{Q(\overline{u}^m)} \phi
$$

$$
= - \int_{\Omega \cap \Omega_h} \frac{u_t^m}{Q(u^m)} \phi - \int_{\Omega_h \setminus \Omega} \nabla \cdot \frac{\nabla \overline{u}^m}{Q(\overline{u}^m)} \phi
$$

$$
= - \frac{1}{\tau} \int_{\Omega_h} \frac{\overline{u}^{m+1} - \overline{u}^m}{Q(\overline{u}^m)} \phi + \int_{\Omega_h} \frac{1}{Q(\overline{u}^m)} \left(\frac{\overline{u}^{m+1} - \overline{u}^m}{\tau} - u_t^m \right) \phi
$$

$$
- \int_{\Omega_h \setminus \Omega} \left(\nabla \cdot \frac{\nabla \overline{u}^m}{Q(\overline{u}^m)} - \frac{u_t^m}{Q(\overline{u}^m)} \right) \phi
$$

and this leads to the equation

$$
\frac{1}{\tau} \int_{\Omega_h} \frac{\overline{u}^{m+1} - \overline{u}^m}{Q(\overline{u}^m)} \phi + \int_{\Omega_h} \frac{\nabla \overline{u}^m}{Q(\overline{u}^m)} \cdot \nabla \phi
$$

$$
= \int_{\Omega_h} \frac{1}{Q(\overline{u}^m)} \left(\frac{\overline{u}^{m+1} - \overline{u}^m}{\tau} - u_t^m \right) \phi - \int_{\Omega_h \setminus \Omega} \left(\nabla \cdot \frac{\nabla \overline{u}^m}{Q(\overline{u}^m)} - \frac{u_t^m}{Q(\overline{u}^m)} \right) \phi. \quad (21)
$$

The difference of equations (13) and (21) gives the error relation (20).

The following Lemma contains the key estimate for the proof of Theorem 2.4. With this result it will be possible to estimate the L^2-norm (on the discrete surface) of the error with respect to the normals, $\nu(u(\cdot, \tau m)) - \nu(u_h^m)$, uniformly in time. For this we provide an estimate for the discrete time derivative of this quantity.

Here and in the following we use the abbreviations

$$
Q^k = Q(\overline{u}^k), Q_h^k = Q(u_h^k), \nu^k = \nu(\overline{u}^k), \nu_h^k = \nu(u_h^k), \quad (22)
$$

for $k = m, m + 1$.

Lemma 2.5.

$$
\int_{\Omega_h} \left(\frac{\nabla \overline{u}^m}{Q(\overline{u}^m)} - \frac{\nabla u_h^m}{Q(u_h^m)} \right) \cdot \nabla ((\overline{u}^{m+1} - u_h^{m+1}) - (\overline{u}^m - u_h^m))
$$

$$
= \frac{1}{2} \int_{\Omega_h} |\nu(\overline{u}^{m+1}) - \nu(u_h^{m+1})|^2 Q(u_h^{m+1}) - \frac{1}{2} \int_{\Omega_h} |\nu(\overline{u}^m) - \nu(u_h^m)|^2 Q(u_h^m)
$$

$$
- \frac{1}{2} \int_{\Omega_h} |(\nu(\overline{u}^{m+1}) - \nu(u_h^{m+1})) - (\nu(\overline{u}^m) - \nu(u_h^m))|^2 Q(u_h^{m+1}) + \int_{\Omega_h} R^m,
$$

$$
(23)
$$

with

$$R^m = (\nu(\overline{u}^{m+1}) - \nu(\overline{u}^m)) \cdot (\nu(\overline{u}^m) - \nu(u_h^m))(Q(\overline{u}^{m+1}) - Q(u_h^{m+1}))$$
$$+ \frac{1}{2}|\nu(\overline{u}^m) - \nu(u_h^m)|^2(Q(\overline{u}^{m+1}) - Q(\overline{u}^m)).$$

Proof. We use the relations

$$(\nabla v, -1) = \nu(v)Q(v), \qquad \nu(v) \cdot (\nu(v) - \nu(w)) = \frac{1}{2}|\nu(v) - \nu(w)|^2$$

and obtain

$$\left(\frac{\nabla \overline{u}^m}{Q^m} - \frac{\nabla u_h^m}{Q_h^m}\right) \cdot \nabla((\overline{u}^{m+1} - u_h^{m+1}) - (\overline{u}^m - u_h^m))$$
$$= (\nu^m - \nu_h^m) \cdot ((\nu^{m+1}Q^{m+1} - \nu_h^{m+1}Q_h^{m+1}) - (\nu^m Q^m - \nu_h^m Q_h^m))$$
$$= (\nu^{m+1} - \nu_h^{m+1}) \cdot (\nu^{m+1}Q^{m+1} - \nu_h^{m+1}Q_h^{m+1}) - (\nu^m - \nu_h^m) \cdot (\nu^m Q^m - \nu_h^m Q_h^m)$$
$$- ((\nu^{m+1} - \nu_h^{m+1}) - (\nu^m - \nu_h^m)) \cdot (\nu^{m+1}Q^{m+1} - \nu_h^{m+1}Q_h^{m+1})$$
$$= \frac{1}{2}|\nu^{m+1} - \nu_h^{m+1}|^2(Q^{m+1} + Q_h^{m+1}) - \frac{1}{2}|\nu^m - \nu_h^m|^2(Q^m + Q_h^m)$$
$$- ((\nu^{m+1} - \nu_h^{m+1}) - (\nu^m - \nu_h^m)) \cdot (\nu^{m+1}Q^{m+1} - \nu_h^{m+1}Q_h^{m+1})$$
$$= \frac{1}{2}|\nu^{m+1} - \nu_h^{m+1}|^2 Q_h^{m+1} - \frac{1}{2}|\nu^m - \nu_h^m|^2 Q_h^m + \tilde{R}^m, \qquad (24)$$

where

$$\tilde{R}^m = \frac{1}{2}|\nu^{m+1} - \nu_h^{m+1}|^2 Q^{m+1} - \frac{1}{2}|\nu^m - \nu_h^m|^2 Q^m$$
$$- ((\nu^{m+1} - \nu_h^{m+1}) - (\nu^m - \nu_h^m)) \cdot (\nu^{m+1}Q^{m+1} - \nu_h^{m+1}Q_h^{m+1}).$$

We rewrite this term as follows

$$\tilde{R}^m = -\frac{1}{2}|\nu^{m+1} - \nu_h^{m+1}|^2 Q_h^{m+1} - \frac{1}{2}|\nu^m - \nu_h^m|^2 Q^m$$
$$+ (\nu^m - \nu_h^m) \cdot (\nu^{m+1} - \nu_h^{m+1})Q_h^{m+1} + (\nu^m - \nu_h^m) \cdot \nu^{m+1}(Q^{m+1} - Q_h^{m+1})$$
$$= -\frac{1}{2}|(\nu^{m+1} - \nu_h^{m+1}) - (\nu^m - \nu_h^m)|^2 Q_h^{m+1} + \frac{1}{2}|\nu^m - \nu_h^m|^2(Q^{m+1} - Q^m)$$
$$+ \frac{1}{2}|\nu^m - \nu_h^m|^2(Q_h^{m+1} - Q^{m+1}) + (\nu^m - \nu_h^m) \cdot \nu^{m+1}(Q^{m+1} - Q_h^{m+1})$$
$$= -\frac{1}{2}|(\nu^{m+1} - \nu_h^{m+1}) - (\nu^m - \nu_h^m)|^2 Q_h^{m+1} + \frac{1}{2}|\nu^m - \nu_h^m|^2(Q^{m+1} - Q^m)$$
$$+ (\nu^m - \nu_h^m) \cdot (\nu^{m+1}(Q^{m+1} - Q_h^{m+1}) + \frac{1}{2}(\nu^m - \nu_h^m)(Q_h^{m+1} - Q^{m+1}))$$
$$= -\frac{1}{2}|(\nu^{m+1} - \nu_h^{m+1}) - (\nu^m - \nu_h^m)|^2 Q_h^{m+1} + \frac{1}{2}|\nu^m - \nu_h^m|^2(Q^{m+1} - Q^m)$$
$$+ (\nu^m - \nu_h^m) \cdot (\nu^{m+1} - \nu^m)(Q^{m+1} - Q_h^{m+1})$$
$$+ (\nu^m - \nu_h^m) \cdot (\nu^m(Q^{m+1} - Q_h^{m+1}) + \frac{1}{2}(\nu^m - \nu_h^m)(Q_h^{m+1} - Q^{m+1})).$$

The last term vanishes because

$$\nu^m(Q^{m+1}-Q_h^{m+1})+\frac{1}{2}(\nu^m-\nu_h^m)(Q_h^{m+1}-Q^{m+1})=(Q^{m+1}-Q_h^{m+1})\frac{1}{2}(\nu^m+\nu_h^m)$$

and we finally have

$$\tilde{R}^m = -\frac{1}{2}|(\nu^{m+1}-\nu_h^{m+1})-(\nu^m-\nu_h^m)|^2 Q_h^{m+1} + \frac{1}{2}|\nu^m-\nu_h^m|^2(Q^{m+1}-Q^m)$$
$$+(\nu^m-\nu_h^m)\cdot(\nu^{m+1}-\nu^m)(Q^{m+1}-Q_h^{m+1}).$$

Together with (24) this proves the Lemma.

Lemma 2.6.

$$\int_{\Omega_h}\left(\frac{\nabla(\overline{u}^{m+1}-\overline{u}^m)}{Q(\overline{u}^m)}-\frac{\nabla(u_h^{m+1}-u_h^m)}{Q(u_h^m)}\right)\cdot\nabla((\overline{u}^{m+1}-u_h^{m+1})-(\overline{u}^m-u_h^m))$$

$$=\int_{\Omega_h}|(\nu(\overline{u}^{m+1})-\nu(u_h^{m+1}))-(\nu(\overline{u}^m)-\nu(u_h^m))|^2 Q(u_h^{m+1})$$

$$+\int_{\Omega_h}\frac{(Q(\overline{u}^{m+1})-Q(\overline{u}^m))^2}{Q(\overline{u}^m)}+\int_{\Omega_h}\frac{(Q(u_h^{m+1})-Q(u_h^m))^2}{Q(u_h^m)}+\int_{\Omega_h}S^m,\ (25)$$

with

$$S^m = |\nu(\overline{u}^{m+1})-\nu(\overline{u}^m)|^2(Q(\overline{u}^{m+1})-Q(u_h^{m+1}))$$
$$+2(\nu(\overline{u}^{m+1})-\nu(\overline{u}^m))\cdot(\nu(u_h^{m+1})-\nu(u_h^m))Q(u_h^{m+1})$$
$$-(\nu(\overline{u}^{m+1})\frac{Q(\overline{u}^{m+1})}{Q(\overline{u}^m)}-\nu(\overline{u}^m))\cdot(\nu(u_h^{m+1})\frac{Q(u_h^{m+1})}{Q(u_h^m)}-\nu(u_h^m))$$
$$\times(Q(\overline{u}^m)+Q(u_h^m)). \tag{26}$$

Proof. We use the relation

$$|\nabla(v-w)|^2 = |\nu(v)-\nu(w)|^2 Q(v)Q(w)+(Q(v)-Q(w))^2 \tag{27}$$

and get

$$\left(\frac{\nabla(\overline{u}^{m+1}-\overline{u}^m)}{Q^m}+\frac{\nabla(u_h^{m+1}-u_h^m)}{Q_h^m}\right)\cdot\nabla((\overline{u}^{m+1}-u_h^{m+1})-(\overline{u}^m-u_h^m))$$

$$=\frac{|\nabla(\overline{u}^{m+1}-\overline{u}^m)|^2}{Q^m}-\frac{|\nabla(u_h^{m+1}-u_h^m)|^2}{Q_h^m}$$

$$-\nabla(\overline{u}^{m+1}-\overline{u}^m)\cdot\nabla(u_h^{m+1}-u_h^m)\left(\frac{1}{Q^m}+\frac{1}{Q_h^m}\right)$$

$$=|\nu^{m+1}-\nu^m|^2 Q^{m+1}+\frac{(Q^{m+1}-Q^m)^2}{Q^m}+|\nu_h^{m+1}-\nu_h^m|^2 Q_h^{m+1}$$

$$+\frac{(Q_h^{m+1}-Q_h^m)^2}{Q_h^m}-(\nu^{m+1}\frac{Q^{m+1}}{Q^m}-\nu^m)\cdot(\nu_h^{m+1}\frac{Q_h^{m+1}}{Q_h^m}-\nu_h^m)(Q^m+Q_h^m)$$

$$= |(\nu^{m+1} - \nu^m) - (\nu_h^{m+1} - \nu_h^m)|^2 Q_h^{m+1} + \frac{(Q^{m+1} - Q^m)^2}{Q^m} + \frac{(Q_h^{m+1} - Q_h^m)^2}{Q_h^m}$$

$$+ |\nu^{m+1} - \nu^m|^2 (Q^{m+1} - Q_h^{m+1}) + \tilde{S}^m,$$

with

$$\tilde{S}^m = 2(\nu^{m+1} - \nu^m) \cdot (\nu_h^{m+1} - \nu_h^m) Q_h^{m+1}$$

$$- (\nu^{m+1} \frac{Q^{m+1}}{Q^m} - \nu^m) \cdot (\nu_h^{m+1} \frac{Q_h^{m+1}}{Q_h^m} - \nu_h^m)(Q^m + Q_h^m).$$

We are now prepared to prove the convergence Theorem for the fully discrete scheme from Algorithm 2.

Proof of Theorem 2.4. We insert

$$\phi_h = I_h(\overline{u}^{m+1} - \overline{u}^m) - (u_h^{m+1} - u_h^m)$$
$$= ((\overline{u}^{m+1} - \overline{u}^m) - (u_h^{m+1} - u_h^m)) - ((\overline{u}^{m+1} - \overline{u}^m) - I_h(\overline{u}^{m+1} - \overline{u}^m))$$
$$= (e^{m+1} - e^m) - (\varepsilon^{m+1} - \varepsilon^m),$$

as a test function into (20), where we set

$$e = \overline{u} - u_h, \quad \varepsilon = \overline{u} - I_h \overline{u},$$

and rephrase the result as

$$A_1 + A_2 + A_3 = A_4 + A_5 + A_6 + A_7 + A_8, \tag{28}$$

with

$$A_1 = \frac{1}{\tau} \int_{\Omega_h} \left(\frac{\overline{u}^{m+1} - \overline{u}^m}{Q(\overline{u}^m)} - \frac{u_h^{m+1} - u_h^m}{Q(u_h^m)} \right) ((e^{m+1} - e^m) - (\varepsilon^{m+1} - \varepsilon^m))$$

$$A_2 = \int_{\Omega_h} \left(\frac{\nabla \overline{u}^{m+1}}{Q^m} - \frac{\nabla u_h^{m+1}}{Q_h^m} \right) \cdot \nabla ((\overline{u}^{m+1} - u_h^{m+1}) - (\overline{u}^m - u_h^m))$$

$$A_3 = - \int_{\Omega_h} \left(\frac{\nabla \overline{u}^{m+1}}{Q^m} - \frac{\nabla u_h^{m+1}}{Q_h^m} \right) \cdot \nabla (\varepsilon^{m+1} - \varepsilon^m)$$

$$A_4 = \int_{\Omega_h} \frac{\nabla (\overline{u}^{m+1} - \overline{u}^m)}{Q^m} \cdot \nabla (e^{m+1} - e^m)$$

$$A_5 = - \int_{\Omega_h} \frac{\nabla (\overline{u}^{m+1} - \overline{u}^m)}{Q^m} \cdot \nabla (\varepsilon^{m+1} - \varepsilon^m)$$

$$A_6 = - \int_{\Omega_h} \left(\frac{\overline{u}^{m+1} - \overline{u}^m}{\tau} - \overline{u}_t^m \right) \frac{1}{Q^m} (\varepsilon^{m+1} - \varepsilon^m)$$

$$A_7 = \int_{\Omega_h} \left(\frac{\overline{u}^{m+1} - \overline{u}^m}{\tau} - \overline{u}_t^m \right) \frac{1}{Q^m} (e^{m+1} - e^m)$$

$$A_8 = - \int_{\Omega_h \setminus \Omega} (H(\overline{u}^m) + V(\overline{u}^m))((e^{m+1} - e^m) - (\varepsilon^{m+1} - \varepsilon^m)).$$

Let us start by estimating the first term A_1 on the left hand side of (28). We use the abbreviation

$$V^m = -\frac{\overline{u}^{m+1} - \overline{u}^m}{\tau Q^m}, \quad V_h^m = -\frac{u_h^{m+1} - u_h^m}{\tau Q_h^m}$$

and have, with Young's inequality and (10),

$$
\begin{aligned}
A_1 &= \tau \int_{\Omega_h} (V^m - V_h^m)^2 Q_h^m + \tau \int_{\Omega_h} V^m (V^m - V_h^m)(Q^m - Q_h^m) \\
&\quad + \int_{\Omega_h} (V^m - V_h^m)(\varepsilon^{m+1} - \varepsilon^m) \\
&\geq \frac{1}{2}\tau \int_{\Omega_h} (V^m - V_h^m)^2 Q_h^m - c\tau \|V^m Q^m\|_{L^\infty(\Omega_h)}^2 \int_{\Omega_h} \left(\frac{1}{Q^m} - \frac{1}{Q_h^m}\right)^2 Q_h^m \\
&\quad - c\frac{1}{\tau} \|\varepsilon^{m+1} - \varepsilon^m\|_{L^2(\Omega_h)}^2 \\
&\geq \frac{1}{2}\tau \int_{\Omega_h} (V^m - V_h^m)^2 Q_h^m - c\tau \int_{\Omega_h} |\nu^m - \nu_h^m|^2 Q_h^m \\
&\quad - ch^4 \int_{m\tau}^{(m+1)\tau} \|\overline{u}_t\|_{H^2(\Omega_h)}^2,
\end{aligned}
\tag{29}
$$

with c depending on $\|\overline{u}_t\|_{L^\infty(\Omega_h \times (0,T))}$.

The most difficult term to estimate is the second one on the right hand side of (28). We use the results of the previous two Lemmas and obtain the identities

$$
\begin{aligned}
A_2 &= \int_{\Omega_h} \left(\frac{\nabla \overline{u}^m}{Q^m} - \frac{\nabla u_h^m}{Q_h^m}\right) \cdot \nabla((\overline{u}^{m+1} - u_h^{m+1}) - (\overline{u}^m - u_h^m)) \\
&\quad + \int_{\Omega_h} \left(\frac{\nabla(\overline{u}^{m+1} - \overline{u}^m)}{Q^m} - \frac{\nabla(u_h^{m+1} - u_h^m)}{Q_h^m}\right) \cdot \nabla((\overline{u}^{m+1} - u_h^{m+1}) - (\overline{u}^m - u_h^m)) \\
&= \frac{1}{2}\int_{\Omega_h} |\nu(\overline{u}^{m+1}) - \nu(u_h^{m+1})|^2 Q(u_h^{m+1}) - \frac{1}{2}\int_{\Omega_h} |\nu(\overline{u}^m) - \nu(u_h^m)|^2 Q(u_h^m) \\
&\quad + \frac{1}{2}\int_{\Omega_h} |(\nu^{m+1} - \nu_h^{m+1}) - (\nu^m - \nu_h^m)|^2 Q_h^{m+1} \\
&\quad + \int_{\Omega_h} \frac{(Q^{m+1} - Q^m)^2}{Q^m} + \int_{\Omega_h} \frac{(Q_h^{m+1} - Q_h^m)^2}{Q_h^m} + \int_{\Omega_h} R^m + \int_{\Omega_h} S^m \\
&= \frac{1}{2}\int_{\Omega_h} |\nu(\overline{u}^{m+1}) - \nu(u_h^{m+1})|^2 Q(u_h^{m+1}) - \frac{1}{2}\int_{\Omega_h} |\nu(\overline{u}^m) - \nu(u_h^m)|^2 Q(u_h^m) \\
&\quad + \frac{1}{2}\int_{\Omega_h} |(\nu^{m+1} - \nu_h^{m+1}) - (\nu^m - \nu_h^m)|^2 Q_h^{m+1} \\
&\quad + \int_{\Omega_h} \frac{(Q^{m+1} - Q_h^{m+1}) - (Q^m - Q_h^m))^2}{Q_h^m} + \int_{\Omega_h} T^m,
\end{aligned}
\tag{30}
$$

with the remainder term

$$T^m = (\nu^{m+1} - \nu^m) \cdot (\nu^m - \nu_h^m)(Q^{m+1} - Q_h^{m+1}) + \frac{1}{2}|\nu^m - \nu_h^m|^2(Q^{m+1} - Q^m)$$

$$+|\nu^{m+1} - \nu^m|^2(Q^{m+1} - Q_h^{m+1}) + (Q^{m+1} - Q^m)^2(\frac{1}{Q^m} - \frac{1}{Q_h^m})$$

$$+2(\nu^{m+1} - \nu^m) \cdot (\nu_h^{m+1} - \nu_h^m)Q_h^{m+1}$$

$$-(\nu^{m+1}\frac{Q^{m+1}}{Q^m} - \nu^m) \cdot (\nu_h^{m+1}\frac{Q_h^{m+1}}{Q_h^m} - \nu_h^m)(Q^m + Q_h^m)$$

$$+2\frac{(Q^{m+1} - Q^m)(Q_h^{m+1} - Q_h^m)}{Q_h^m}.$$

Elementary calculations lead to the following form of T^m with the special and important feature, that it contains triple products of differences only.

$$T^m = (\nu^{m+1} - \nu^m) \cdot (\nu^m - \nu_h^m)(Q^{m+1} - Q_h^{m+1}) + \frac{1}{2}|\nu^m - \nu_h^m|^2(Q^{m+1} - Q^m)$$

$$+|\nu^{m+1} - \nu^m|^2(Q^{m+1} - Q_h^{m+1}) + (Q^{m+1} - Q^m)^2(\frac{1}{Q^m} - \frac{1}{Q_h^m})$$

$$+(\nu_h^{m+1} - \nu_h^m) \cdot (\nu^{m+1} - \nu^m)(Q_h^{m+1} - Q^{m+1})$$

$$+(\nu^{m+1} - \nu_h^{m+1}) \cdot (\nu^{m+1}Q^{m+1} - \nu^m Q^m)\frac{Q_h^{m+1} - Q_h^m}{Q_h^m}$$

$$-\frac{1}{2}|\nu^{m+1} - \nu^m|^2 Q^m\frac{Q_h^{m+1} - Q_h^m}{Q_h^m} - \frac{1}{2}|\nu_h^{m+1} - \nu_h^m|^2 Q_h^m\frac{Q^{m+1} - Q^m}{Q^m}$$

$$+(\nu_h^{m+1} - \nu^{m+1}) \cdot (\nu_h^{m+1}Q_h^{m+1} - \nu_h^m Q_h^m)\frac{Q^{m+1} - Q^m}{Q^m}$$

$$+(Q^{m+1} - Q^m)(Q_h^{m+1} - Q_h^m)(\frac{1}{Q_h^m} - \frac{1}{Q^m})$$

$$+(\nu^{m+1} - \nu^m) \cdot (\nu_h^m - \nu^m)(Q_h^m - Q_h^{m+1})$$

$$+(\nu_h^{m+1} - \nu_h^m) \cdot (\nu^m - \nu_h^m)(Q^m - Q^{m+1})$$

$$-\frac{1}{2}|\nu^{m+1} - \nu^m|^2(Q_h^m - Q_h^{m+1}) - \frac{1}{2}|\nu_h^{m+1} - \nu_h^m|^2(Q^m - Q^{m+1}).$$

In the following we provide a first estimate for the remainder term T^m. The constants depend on the norms $\|\nabla u_t\|_{L^\infty(\Omega \times (0,T))}$ and $\|\nabla u\|_{L^\infty(\Omega \times (0,T))}$.

$$c|T^m| \leq \tau|\nu^m - \nu_h^m||Q^{m+1} - Q_h^{m+1}| + \tau|\nu^m - \nu_h^m|^2 + \tau^2|Q^{m+1} - Q_h^{m+1}|$$

$$+\tau^2|\nu^m - \nu_h^m| + \tau|\nu_h^{m+1} - \nu_h^m||Q^{m+1} - Q_h^{m+1}|$$

$$+\tau|\nu^{m+1} - \nu_h^{m+1}||\nu_h^{m+1} - \nu_h^m|Q^{m+1} - Q_h^m| + \tau^2|\nu_h^{m+1} - \nu_h^m|Q_h^{m+1}$$

$$+\tau|\nu^{m+1} - \nu_h^{m+1}|(|\nu_h^{m+1} - \nu_h^m|Q_h^{m+1}| + |Q_h^{m+1} - Q_h^m|)$$

$$+\tau|\nu_h^{m+1} - \nu^m|^2 Q^m + \tau|\nu^m - \nu_h^m||Q^{m+1} - Q_h^m|$$

$$+\tau|\nu^m - \nu_h^m||\nu_h^{m+1} - \nu_h^m| + \tau^2|Q_h^{m+1} - Q_h^m| + \tau|\nu_h^{m+1} - \nu_h^m|^2.$$

We use the simple estimates

$$|Q^{m+1} - Q_h^{m+1}| \leq |(Q^{m+1} - Q_h^{m+1}) - (Q^m - Q_h^m)| + |\nu^m - \nu_h^m|Q_h^m Q^m,$$
$$|\nu^{m+1} - \nu_h^{m+1}| \leq |(\nu^{m+1} - \nu_h^{m+1}) - (\nu^m - \nu_h^m)| + |\nu^m - \nu_h^m|,$$
$$|Q_h^{m+1} - Q_h^m| \leq |(Q^{m+1} - Q_h^{m+1}) - (Q^m - Q_h^m)| + c\tau,$$
$$|\nu_h^{m+1} - \nu_h^m| \leq |(\nu^{m+1} - \nu_h^{m+1}) - (\nu^m - \nu_h^m)| + c\tau,$$
$$Q_h^m \leq c\frac{|(Q^{m+1} - Q_h^{m+1}) - (Q^m - Q_h^m)|^2}{Q_h^m} + Q_h^{m+1} + c\tau,$$
$$\tau^2|\nu^m - \nu_h^m| \leq c\tau|\nu^m - \nu_h^m|^2 + c\tau^3,$$

(31)

together with Young's inequality and get for every positive number δ, by standard techniques, the estimate

$$|T^m| \leq (\delta + c_\delta \tau)\frac{|(Q^{m+1} - Q_h^{m+1}) - (Q^m - Q_h^m)|^2}{Q_h^m}$$
$$+ (\delta + c_\delta \tau)|(\nu^{m+1} - \nu_h^{m+1}) - (\nu^m - \nu_h^m)|^2 Q_h^{m+1}$$
$$+ c_\delta \tau|\nu^m - \nu_h^m|^2 Q_h^m + c_\delta \tau^3(Q_h^m + Q_h^{m+1}).$$

(32)

Thus from (30) and (32) we get, after a suitable choice of δ and τ_0 for $\tau \leq \tau_0$, the estimate

$$A_2 = \int_{\Omega_h} \left(\frac{\nabla \overline{u}^{m+1}}{Q^m} - \frac{\nabla u_h^{m+1}}{Q_h^m}\right) \cdot \nabla((\overline{u}^{m+1} - u_h^{m+1}) - (\overline{u}^m - u_h^m))$$
$$\geq \frac{1}{2}\int_{\Omega_h} |\nu^{m+1} - \nu_h^{m+1}|^2 Q_h^{m+1} - \frac{1}{2}\int_{\Omega_h} |\nu^m - \nu_h^m|^2 Q_h^m$$
$$+ \frac{1}{4}\int_{\Omega_h} |(\nu^{m+1} - \nu_h^{m+1}) - (\nu^m - \nu_h^m)|^2 Q_h^{m+1}$$
$$+ \frac{1}{4}\int_{\Omega_h} \frac{((Q^{m+1} - Q_h^{m+1}) - (Q^m - Q_h^m))^2}{Q_h^m}$$
$$- c\tau \int_{\Omega_h} |\nu^m - \nu_h^m|^2 Q_h^m - c\tau^3 \int_{\Omega_h} (Q_h^m + Q_h^{m+1}).$$

(33)

From (31) we infer

$$\left|\left(\frac{\nabla \overline{u}^{m+1}}{Q^m} - \frac{\nabla u_h^{m+1}}{Q_h^m}\right)\right|$$
$$= |\nu^{m+1}\frac{Q^{m+1}}{Q^m} - \nu_h^{m+1}\frac{Q_h^{m+1}}{Q_h^m} + \left(\frac{1}{Q^m} - \frac{1}{Q_h^m}\right)e_{n+1}|$$
$$\leq c|\nu^{m+1} - \nu_h^{m+1}|\frac{Q^{m+1}}{Q^m} + |\frac{Q^{m+1}}{Q^m} - \frac{Q_h^{m+1}}{Q_h^m}| + |\nu^m - \nu_h^m|$$

$$\leq \frac{|(Q^{m+1} - Q_h^{m+1}) - (Q^m - Q_h^m)|}{Q_h^m} + c|\nu^m - \nu_h^m|$$

$$+c|(\nu^{m+1} - \nu_h^{m+1}) - (\nu^m - \nu_h^m)|$$

$$\leq \frac{|(Q^{m+1} - Q_h^{m+1}) - (Q^m - Q_h^m)|}{\sqrt{Q_h^m}}$$

$$+c|\nu^m - \nu_h^m|\sqrt{Q_h^m} + c|(\nu^{m+1} - \nu_h^{m+1}) - (\nu^m - \nu_h^m)|\sqrt{Q_h^{m+1}}.$$

This then leads to an estimate for A_3

$$A_3 = -\int_{\Omega_h} \left(\frac{\nabla \overline{u}^{m+1}}{Q^m} - \frac{\nabla u_h^{m+1}}{Q_h^m} \right) \cdot \nabla(\varepsilon^{m+1} - \varepsilon^m)$$

$$\geq -\delta \int_{\Omega_h} \frac{((Q^{m+1} - Q_h^{m+1}) - (Q^m - Q_h^m))^2}{Q_h^m} - c_\delta \tau \int_{\Omega_h} |\nu^m - \nu_h^m|^2 Q_h^m$$

$$-\delta \int_{\Omega_h} |(\nu^{m+1} - \nu_h^{m+1}) - (\nu^m - \nu_h^m)|^2 Q_h^{m+1} - c_\delta \tau h^2 \int_{m\tau}^{(m+1)\tau} \|\overline{u}_t\|_{H^2(\Omega_h)}^2.$$

$$(34)$$

Clearly for A_5,

$$A_5 = -\int_{\Omega_h} \frac{\nabla(\overline{u}^{m+1} - \overline{u}^m)}{Q^m} \cdot \nabla(\varepsilon^{m+1} - \varepsilon^m)$$

$$\leq c\tau^{\frac{3}{2}} h \|\nabla \overline{u}_t\|_{L^\infty(\Omega_h \times (0,T))} \left(\int_{m\tau}^{(m+1)\tau} \|\overline{u}_t\|_{H^2(\Omega_h)}^2 \right)^{\frac{1}{2}}$$

$$\leq ch^2 \int_{m\tau}^{(m+1)\tau} \|\overline{u}_t\|_{H^2(\Omega_h)}^2 + c\tau^3.$$

$$(35)$$

Let us now estimate the first term A_4 on the right hand side.

$$\int_{\Omega_h} \frac{\nabla(\overline{u}^{m+1} - \overline{u}^m)}{Q^m} \cdot \nabla(e^{m+1} - e^m) = -\int_{\Omega_h} \nabla \cdot \frac{\nabla(\overline{u}^{m+1} - \overline{u}^m)}{Q^m} (e^{m+1} - e^m)$$

$$= -\int_{\Omega_h} \frac{1}{Q_h^m} \Delta(\overline{u}^{m+1} - \overline{u}^m)(e^{m+1} - e^m)$$

$$-\int_{\Omega_h} \left(\frac{1}{Q^m} - \frac{1}{Q_h^m} \right) \Delta(\overline{u}^{m+1} - \overline{u}^m)(e^{m+1} - e^m)$$

$$-\int_{\Omega_h} \nabla \frac{1}{Q^m} \cdot \nabla(\overline{u}^{m+1} - \overline{u}^m)(e^{m+1} - e^m)$$

$$\leq \int_{\Omega_h} |\Delta((\overline{u}^{m+1} - \overline{u}^m)| \frac{|e^{m+1} - e^m|}{Q_h^m}$$

$$+ \int_{\Omega_h} |\nu^m - \nu_h^m||\Delta((\overline{u}^{m+1} - \overline{u}^m)||e^{m+1} - e^m|$$

$$+ c \int_{\Omega_h} |\nabla(\overline{u}^{m+1} - \overline{u}^m)||e^{m+1} - e^m|$$

$$=: I_1 + I_2 + I_3.$$

Let us estimate these three terms separately. For all of them we use the inequality

$$|e^{m+1} - e^m| \leq c\tau |V^m - V_h^m|Q_h^m + c\tau |\nu^m - \nu_h^m|Q_h^m. \tag{36}$$

Thus, we have

$$I_1 \leq c\tau \int_{m\tau}^{(m+1)\tau} \|\overline{u}_t\|_{H^2(\Omega_h)} \left(\|V^m - V_h^m\|_{L^2(\Omega_h)} + \|\nu^m - \nu_h^m\|_{L^2(\Omega_h)} \right)$$

$$\leq \delta\tau \int_{\Omega_h} (V^m - V_h^m)^2 Q_h^m + c\tau \int_{\Omega_h} |\nu^m - \nu_h^m|^2 Q_h^m + c_\delta \tau^2 \int_{m\tau}^{(m+1)\tau} \|\overline{u}_t\|_{H^2(\Omega_h)}^2.$$

With (36) we get

$$I_2 \leq c\|\overline{u}\|_{L^\infty((0,T),H^2(\Omega_h))} \left(\int_{\Omega_h} |\nu^m - \nu_h^m|^2 Q_h^m \right)^{\frac{1}{2}} \left(\int_{\Omega_h} \frac{(e^{m+1} - e^m)^2}{Q_h^m} \right)^{\frac{1}{2}}$$

$$\leq \delta\tau \int_{\Omega_h} (V^m - V_h^m)^2 Q_h^m + c_\delta \tau \int_{\Omega_h} |\nu^m - \nu_h^m|^2 Q_h^m,$$

$$I_3 \leq \delta\tau \int_{\Omega_h} (V^m - V_h^m)^2 Q_h^m + c\tau \int_{\Omega_h} |\nu^m - \nu_h^m|^2 Q_h^m + c_\delta \tau^3 \int_{\Omega_h} Q_h^m.$$

We collect the estimates for I_1, I_2, I_3 and arrive at the estimate

$$A_4 = \int_{\Omega_h} \frac{\nabla(\overline{u}^{m+1} - \overline{u}^m)}{Q^m} \cdot \nabla(e^{m+1} - e^m) \leq 3\delta\tau \int_{\Omega_h} (V^m - V_h^m)^2 Q_h^m$$

$$+ c_\delta \tau \int_{\Omega_h} |\nu^m - \nu_h^m|^2 Q_h^m + c_\delta \tau^2 \int_{m\tau}^{(m+1)\tau} \|\overline{u}_t\|_{H^2(\Omega_h)}^2 + c_\delta \tau^3 \int_{\Omega_h} Q_h^m. \tag{37}$$

For the third term on the right hand side of (28) we observe

$$A_6 = - \int_{\Omega_h} \left(\frac{\overline{u}^{m+1} - \overline{u}^m}{\tau} - \overline{u}_t^m \right) \frac{1}{Q^m} (\varepsilon^{m+1} - \varepsilon^m)$$

$$\leq \sqrt{\tau} \left(\int_{m\tau}^{(m+1)\tau} \|\overline{u}_{tt}\|_{L^2(\Omega_h)}^2 \right)^{\frac{1}{2}} \|\varepsilon^{m+1} - \varepsilon^m\|_{L^2(\Omega_h)}$$

$$\leq c\tau h^2 \left(\int_{m\tau}^{(m+1)\tau} \|\overline{u}_{tt}\|_{L^2(\Omega_h)}^2 + \int_{m\tau}^{(m+1)\tau} \|\overline{u}_t\|_{H^2(\Omega_h)}^2 \right) \tag{38}$$

and similarly, with the use of (36),

$$A_7 = \int_{\Omega_h} \left(\frac{\overline{u}^{m+1} - \overline{u}^m}{\tau} - \overline{u}_t^m \right) \frac{1}{Q^m} (e^{m+1} - e^m)$$

$$\leq \delta \tau \int_{\Omega_h} |V^m - V_h^m|^2 Q_h^m + c_\delta \tau \int_{\Omega_h} |\nu^m - \nu_h^m|^2 Q_h^m + c_\delta \tau^2 \int_{m\tau}^{(m+1)\tau} \|\overline{u}_{tt}\|_{L^2(\Omega_h)}^2.$$

$$(39)$$

For the last term in (28) we use the fact that $H(\overline{u}^m)$ and $V(\overline{u}^m)$ are bounded uniformly and that $|\Omega_h \setminus \Omega| \leq ch^2$. Thus

$$-\int_{\Omega_h \setminus \Omega} (H(\overline{u}^m) + V(\overline{u}^m))((e^{m+1} - e^m) - (\varepsilon^{m+1} - \varepsilon^m))$$

$$\leq c\|e^{m+1} - e^m\|_{L^1(\Omega_h \setminus \Omega)} + ch\|\varepsilon^{m+1} - \varepsilon^m\|_{L^2(\Omega_h)}$$

$$\leq c\|e^{m+1} - e^m\|_{L^1(\Omega_h \setminus \Omega)} + c\tau h^2 + ch^4 \int_{m\tau}^{(m+1)\tau} \|\overline{u}_t\|_{H^2(\Omega_h)}^2. \quad (40)$$

Note further that

$$\|e^{m+1} - e^m\|_{L^1(\Omega_h \setminus \Omega)} \leq \int_{\Omega_h \setminus \Omega} \frac{|e^{m+1} - e^m|}{\sqrt{Q_h^m}} \left(|\sqrt{Q_h^m} - \sqrt{Q^m}| + c \right)$$

$$\leq \left(\int_{\Omega_h} \frac{|e^{m+1} - e^m|^2}{Q_h^m} \right)^{\frac{1}{2}} \left(\int_{\Omega_h \setminus \Omega} |\sqrt{Q_h^m} - \sqrt{Q^m}|^2 + ch^2 \right)^{\frac{1}{2}}$$

$$\leq c\tau \left(\int_{\Omega_h} |V^m - V_h^m|^2 Q_h^m + \int_{\Omega_h} |\nu^m - \nu_h^m|^2 Q_h^m \right)^{\frac{1}{2}} \left(\int_{\Omega_h} |\nu^m - \nu_h^m|^2 Q_h^m + ch^2 \right)^{\frac{1}{2}},$$

where we used (36) together with the estimate

$$\int_{\Omega_h \setminus \Omega} |\sqrt{Q_h^m} - \sqrt{Q^m}|^2 \leq \int_{\Omega_h} \frac{|Q_h^m - Q^m|^2}{Q_h^m} \leq c \int_{\Omega_h} |\nu_h^m - \nu^m|^2 Q_h^m.$$

So, for arbitrary positive δ,

$$A_8 = -\int_{\Omega_h \setminus \Omega} (H(\overline{u}^m) + V(\overline{u}^m))((e^{m+1} - e^m) - (\varepsilon^{m+1} - \varepsilon^m))$$

$$\leq \delta \tau \int_{\Omega_h} (V^m - V_h^m)^2 Q_h^m + c_\delta \tau \int_{\Omega_h} |\nu^m - \nu_h^m|^2 Q_h^m$$

$$+ c(\tau^2 + h^4) \int_{m\tau}^{(m+1)\tau} \|\overline{u}_t^m\|_{H^2(\Omega_h)}^2 + c_\delta \tau h^2. \quad (41)$$

We collect the estimates (28,29,33,34–37,38–41), choose δ in a suitable way, and finally get the following result

$$\frac{1}{4}\tau \int_{\Omega_h} (V^m - V_h^m)^2 Q_h^m$$

$$+ \int_{\Omega_h} |\nu^{m+1} - \nu_h^{m+1}|^2 Q_h^{m+1} - \int_{\Omega_h} |\nu^m - \nu_h^m|^2 Q_h^m$$

$$+ \frac{1}{8} \int_{\Omega_h} |(\nu^{m+1} - \nu_h^{m+1}) - (\nu^m - \nu_h^m)|^2 Q_h^{m+1}$$

$$+ \frac{1}{8} \int_{\Omega_h} ((Q^{m+1} - Q_h^{m+1}) - (Q^m - Q_h^m))^2 / Q_h^m$$

$$\leq c\tau \int_{\Omega_h} |\nu^m - \nu_h^m|^2 Q_h^m + c(\tau^2 + h^2) \int_{m\tau}^{(m+1)\tau} (\|\bar{u}_t\|_{H^2(\Omega_h)}^2 + \|\bar{u}_{tt}\|_{L^2(\Omega_h)}^2)$$

$$+ c\tau^3 \int_{\Omega_h} (Q_h^m + Q_h^{m+1}) + c\tau h^2. \tag{42}$$

We know, from Theorem 2.2, that $\int_{\Omega_h} Q_h^m$ is bounded and so from the previous estimate it follows that

$$\int_{\Omega_h} |\nu^{m+1} - \nu_h^{m+1}|^2 Q_h^{m+1} - \int_{\Omega_h} |\nu^m - \nu_h^m|^2 Q_h^m$$

$$\leq c\tau \int_{\Omega_h} |\nu^m - \nu_h^m|^2 Q_h^m + c(\tau^2 + h^2) \int_{m\tau}^{(m+1)\tau} (\|\bar{u}_t\|_{H^2(\Omega_h)}^2 + \|\bar{u}_{tt}\|_{L^2(\Omega_h)}^2)$$

$$+ c\tau^3 + c\tau h^2. \tag{43}$$

Summation over $m = 0, \ldots, l - 1$, with $l \leq M - 1$, leads to

$$\int_{\Omega_h} |\nu^l - \nu_h^l|^2 Q_h^l \leq c(\tau^2 + h^2) \int_0^T (\|\bar{u}_t\|_{H^2(\Omega_h)}^2 + \|\bar{u}_{tt}\|_{L^2(\Omega_h)}^2)$$

$$+ c\tau^2 + ch^2 + c\tau \sum_{m=0}^{l-1} \int_{\Omega_h} |\nu^m - \nu_h^m|^2 Q_h^m$$

$$\leq c(\tau^2 + h^2) + c\tau \sum_{m=0}^{l-1} \int_{\Omega_h} |\nu^m - \nu_h^m|^2 Q_h^m, \tag{44}$$

or with

$$E^l = \int_{\Omega_h} |\nu^l - \nu_h^l|^2 Q_h^l,$$

$$E^l \leq c\tau \sum_{m=0}^{l-1} E^m + c(\tau^2 + h^2).$$

The discrete Gronwall Lemma then implies

$$E^l \leq c(\tau^2 + h^2),$$

which is the final error estimate for the approximation of the normal,

$$\sup_{m=0,\ldots,M} \int_{\Omega_h} |\nu(\overline{u}^m) - \nu(u_h^m)|^2 Q(u_h^m) \le c(\tau^2 + h^2).$$ (45)

This implies the first statement of the Theorem. The second statement concerning the numerical approximation of the normal velocity then follows from (42) if we use (45), i.e.

$$\tau \sum_{m=0}^{M-1} \int_{\Omega_h} (V^m - V_h^m)^2 Q_h^m \le c(\tau^2 + h^2).$$ (46)

Thus the Theorem is proved.

3 Mean Curvature Flow of Level Sets

3.1 Viscosity Solutions

The level set approach to mean curvature flow allows the modeling of topological changes during the flow. The interface is written as the zero level set of a scalar function,

$$\Gamma(t) = \{x \in \mathbb{R}^n \,|\, u(x,t) = 0\} \quad (t \ge 0).$$ (47)

As in Section 2, we derive the differential equation for u from the law of motion. Here,

$$V(u) = -\frac{u_t}{|\nabla u|}, \quad H(u) = \nabla \cdot \frac{\nabla u}{|\nabla u|},$$

so that (1) is equivalent to

$$u_t - |\nabla u| \nabla \cdot \frac{\nabla u}{|\nabla u|} = 0.$$ (48)

This differential equation is degenerate in two ways: it is not defined where the gradient of u vanishes – this is the case at points where the topology of Γ changes – and the equation is not uniformly parabolic. We have to solve the following initial value problem

$$u_t - \sum_{i,j=1}^{n} \left(\delta_{ij} - \frac{u_{x_i} u_{x_j}}{|\nabla u|^2}\right) u_{x_i x_j} = 0 \quad \text{in } \mathbb{R}^n \times (0, \infty),$$ (49)

$$u(\cdot, 0) = u_0 \quad \text{in } \mathbb{R}^n.$$ (50)

The initial interface $\Gamma_0 = \{x \in \mathbb{R}^n \,|\, u_0(x) = 0\}$ is given by a continuous function $u_0 : \mathbb{R}^n \to \mathbb{R}$. If $u : \mathbb{R}^n \times [0, \infty) \to \mathbb{R}$ is the unique viscosity solution of (48), then we call (47) the generalized solution of the mean curvature flow problem. As $\Gamma(t)$ exists for all times, it provides a notion of solution beyond singularities. For this reason, the level set approach has also become very important in the numerical approximation of mean curvature flow and related problems. We shall show how to use the results from the previous chapter to solve (49), (50) numerically in a reliable way.

We briefly repeat the analytical framework.

Definition 3.1. *A function $u \in C^0(\mathbb{R}^n \times [0, \infty))$ is called a viscosity sub-solution of (49) provided that for each $\phi \in C^\infty(\mathbb{R}^{n+1})$, if $u - \phi$ has a local maximum at $(x_0, t_0) \in \mathbb{R}^n \times (0, \infty)$, then*

$$\phi_t - \sum_{i,j=1}^n \left(\delta_{ij} - \frac{\phi_{x_i} \phi_{x_j}}{|\nabla \phi|^2} \right) \phi_{x_i x_j} \le 0 \text{ at } (x_0, t_0), \text{ if } \nabla \phi(x_0, t_0) \ne 0,$$

$$\phi_t - \sum_{i,j=1}^n \left(\delta_{ij} - p_i p_j \right) \phi_{x_i x_j} \le 0 \text{ at } (x_0, t_0) \text{ for some } |p| \le 1, \text{ if } \nabla \phi(x_0, t_0) = 0.$$

$$(51)$$

A viscosity supersolution is defined analogously: maximum is replaced by minimum and $\le$ by $\ge$. A viscosity solution of (49) is a function $u \in C^0(\mathbb{R}^n \times [0, \infty))$ that is both a subsolution and a supersolution.

We assume that the initial function u_0 is smooth. The following existence and uniqueness theorem is a special case of results proved independently by Evans, Spruck [10] and Chen, Giga, Goto [1].

Theorem 3.1. *Assume $u_0 : \mathbb{R}^n \to \mathbb{R}$ with $u_0(x) = 1$ for $|x| \ge S$ for some $S > 0$. Then there exists a unique viscosity solution of (49), (50), such that $u(x, t) = 1$ for $|x| + t \ge R$, for some $R > 0$ depending only on S.*

The authors also proved that the sets $\Gamma(t)$ are independent of the particular choice of u_0 which has Γ_0 as its zero level set, so that the generalized evolution $(\Gamma(t))_{t \ge 0}$ is well defined for a given Γ_0. The level set solution has been investigated further in several papers, among which we mention in particular [11], [12], [13] and [20].

3.2 Regularization

Evans and Spruck proved in [10] that the solutions u^ε of

$$u_t^\varepsilon - \sqrt{\varepsilon^2 + |\nabla u^\varepsilon|^2} \, \nabla \cdot \frac{\nabla u^\varepsilon}{\sqrt{\varepsilon^2 + |\nabla u^\varepsilon|^2}} = 0 \text{ in } \mathbb{R}^n \times (0, \infty), \quad (52)$$

$$u^\varepsilon(\cdot, 0) = u_0 \text{ in } \mathbb{R}^n \quad (53)$$

converge locally uniformly for $\varepsilon \to 0$ to the unique viscosity solution u of (49), (50). We shall use this regularised form of (48) as a basis for the computation of the viscosity solution. Then it is important to know the asymptotic error between the viscosity solution and the solution of the regularised problem quantitatively for $\epsilon \to 0$. [2] contains the proof of the following theorem together with several a priori estimates and their dependence on ϵ.

Theorem 3.2. *For every $\alpha \in (0, \frac{1}{2})$, $0 < T < \infty$, there is a constant $c = c(u_0, T, \alpha)$ such that*

$$\|u - u^\varepsilon\|_{L^\infty(\mathbb{R}^n \times (0,T))} \leq c\varepsilon^\alpha$$

for all $\varepsilon > 0$.

If one wants to compute approximations to the viscosity solution u of (49), (50) then, according to Theorem 3.2, it is sufficient to solve the regularised problem (52), (53). But one has to restrict the computations to some finite domain. This means that one has to solve the problem

$$u_{\varepsilon t} - \sqrt{\varepsilon^2 + |\nabla u_\varepsilon|^2} \; \nabla \cdot \frac{\nabla u_\varepsilon}{\sqrt{\varepsilon^2 + |\nabla u_\varepsilon|^2}} = 0 \text{ in } \Omega \times (0,T), \qquad (54)$$

$$u_\varepsilon(\cdot, 0) = u_0 \text{ on } \partial\Omega \times (0,T) \cup \Omega \times \{0\}, \qquad (55)$$

on a domain Ω which is large enough.

Corollary 3.1. *Let $\Omega = B_{\tilde{S}}(0)$ with $\tilde{S} > R = R(S)$, where R is the radius from Theorem 3.1. Then for every $\alpha \in (0, \frac{1}{2})$, $0 < T < \infty$ there is a constant $c = c(u_0, T, \alpha)$ such that*

$$\|u - u_\varepsilon\|_{L^\infty(\Omega \times (0,T))} \leq c\varepsilon^\alpha \qquad (56)$$

for all $\varepsilon > 0$.

Proof. The functions u^ε and u_ε are solutions of the same parabolic differential equation in the domain Ω with the same initial data. The comparison theorem then implies that

$$\|u^\varepsilon - u_\varepsilon\|_{L^\infty(\Omega \times (0,T))} \leq \|u^\varepsilon - u_\varepsilon\|_{L^\infty(\partial\Omega \times (0,T))}.$$

On $\partial\Omega$ on has that $u_\varepsilon = 1$ and $u = 1$. This means that one can continue the estimate by

$$\leq \|u^\varepsilon - u\|_{L^\infty(\partial\Omega \times (0,T))}.$$

But then the result follows from Theorem 3.2.

We are now in the position to use the results which were obtained for the mean curvature flow of graphs as a tool for the computation of the viscosity solution to the mean curvature flow of level sets. If we scale

$$U := \frac{u_\varepsilon}{\varepsilon}, \qquad (57)$$

then U is a solution of the mean curvature flow problem for graphs (see (4))

$$U_t - \sqrt{1 + |\nabla U|^2} \; \nabla \cdot \frac{\nabla U}{\sqrt{1 + |\nabla U|^2}} = 0 \text{ in } \Omega \times (0,T),$$

$$U = \frac{u_0}{\varepsilon} \text{ on } \partial\Omega \times (0,T) \cup \Omega \times \{0\}. \qquad (58)$$

This is a theoretical observation and implies that we can apply the results concerning the mean curvature flow of graphs to the mean curvature flow of level sets. But computationally we shall not use (58), but directly the unscaled version for u_ε.

3.3 Convergence of the numerical scheme for the level set problem

The semi-discrete scheme In the following we use the abbreviations

$$\nu_\varepsilon(v) = \frac{(\nabla v, -\varepsilon)}{Q_\varepsilon(v)}, \quad Q_\varepsilon(v) = \sqrt{\varepsilon^2 + |\nabla v|^2}, \quad V_\varepsilon(v) = -\frac{v_t}{Q_\varepsilon(v)}.$$

Our results for the mean curvature flow of a graph can directly be transformed into a convergence result for the regularised level set problem.

Theorem 3.3. *Let u_ε be the solution of (54), (55) and $u_{\varepsilon h}$ be the solution of the semi-discrete problem $u_{\varepsilon h}(\cdot, t) \in X_h$ with $u_{\varepsilon h}(\cdot, t) - I_h u_0 \in X_{h0}$, $u_{\varepsilon h}(\cdot, 0) = u_{h0} = I_h u_0$,*

$$\int_{\Omega_h} \frac{u_{\varepsilon h t} \phi_h}{Q_\varepsilon(u_{\varepsilon h})} + \int_{\Omega_h} \frac{\nabla u_{\varepsilon h} \cdot \nabla \phi_h}{Q_\varepsilon(u_{\varepsilon h})} = 0 \qquad (59)$$

for all $t \in (0, T)$ and all discrete test functions $\phi_h \in X_{h0}$. Then

$$\int_0^T \int_{\Omega \cap \Omega_h} (V_\varepsilon(u_\varepsilon) - V_\varepsilon(u_{\varepsilon h}))^2 Q_\varepsilon(u_{\varepsilon h}) \le c_\varepsilon h^2,$$

$$\sup_{(0,T)} \int_{\Omega \cap \Omega_h} |\nu_\varepsilon(u_\varepsilon) - \nu_\varepsilon(u_{\varepsilon h})|^2 Q_\varepsilon(u_{\varepsilon h}) \le c_\varepsilon h^2.$$

The *proof* is based on the scaling argument, which we shall use for the fully discrete scheme below and therefore we omit it here. For a detailed discussion of the dependency on ε see [2].

In two space dimensions we can prove that the computed solutions $u_{\varepsilon h}$ converge in L^∞ to the viscosity solution. The proof is contained in [7].

Theorem 3.4. *Let u be the viscosity solution of (49), (50) and let $u_{\varepsilon h}$ be the solution of the problem 59. with $\Omega \subset \mathbb{R}^2$ as in Corollary 3.1. Then there exist a function $h = h(\varepsilon) \to 0$, $\varepsilon \to 0$ such that*

$$\lim_{\varepsilon \to 0} \|u - u_{\varepsilon h(\varepsilon)}\|_{L^\infty(\Omega \times (0,T))} = 0.$$

The fully discrete scheme A similar transformation of the graph problem into the level set problem is available for the fully discrete scheme. Let us formulate the algorithm for this case. It can be implemented into an available solver for linear parabolic problems easily.

Algorithm 3 *Let $u_{\varepsilon h}^0 = I_h u_0$. For $m = 0, \dots M - 1$ compute $u_{\varepsilon h}^{m+1} \in X_h$ such that $u_{\varepsilon h} - u_{\varepsilon h}^0 \in X_{h0}$ and for every $\phi_h \in X_{h0}$*

$$\frac{1}{\tau} \int_{\Omega_h} \frac{u_{\varepsilon h}^{m+1} \phi_h}{Q(u_{\varepsilon h}^m)} + \int_{\Omega_h} \frac{\nabla u_{\varepsilon h}^{m+1} \cdot \nabla \phi_h}{Q(u_{\varepsilon h}^m)} = \frac{1}{\tau} \int_{\Omega_h} \frac{u_{\varepsilon h}^m \phi_h}{Q(u_{\varepsilon h}^m)}. \qquad (60)$$

We now have the following convergence theorem for the fully discrete regularised level set problem. The required regularity of the continuous solution is available because the domain is a ball and thus mean convex.

Theorem 3.5. *Let u_ε be the solution of (54), (55) and let $u_{\varepsilon h}^m$, $(m = 1, \ldots, M)$ be the solution of (60). Then there exists a $\tau_0 > 0$ such that, for all $0 < \tau \le \tau_0$,*

$$\tau \sum_{m=0}^{M} \int_{\Omega \cap \Omega_h} (V_\varepsilon(u_\varepsilon^m) - V_{\varepsilon h}^m)^2 Q_\varepsilon(u_{\varepsilon h}^m) \le c_\varepsilon (\tau^2 + h^2), \tag{61}$$

$$\sup_{m=0,\ldots,M} \int_{\Omega \cap \Omega_h} |\nu_\varepsilon(u_\varepsilon^m) - \nu_\varepsilon(u_{\varepsilon h}^m)|^2 Q_\varepsilon(u_{\varepsilon h}^m) \le c_\varepsilon (\tau^2 + h^2), \tag{62}$$

with $M = [\frac{T}{\tau}]$. $V_{\varepsilon h}^m = -(u_{\varepsilon h}^{m+1} - u_{\varepsilon h}^m)/(\tau Q_\varepsilon(u_{\varepsilon h}^m))$ is the regularised discrete normal velocity.

Proof. We scale as in (57). Then U is a solution of the equation for the mean curvature flow of a graph (4) with initial and boundary data $U_0 = u_0/\varepsilon$. Let $U_h^m = u_{\varepsilon h}^m/\varepsilon$ be the scaled solution of the fully discrete scheme (60). U_h^m then is the solution of the fully discrete scheme (13) for graphs and we can apply Theorem 2.4 and get the error estimates (18), (19). The constants on the right hand side of these estimates now depend on ε because the discrete boundary and initial data depend on this parameter. We scale back the error estimates, observe that

$$\nu(U) = \nu_\varepsilon(u_\varepsilon), \ V(U) = V_\varepsilon(u_\varepsilon), \ \nu(U_h^m) = \nu_\varepsilon(u_{\varepsilon h}^m), \ V_h^m = V_{\varepsilon h}^m$$

and

$$\varepsilon Q(U) = Q_\varepsilon(u_\varepsilon), \ \varepsilon Q(U_h^m) = Q_\varepsilon(u_{\varepsilon h}^m),$$

so the Theorem is proved.

We now dovetail the results from Corollary 3.1 and Theorem 3.5 to get a convergence result for viscosity solutions of the level set problem in $\mathbb{R}^2$. The gradient of the discrete solution is bounded. The gradient of the continuous solution is bounded independently of ε, since the domain is a ball. Let $S \in \mathcal{T}_h$ be a simplex of the triangulation. Then

$$\int_S Q(u_{\varepsilon h}^m) \le \int_S |Q(u_\varepsilon^m) - Q(u_{\varepsilon h}^m)| + ch^2$$

$$\le c \int_S |\nu(u_\varepsilon^m) - \nu(u_{\varepsilon h}^m)| Q(u_{\varepsilon h}^m) + ch^2$$

$$\le c \left(\int_S |\nu(u_\varepsilon^m) - \nu(u_{\varepsilon h}^m)|^2 Q(u_{\varepsilon h}^m) \right)^{\frac{1}{2}} \left(\int_S Q(u_{\varepsilon h}^m) \right)^{\frac{1}{2}} + ch^2$$

$$\le \delta \int_S Q(u_{\varepsilon h}^m) + \frac{c}{\delta} \int_S |\nu(u_\varepsilon^m) - \nu(u_{\varepsilon h}^m)|^2 Q(u_{\varepsilon h}^m) + ch^2$$

$$\le \delta \int_S Q(u_{\varepsilon h}^m) + \frac{c_\varepsilon}{\delta}(\tau^2 + h^2) + ch^2,$$

and this implies that

$$\int_S Q(u_{\varepsilon h}^m) \le c_\varepsilon(\tau^2 + h^2).$$

But then

$$\sup_S Q(u_{\varepsilon h}^m) \le c_\varepsilon(\frac{\tau^2}{h^2} + 1),$$

and so

$$\sup_{\Omega_h} Q(u_{\varepsilon h}^m) \le c_\varepsilon(\frac{\tau^2}{h^2} + 1).$$

From (27) we then can conclude that

$$\begin{aligned}
|\nabla(u_\varepsilon^m - u_{\varepsilon h}^m)|^2 &= |\nu(u_\varepsilon^m) - \nu(u_{\varepsilon h}^m)|^2 Q(u_{\varepsilon h}^m) Q(u_\varepsilon^m) + (Q(u_\varepsilon^m) - Q(u_{\varepsilon h}^m))^2 \\
&\le c|\nu(u_\varepsilon^m) - \nu(u_{\varepsilon h}^m)|^2 Q(u_{\varepsilon h}^m)(1 + Q(u_{\varepsilon h}^m)) \\
&\le c_\varepsilon|\nu(u_\varepsilon^m) - \nu(u_{\varepsilon h}^m)|^2 Q(u_{\varepsilon h}^m)(1 + \frac{\tau^2}{h^2}).
\end{aligned}$$

With the error estimate from Theorem 3.5 we get

$$\|\nabla(u_\varepsilon^m - u_{\varepsilon h}^m)\|_{L^2(\Omega_h)} \le c_\varepsilon(\tau + h)(1 + \frac{\tau}{h}) \le c_\varepsilon(h + \tau) \le c_\varepsilon(h + \tau),$$

under the assumption that $\tau \le ch$. With the well known estimate [22]

$$\|w_h\|_{L^\infty(\Omega_h)} \le c|\log h|^{\frac{1}{2}} \|\nabla w_h\|_{L^2(\Omega_h)},$$

for functions $w_h \in X_{h0}$ (in $\mathbb{R}^2$), we can conclude that

$$\begin{aligned}
\|u_\varepsilon^m - u_{\varepsilon h}^m\|_{L^\infty(\Omega_h)} &\le \|u_\varepsilon^m - I_h u_\varepsilon^m\|_{L^\infty(\Omega_h)} + \|I_h u_\varepsilon^m - u_{\varepsilon h}^m\|_{L^\infty(\Omega_h)} \\
&\le c_\varepsilon h^2 + c|\log h|^{\frac{1}{2}} \|\nabla(I_h u_\varepsilon^m - u_{\varepsilon h}^m)\|_{L^2(\Omega_h)} \\
&\le c_\varepsilon h^2 + c|\log h|^{\frac{1}{2}} \\
&\quad \times (\|\nabla(I_h u_\varepsilon^m - u_\varepsilon^m)\|_{L^2(\Omega_h)} + \|\nabla(u_\varepsilon^m - u_{h\varepsilon}^m)\|_{L^2(\Omega_h)}) \\
&\le c_\varepsilon h^2 + c_\varepsilon|\log h|^{\frac{1}{2}}(h + \tau) \\
&\le c_\varepsilon|\log h|^{\frac{1}{2}}(h + \tau).
\end{aligned}$$

This estimate, together with the estimate from Corollary 3.1, implies the following bound for the difference between the viscosity solution u and the fully discrete regularised solution $u_{\varepsilon h}^m$

$$\begin{aligned}
\|u^m - u_{\varepsilon h}^m\|_{L^\infty(\Omega_h)} &\le \|u^m - u_\varepsilon^m\|_{L^\infty(\Omega_h)} + \|u_\varepsilon^m - u_{\varepsilon h}^m\|_{L^\infty(\Omega_h)} \\
&\le c\varepsilon^\alpha + c_\varepsilon|\log h|^{\frac{1}{2}}(h + \tau).
\end{aligned}$$

This is true for arbitrary fixed $\alpha \in (0, \frac{1}{2})$. Now for $\varepsilon \to 0$ we can find $h = h(\varepsilon) \to 0$ and $\tau = \tau(\varepsilon) \to 0$ such that with the piecewise linear interpolant $u_{\varepsilon h\tau}$ of $u_{\varepsilon h}^m$ with respect to time we have proved the final Theorem.

Theorem 3.6. *Let u be the viscosity solution from Theorem 3.1 and let Ω be the domain from Corollary 3.1. Denote by $u_{\varepsilon h\tau}$ the time interpolated solution of the fully discrete scheme (60). Then there exist functions $h = h(\varepsilon) \to 0$ and $\tau = \tau(\varepsilon) \to 0$ for $\varepsilon \to 0$ such that*

$$\|u - u_{\varepsilon h(\varepsilon)\tau(\varepsilon)}\|_{L^\infty(\Omega_h \times (0,T))} \to 0.$$

3.4 Numerical Tests for the Level Set Algorithm

The regular case. We first test the algorithm for a regular case, i.e. for a case in which the gradient of the solution to (48) does not vanish inside the domain Ω. In order to be able to compute the error between continuous and fully discrete solution we use the exact solution for a shrinking circle,

$$u(x,t) = 8 - 2t - |x|^2,$$

on the domain $\Omega = \{x \in \mathbb{R}^2 \mid 1 < |x| < 3\}$ for the time interval $(0,T)$ with $T = 2$. We use the exact solution as boundary data. Since the gradient of u is nonzero on Ω, we choose the regularization parameter

$$\varepsilon = 0.$$

In Table 1 we show the results. For each refinement level of the globally refined grid we provide the grid parameter h, the error of the normal,

$$E_\nu = \Big(\sup_{m\in\{0,\ldots,M\}} \int_{\Omega_h} |\nu(u(\cdot,m\tau)) - \nu(u_{\varepsilon h}^m)|^2 |\nabla u_{\varepsilon h}^m| \Big)^{\frac{1}{2}},$$

the error for the normal velocity,

$$E_V = \Big(\tau \sum_{m=0}^{M-1} \int_{\Omega_h} (V(u(\cdot,m\tau)) - V(u_{\varepsilon h}^m))^2 |\nabla u_{\varepsilon h}^m| \Big)^{\frac{1}{2}},$$

where we have set

$$\nu(w) = \frac{\nabla w}{|\nabla w|}, \qquad V(w) = -\frac{w_t}{|\nabla w|},$$

for a generic function w, and the error for the approximation of u by $u_{\varepsilon h\tau}$ in the norms

$$E_{L^\infty(L^2)} = \|u - u_{\varepsilon h\tau}\|_{L^\infty((0,T),L^2(\Omega_h))}$$

and

$$E_{L^2(H^1)} = \|u - u_{\varepsilon h\tau}\|_{L^2((0,T),H^1(\Omega_h))},$$

where we denote by $u_{\varepsilon h\tau}$ the piecewise linear interpolation of $u_{\varepsilon h}^m$ ($m = 0,\ldots,M$) with respect to time. Between two spatial discretization levels h_1 and h_2 we compute the experimental order of convergence

$$eoc(h_1, h_2) = \log \frac{E(h_1)}{E(h_2)} / \log \frac{h_1}{h_2},$$

for the errors $E(h_1)$ and $E(h_2)$, and for each of the error norms E_ν, E_V, $E_{L^\infty(L^2)}$ and $E_{L^2(H^1)}$. For the time step size we have used $\tau = 0.01h$. Obviously the results of the Theorem are reproduced in practical computations. We also observe second order convergence for the L^2-norm and linear convergence for the H^1-norm.

Table 1. The regular case: errors and experimental orders of convergence for the test problem

lev.	h	E_ν	eoc	E_V	eoc	$E_{L^\infty(L^2)}$	eoc	$E_{L^2(H^1)}$	eoc
2	2.1248	1.2140	-	1.3400	-	3.1444	-	4.7184	-
4	1.1798	0.5748	1.27	0.8245	0.83	0.9004	2.13	2.5803	1.03
6	0.6161	0.2730	1.15	0.4063	1.09	0.2234	2.15	1.2476	1.12
8	0.3142	0.1318	1.08	0.1962	1.08	0.05678	2.03	0.6118	1.06
10	0.1585	0.06454	1.04	0.09469	1.07	0.01426	2.02	0.3017	1.03

Mildly singular case. As a first test for the singular case we use the same exact solution as in the previous section, but now we choose the domain to be

$$\Omega = \{ x \in \mathbb{R}^2 \, | \, |x| < 3 \}.$$

There is exactly one singular point within the domain now, namely $x = 0$. The results are shown in Table 2.

Table 2. The mildly singular case: errors and experimental orders of convergence for the test problem with the choice $\varepsilon = h$

lev.	h	E_ν	eoc	E_V	eoc	$E_{L^\infty(L^2)}$	eoc	$E_{L^2(H^1)}$	eoc
2	3.0000	2.8489	-	4.6263	-	15.7814	-	7.5245	-
4	1.7132	1.4571	1.19	3.9263	0.29	6.3723	1.62	4.2222	1.03
6	0.9137	0.6733	1.24	2.9242	0.47	1.8536	1.96	2.0767	1.13
8	0.4713	0.3015	1.21	2.1304	0.48	0.4845	2.03	0.8617	1.33
10	0.2393	0.1382	1.15	1.5250	0.49	0.1219	2.04	0.3822	1.20

We have chosen $\varepsilon = h$ and $\tau = 0.01h$ and we have set $\nu(u) := 0$ where $\nabla u = 0$. The results for similar parameters but with $\varepsilon = h^2$ are shown in Table 3. We observe convergence for $\varepsilon = h$ but the order of convergence for the normal velocities may break down for the choice $\varepsilon = h^2$. Recall, that we did not prove convergence of the discrete solution to the viscosity solution in stronger norms than the $L^\infty(\Omega_h \times (0,T))$-norm. Thus we only expect convergence for the $L^\infty((0,T), L^2(\Omega))$-norm. But the fact that the set of singular points is very small in this case may be responsible for convergence of the computed solution to the viscosity solution for higher norms in this case.

Table 3. The mildly singular case: errors and experimental orders of convergence for the test problem with the choice $\varepsilon = h^2$

lev.	h	E_ν	eoc	E_V	eoc	$E_{L^\infty(L^2)}$	eoc	$E_{L^2(H^1)}$	eoc
2	3.0000	2.7983	-	4.9629	-	16.1827	-	7.8973	-
4	1.7132	1.4510	1.17	4.7450	0.08	8.9268	1.06	6.0774	0.47
6	0.9137	0.6740	1.22	2.8106	0.83	1.6596	2.68	2.0154	1.76
8	0.4713	0.3030	1.21	1.6609	0.80	0.2517	2.85	0.7483	1.50
10	0.2393	0.1389	1.15	1.4901	0.16	0.05850	2.15	0.4523	0.74

The highly singular case. As a computational example for a highly singular case with a large singular set $\{x \in \Omega | \nabla u(x) = 0\}$, we take a mathematically trivial viscosity solution of the mean curvature flow for level sets. But it is numerically very interesting. We choose the piecewise linear function

$$u_0(x_1, x_2) = \begin{cases} x_1 + 1 & : & x_1 \in [-3, -1] \\ 0 & : & x_1 \in [-1, 1] \\ x_1 - 1 & : & x_1 \in [1, 3]. \end{cases}$$

The level lines of u_0 are straight lines, i.e., their curvature is zero. Therefore they do not move by mean curvature flow. The viscosity solution u is given by $u = u_0$. Nevertheless the singular set is the strip $\{x \in \mathbb{R}^2 | |x| < 3, x_1 \in (-1,1)\}$ and so is a large set. We only expect convergence for the $L^\infty((0,T), L^2(\Omega))$-norm. The results are shown in Table 4. The parameters are $\tau = 0.01h$, $\varepsilon = 0.01\sqrt{h}$.

The dumb-bell problem. In Figure 4 we show the results of the computation of the mean curvature flow for a dumb-bell type surface. The computation uses the numerical method from [8] and naturally breaks down when the singularity appears. We also used the numerical methods for the computation of viscosity solutions for the level set formulation of mean curvature flow from

Table 4. The highly singular case: errors and experimental orders of convergence for the test problem with the choice $\varepsilon = 0.01\sqrt{h}$

lev.	h	E_ν	eoc	E_V	eoc	$E_{L^\infty(L^2)}$	eoc	$E_{L^2(H^1)}$	eoc
2	3.0000	2.7027	-	0.4385	-	1.7160	-	2.5481	-
4	1.7132	1.1996	1.45	0.3454	0.43	0.2753	3.27	0.7667	2.14
6	0.9137	1.2003	0.00	0.3553	-0.05	0.1955	0.54	0.7988	-0.07
8	0.4713	0.9483	0.36	0.3043	0.23	0.1024	1.16	0.7592	0.08
10	0.2393	0.6688	0.52	0.2602	0.23	0.04739	1.14	0.5688	0.43

this paper to compute the evolution across the singularity for the same initial surface. We have used the oriented distance function of the initial dumb-bell as initial value for the level set problem. We kept the boundary data fixed. The problem is axially symmetric and so we have solved the corresponding two dimensional problem in the x_1-r-plane, where $r = \sqrt{x_2^2 + x_3^2}$. The computational domain was $\Omega = (-2, 2) \times (-2, 2)$. The differential equation for the axially symmetric problem is similar to the twodimensional problem except for a weight r in the integrals. Figure 2 shows level lines of the initial function and level lines of the solution at the time when the surface breaks apart. Figure 3 shows the results for several time steps.

Fig. 2. Level lines of the initial values (left) and of the solution at the 325-th time step for the axially symmetric dumb-bell

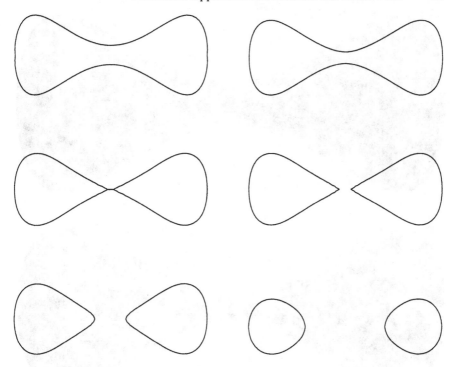

Fig. 3. The 0-level line for the axially symmetric dumb-bell problem at time steps 0, 200, 314, 325, 400 and 1000

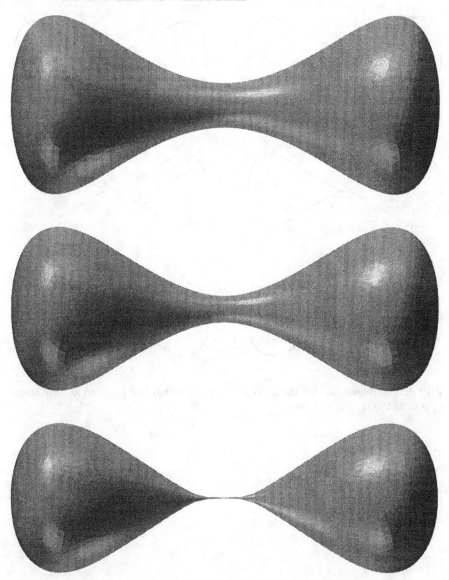

Fig. 4. Formation of a singularity in mean curvature flow

References

[1]   CHEN, Y-G., GIGA, Y. & GOTO, S. Uniqueness and existence of viscosity solutions of generalized mean curvature flow equations. *J. Diff. Geom* **33** (1991), 749–786

[2] DECKELNICK, K. Error analysis for a difference scheme approximating mean curvature flow. *Interfaces and Free Boundaries* **2** (2000), 117–142

[3] DECKELNICK, K. & DZIUK, G. Convergence of a finite element method for non–parametric mean curvature flow. *Numer. Math.* **72** (1995), 197–222

[4] DECKELNICK, K. & DZIUK, G. Discrete anisotropic curvature flow of graphs. *M2AN Math. Model. Numer. Anal.* **33** (1999), 1203–1222

[5] DECKELNICK, K. & DZIUK, G. Error estimates for a semi implicit fully discrete finite element scheme for the mean curvature flow of graphs. *Interfaces and Free Boundaries* **2** (2000), 341–359

[6] DECKELNICK, K. & DZIUK, G. A fully discrete numerical scheme for weighted mean curvature flow. *Numer. Math.* **91** (2002), 423–452

[7] DECKELNICK, K. & DZIUK, G. Convergence of numerical schemes for the approximation of level set solutions to mean curvature flow. *Preprint Mathematische Fakultät Freiburg* (2000), 1–17

[8] DZIUK, G. An algorithm for evolutionary surfaces. *Numer. Math.* **58** (1991), 603–611

[9] DZIUK, G. Numerical schemes for the mean curvature flow of graphs. *In P. Argoul, M. Frémond, Q. S. Nguyen (Eds.): IUTAM Symposium on Variations of Domains and Free-Boundary Problems in Solid Mechanics. 63–70, Kluwer Academic Publishers, Dordrecht-Boston-London 1999*

[10] EVANS, L.C. & SPRUCK, J. Motion of level sets by mean curvature I. *J. Diff. Geom.* **33** (1991), 636–681

[11] EVANS, L.C. & SPRUCK, J. Motion of level sets by mean curvature II. *Trans. Amer. Math. Soc.* **330** (1992), 321–332

[12] EVANS, L.C. & SPRUCK, J. Motion of level sets by mean curvature III. *J. Geom. Anal.* **2** (1992), 121–150

[13] EVANS, L.C. & SPRUCK, J. Motion of level sets by mean curvature IV. *J. Geom. Anal.* **5** (1995), 77–114

[14] FRIED, M. Niveauflächen zur Berechnung zweidimensionaler Dendrite. *Dissertation Freiburg* 1999

[15] HUISKEN, G. Non–parametric mean curvature evolution with boundary conditions. *J. Differential Equations* **77** (1989), 369–378

[16] LIEBERMAN, G. A. The first initial–boundary value problem for quasilinear second order parabolic equations *Ann. Sci. Norm. Sup. Pisa Ser. IV* **8** (1986), 347–387

[17] OLIKER, V.I. & URALTSEVA, N.N. Evolution of nonparametric surfaces with speed depending on curvature II. The mean curvature case. *Comm. Pure Appl. Math.* **46** (1993), 97–135

[18] OSHER, S. & SETHIAN, J.A. Fronts propagating with curvature dependent speed: Algorithms based on Hamilton–Jacobi formulations. *J. Comp. Phys.* **79** (1988), 12–49

[19] SETHIAN, J.A. *Level set methods.* Cambridge Monographs on Applied and Computational Mathematics **3**, Cambridge University Press (1996)

[20] SONER, H.M. Motion of a set by the curvature of its boundary. *J. Differ. Equations* **101** (1993), 313–372

[21] WALKINGTON, N.J. Algorithms for computing motion by mean curvature. *SIAM J. Numer. Anal.* **33** (1996), 2215–2238

[22] YSERENTANT, H. On the multi–level splitting of finite element spaces. *Numer. Math.* **49** (1986), 379–412

Reaction-Diffusion Systems Arising in Biological and Chemical Systems: Application of Singular Limit Procedures

Masayasu Mimura

Department of Mathematical and Life Sciences
Institute of Nonlinear Sciences and Applied Mathematics
Graduate School of Science, Hiroshima University, Japan
mimura@math.sci.hiroshima-u.ac.jp

Contents

1 What is diffusion?

Diffusion enhances spatial homogenization. Nonlinear phenomena related to diffusion appear in chemistry, biology and other scientific fields. In order to theoretically understand these phenomena, many discrete and continuous diffusion models have been proposed.

1.1 Discrete diffusion models

Two cells model We consider the simple situation where two cells C_1 and C_2 are connected by a diffusive membrane and $u_1(t)$ and $u_2(t)$ be the concentrations of a single chemical species at time t in C_1 and C_2. Assuming that each species flows from one cell to the other, in proportional to the difference of their concentrations, the time evolution of $u_1(t)$ and $u_2(t)$ is simply given by

$$\begin{aligned} u_1'(t) &= \underline{d}[u_2(t) - u_1(t)] \\ u_2'(t) &= \underline{d}[u_1(t) - u_2(t)] \end{aligned} \quad , \quad t > 0, \tag{1}$$

where $'$ indicates the differentiation with respect to time and d is the diffusion coefficient of the membrane. By using the property $u_1(t) + u_2(t) = u_1(0) + u_2(0)$, one can easily obtain

$$\lim_{t\to\infty} u_1(t) = \lim_{t\to\infty} u_2(t) = \frac{u_1(0) + u_2(0)}{2}, \tag{2}$$

that is, the concentration of the species in the two cells becomes equal. This clearly implies that (1) possesses the effect of diffusion.

N cells model Let $u_1(t)$, $u_2(t)$, ..., $u_N(t)$ be the concentrations of a single chemical species in N cells C_1, C_2, ..., C_N, which are one-dimensionally connected by diffusive membranes. A diffusion model corresponding to (1) is

$$u_1'(t) = \underline{d}[u_2(t) - u_1(t)] = \underline{d}\nabla + u_1(t)$$
$$u_2'(t) = \underline{d}[u_1(t) - u_2(t)] + d[u_3(t) - u_2(t)] = \underline{d}\nabla + \nabla_- u_2(t) = \underline{d}\Delta_{+-} u_2(t)$$

$$\cdots$$

$$u_i'(t) = \underline{d}[u_{i-1}(t) - u_i(t)] + d[u_{i+1}(t) - u_i(t)] = \underline{d}\Delta_{+-} u_i(t)$$

$$\cdots$$

$$u_{N-1}'(t) = \underline{d}[u_{N-2}(t) - u_{N-1}(t)] + d[u_N(t) - u_{N-1}(t)] = \underline{d}\Delta_{+-} u_{N-1}(t)$$
$$u_N'(t) = \underline{d}[u_{N-1}(t) - u_N(t)] = -\underline{d}\nabla - u_N(t)$$

$$\tag{3}$$

for $t > 0$. Similarly to (2), we can also show

$$\lim_{t\to\infty} u_i(t) = \frac{u_1(0) + u_2(0) + \cdots + u_N(0)}{N} \qquad (i = 1, 2, \ldots, N), \tag{4}$$

that is, $u_i(t)$ $(i = 1, 2, \ldots, N)$ tends to the average of $u_1(0)$, $u_2(0)$, ... $u_N(0)$. System (3) can be easily extended to general two-dimensional discrete model for $u_{ij}(t)$ in N^2 cells C_{ij} $(i, j = 1, 2, \ldots, N)$.

1.2 Continuous models

We point out that (3) can be naturally represented in the continuous version. Let $u(t, x)$ be the concentration or the density of a single species in isotropic diffusive medium at time t and position x. Then, it is described by

$$u_t = d\Delta u, \qquad t > 0, \; x \in \Omega, \tag{5}$$

where d is the diffusion rate, Δ is the Laplacian operator in $\mathbb{R}^n$. Consider (5) in a bounded domain Ω in $\mathbb{R}^n$ under the following initial and boundary conditions

$$u(0, x) = a(x), \qquad x \in \Omega, \tag{6}$$

$$\frac{\partial u}{\partial \nu} = 0, \qquad t > 0, \; x \in \partial\Omega, \tag{7}$$

where ν is the outerward normal vector on the boundary $\partial\Omega$. One knows

$$\lim_{t \to \infty} u(t, x) = <a>, \tag{8}$$

where $<f>$ means the spatial average of $f(x)$ in Ω. It turns out that the concentration becomes homogeneous in space, which is obviously a continuous version of (4). Furthermore, (5) is extended to

$$u_t = d\Delta u + f(u) , \qquad t > 0, \; x \in \Omega, \tag{9}$$

where $f(u)$ is a reaction term (9) is simply called a scalar *reaction-diffusion (RD) equation*. In the fields of chemistry, ecollogy and biology, a lot of models described by the form (9) have been proposed so far.

For the asymptotic behavior of solutions of (9) with (6) and (7), the following results are already known:

(i) w-limit set consists of equilibrium solutions only [1];
(ii) when the domain Ω is convex, any non-constant equilibrium solutions are unstable, even if they exist [2];
(iii) there are suitable non-convex domain Ω and $f(u)$ such that there are stable non-constant equilibrium solutions [3].

These properties indicate that if the domain Ω is convex, any bounded solution generically approach one of the stable constant equilibrium solutions which are given by stable critical points of the diffusionless equation of (9)

$$u_t = f(u) , \qquad t > 0. \tag{10}$$

This result seems to be obvious, because we already know that diffusion enhances spatial homogeneization. Therefore, it was long believed that RD equations could not generate any spatial patterns and waves and therefore these were not interested from pattern formation viewpoints.

2 Paradox of diffusion

In 1952, two surprising paradoxical evidences of diffusion were demonstrated by using mathematical models. The first contributor is a mathemacian, A. Turing, who stated that *diffusion enhances spatial inhomogeneities* [4]. In order to explain this paradox, he proposed a discrete two cells model, in which two different types of chemical species, one activator and its inhibitor, are included. Let $A_i(t)$ and $I_i(t)$ be respectively the concentrations of the activator and its inhibitor at time t in the i-th cell ($i = 1, 2$). The resulting system for $(u_i(t), v_i(t))$ ($i = 1, 2$) is described by the following four-dimensional ODEs

$$\begin{aligned}
A_1' &= \underline{d}_A[A_2 - A_1] + f(A_1, I_1) \\
I_1' &= \underline{d}_I[I_2 - I_1] + g(A_1, I_1) \\
A_2' &= \underline{d}_A[A_1 - A_2] + f(A_2, I_2) \\
I_2' &= \underline{d}_I[I_1 - I_2] + g(A_2, I_2)
\end{aligned} \qquad , \qquad t > 0, \tag{11}$$

where $\underline{d}_A$ and $\underline{d}_I$ are respectively the diffusion coefficients of the activator and its inhibitor. f and g describe the interaction of the activator and its inhibitor. The interaction is given as follows

(i) Activator grows in an autocatalytic way;
(ii) Activator produces inhibito, by which its growth is inhibited;
(iii) Inhibitor decays in a constant rate.

When $A_1 = A_2 = A$ and $I_1 = I_2 = I$, (11) reduces to

$$\begin{array}{l} A' = f(A, I) \\ I' = g(A, I) \end{array}, \quad t > 0. \tag{12}$$

Suppose that there is a critical point, say (A, I), of (12). Then

$$(A_1, I_1; A_2, I_2) = (A, I, ; A, I) \quad \text{is a critical point of (11).}$$

We call it a homogeneous critical point, because the states in two cells C_1 and C_2 are equal. Assume that (A, I) is stable in (12). Turing claimed that the critical point $(A, I; A, I)$ is not necessarily stable in (11). In fact, if $\underline{d}_I$ is suitably larger than $\underline{d}_A$, the critical point is *unstable* in (11), in other words, the difference of $A_1(t)$ and $A_2(t)$ (and also $I_1(t)$ and $I_2(t)$) increases with time. This is called *diffusion-induced instability*. In order to demonstrate this instability explicitly, we propose a linearized system of (12) with a critical point $(A, I) = (0, 0)$ for simplicity,

$$\begin{array}{l} A' = 5A - 6I \\ I' = 6A - 7I \end{array}, \quad t > 0. \tag{13}$$

It is obvious to see that $(0, 0)$ is asymptotically stable, that is,

$$\lim_{t \to \infty} (A(t), I(t)) = (0, 0).$$

We now consider the two cell systems for $(A_1, I_1; A_2, I_2)$ corresponding to (11)

$$\begin{array}{l} A_1' = \underline{d}_A[A_2 - A_1] + 5A_1 - 6I_1 \\ I_1' = \underline{d}_I[I_2 - I_1] + 6A_1 - 7I_1 \\ A_2' = \underline{d}_A[A_1 - A_2] + 5A_2 - 6I_2 \\ I_2' = \underline{d}_I[I_1 - I_2] + 6A_2 - 7I_2 \end{array}, \quad t > 0. \tag{14}$$

Of course, $(A_1, I_1; A_2, I_2) = (0, 0; 0, 0)$ is the homogeneous critical point which is independent of the diffusion rates $\underline{d}_A$ and $\underline{d}_I$. In order to study the stability of this critical point, we introduce four new variables P, Q, R and S by

$$\begin{array}{ll} A_1 + A_2 = P, & A_1 - A_2 = R, \\ I_1 + I_2 = Q, & I_1 - I_2 = S. \end{array} \tag{15}$$

Then (14) is rewritten as

$$P' = P - 6Q$$
$$Q' = 6P - 7Q \tag{16}$$

and

$$R' = 5R - 6S - 2\underline{d}_A R$$
$$S' = 6R - 7S - 2\underline{d}_I S \tag{17}$$

One find that (16), (17) are decoupled. Since (16) is the same as (13), we see that the critical point $(P, Q) = (0, 0)$ is asymptotically stable, that is,

$$\lim_{t \to \infty} (A_1(t) + A_2(t), I_1(t) + I_2(t)) = (0, 0). \tag{18}$$

We next consider (17). The stability analysis indicates the following: if $\underline{d}_A$ and $\underline{d}_I$ satisfy $(5 - 2\underline{d}_A)(7 + 2\underline{d}_I) < 36$, the critical point $(R, S) = (0, 0)$ is asymptotically stable, while if $(5 - 2\underline{d}_A)(7 + 2\underline{d}_I) > 36$, it is unstable in $(\underline{d}_A, \underline{d}_I)$ - plane, as shown in Fig. 1.

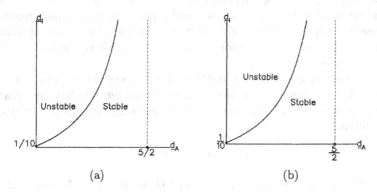

(a)   (b)

Fig. 1. Bifurcation curve in $(\underline{d}_A, \underline{d}_I)$-plane

Consequently, if $(R, S) = (0, 0)$ is asymptotically stable, we infer

$$\lim_{t \to \infty} (A_1(t) - A_2(t), I_1(t) - I_2(t)) = (0, 0), \tag{19}$$

so that, by (18),

$$\lim_{t \to \infty} A_1(t) = \lim_{t \to \infty} A_2(t) = 0 \quad \text{and} \quad \lim_{t \to \infty} I_1(t) = \lim_{t \to \infty} I_2(t) = 0,$$

that is, the homogeneous critical point $(0, 0; 0, 0)$ is asymptotically stable. On the other hand, if $(R, S) = (0, 0)$ is unstable, even if the difference of $A_1(0)$ and $A_2(0)$ or $I_1(0)$ and $I_2(0)$ is small, the differences of $A_1(t)$ and

$A_2(t)$ or $I_1(t)$ and $I_2(t)$ are exponentially increasing. In fact, suppose $\underline{d}_A$ is arbitrarily fixed in order to satisfy $0 < \underline{d}_A < 2/5$. By Fig. 1, one finds that, if $\underline{d}_I$ is small, $(0,0;0,0)$ is asymptotically stable; however, if $\underline{d}_I$ is increasing, it becomes unstable. This is the diffusion-induced instabilty which Turing originally stated in 1952. The idea of the above is naturally extended to an N-cells model and furthermore, to the following two component RD equations

$$\begin{array}{c} u_t = d_u \Delta u + f(u,v) \\ v_t = d_v \Delta v + g(u,v) \end{array} \quad , \quad t > 0, \ x \in \Omega \subset \mathbb{R}^n, \tag{20}$$

in a bounded domain Ω, where d_u and d_v are the diffusion coefficients of u and v, respectively. Under the zero flux boundary conditions, numerical calculations reveal that RD systems of the form (20) generate a surprisingly variety of spatial and/or spatio-temporal patterns, when f and g fall into the framework of activator-inhibitor interaction.

The second contributors are two neurophysiologists, A. L. Hodgkin and A. F. Huxley who investigated the mechanism of impulses with constant shape and velocity propagating along nerve fiber. They proposed a RD model to describe the propagation of impulses along the fiber [5]. The model is given by the coupling of a single RD equation and three ODEs. Numerical simulation showed that their model generated a traveling pulse which represents propagating impulse. This is also another paradoxical evidence of diffusion in a sense that *diffusion generate a localized wave*, if it is coupled with some reaction process.

These two evidences arising in RD equations seem to contradict the result stated in the previous section. However, we should remark that the equations which generate such patterns and waves are not single equations but systems of the following form

$$U_t = D\Delta U + F(U), \tag{21}$$

where $U(t,x) = (u_1, u_2, \ldots, u_N)(t,x)$, D is an $N \times N$ diagonal matrix with non-negative elements $\{d_i\}(i = 1, 2, \ldots, N)$. Moreover, $F(U) = (f_1(U), f_2(U), \ldots, f_N(U))$ describes suitable reaction terms which represents the interaction among U.

In the following section, as an example, we will introduce bistable RD equations and demonstrate how the interplay of diffusion and reaction generates complex spatio-temporal patterns which can not be expected by the diffusionless system of (21).

3 Diffusive patterns and waves - bistable RD equations

In general, RD equations of the form (21) consist of two well studies equations. One is the heat equation to describe "diffusion process"

$$u_t = d\Delta u \qquad (22)$$

and the other is a system of ODEs to describe "reaction process"

$$U_t = F(U). \qquad (23)$$

Then, a naive question arises: can qualitative properties of solutions to RD equations (21) be known by the information of the diffusionless equations (23)? In the next section, we will show that this can not be necessarily expected.

3.1 Scalar bistable RD equation

Let us first consider a scalar RD equation with cubic nonlinearity $f(u) = u(1 - u)(u - a)$ satisfying $0 < a < 1$

$$u_t = d\Delta u + f(u), \qquad t > 0, \ x \in \Omega \subset \mathbb{R}^n, \qquad (24)$$

where d is the diffusion coefficient. We simply assume that Ω is a bounded and convex domain. One easily knows that stable equilibrium solutions are spatially constant equilibria $u = 0$ and $u = 1$ only, while the rest $u = a$ is unstable. Therefore, we call that (22) is a scalar bistable RD equation. First consider the one dimensional problem of (24)

$$u_t = du_{xx} + f(u), \qquad t > 0, \ x \in (-L, L), \qquad (25)$$

under the zero flux boundary conditions $x = \pm L$ and the initial conditions

$$\begin{aligned}
u(0, x) = 1 \quad &\text{for} \quad -L < x < 0 \\
u(0, x) = 0 \quad &\text{for} \quad 0 < x < L
\end{aligned} \qquad (26)$$

Then, because of bistability, one can expect the occurrence of a single transition layer, connecting two states where u takes nearly 1 at the left hand side and 0 in the right. After that, the layer generally propagates to either the wall $x = L$ or $x = -L$, as if its shape and velocity are both constant, as in Fig. 2, and eventually it approaches to the wall and then the solution u becomes one of the constant states: either $u = 0$ or $u = 1$. This expectation arrives at that the final state would be either $u = 0$ or $u = 1$. It turns out that the result is the same as the one of the corresponding ODE

$$u_t = f(u), \qquad t > 0. \qquad (27)$$

Only the difference is the transient behavior of solutions. Here we emphasize that the information on travelling front solutions play an important role on the study of this behavior.

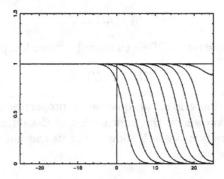

Fig. 2. Layer propagation in (25)-(26)

(i) Travelling front solutions

We consider a special solution of the form $u(t, x) = u(z)$ with $z = x - ct$ in $-\infty < z < \infty$, where c is called the velocity. It satisfies

$$-cu' = du'' + f(u), \qquad -\infty < z < \infty, \qquad (28)$$

with the boundary conditions

$$u(-\infty) = 1 \quad \text{and} \quad u(\infty) = 0. \qquad (29)$$

For the problem (28), (29), the velocity c is explicitly given by $c = c(a) = (1-2a)/\sqrt{2}$ for which there is a unique solution $\phi(z; a)$ except for translational free. If $0 < a < 1/2$ (resp. $1/2 < a < 1$), $c > 0$ (resp. $c < 0$), $\phi(x - c(a)t; a)$ propagates to the right (resp. left) direction. We therefore call it a *travelling front*. In particular, for the equi-stable case when $a = 1/2$, $c = 0$ so that $\phi(x; 1/2)$ is a nonconstant equilibrium solution, which is called a *standing front*. By this information on travelling velocity, one can expect that if $c > 0$, the layer in (25), (26) propagates to the right direction so that the final state solutions could be $u = 1$, while if $c < 0$, the situation is opposite. The equi-stable case when $c = 0$ is exceptional. The standing wave solution indicates that the layer does not move, as if the solution u is an equilibrium solution of (25). However, we already know that any nonconstant equilibrium solutions are unstable so that the layer should move under small perturbations. Thus, the following question naturally arises: "how does the transition layer move?". Unfortunately, the standing wave solution does not answer to this question. It will be explained later.

Except for the equi-stable case, the transient as well as asymptotic behaviors of the one-dimensional case of (24) are inherited to the two-dimensional case where the initial function $u(0, x, y)$ is a suitably large delta-like function. If $c > 0$ ($0 < a < 1/2$), $u(t, x)$ grows and expands uniformly so that the final state is $u = 1$, while if $c < 0$ ($1/2 < a < 1$), it shrinks so that the final state is $u = 0$. The equi-stable case when $a = 1/2$ is very delicate similarly to the one-dimesional case.

(ii) Slow diffusion and fast reaction limit

In order to know the time evolution of transition layer, we consider the special situation where diffusion is very slow and reaction is very fast so that (24) is rewritten as the following RD equation, with a small parameter $\varepsilon > 0$,

$$u_t = \varepsilon \Delta u + \varepsilon^{-1} f(u), \qquad t > 0, \; (x, y) \in \Omega \subset \mathbb{R}^2. \tag{30}$$

Because ε is sufficiently small, there generally appear transition sharp layer regions with narrow width in a short time, which seperate two qualitatively different regions where u takes nearly either 1 or 0. Singular perturbation arguments say that such layer regions become interfacial curves, say $\Gamma(t)$, in the limit $\varepsilon \downarrow 0$. For sufficiently small $\varepsilon > 0$, the time evolution of such layer regions is described well approximately by the following evolution equation for the interface $\Gamma(t)$

$$V(t) = c(a) - \varepsilon \kappa, \qquad t > 0, \; (x, y) \in \Gamma(t), \tag{31}$$

where $V(t)$ is the normal velocity of the interface $\Gamma(t)$ and κ is the curvature of the interfacial curve [6]. If the initial interface $\Gamma(0)$ is given by a circle with radius r_0, for instance, then the resulting interface $\Gamma(t)$ is also given by a circle with radius $r(t)$ which is a solution of

$$
\begin{aligned}
r' &= c(a) - \varepsilon/r, \qquad t > 0, \\
r(0) &= r_0.
\end{aligned}
\tag{32}
$$

When $c(a) > 0$, one ovbiously finds that, if $r_0 > \varepsilon/c(a)$, the circular interface expands, while when $c(a) < 0$, it shrinks to zero in finite time. This suggests us that $u(t, x)$ of (30) eventually tends to the constant state $u = 1$ for the former case, and $u = 0$ for the latter case.

In particular, for the equi-stable case when $a = 1/2$ or $c = 0$, (31) reduces to an equation for motion of curvature

$$V(t) = -\varepsilon \kappa, \qquad t > 0, \tag{33}$$

which has been intensively investigated in the fields of differential geometry and nonlinear PDEs ([7], for instance). For the equi-stable case, we find that the curvature effect generically plays an important role on the dynamics of layers in higher dimensions.

A special case is the one dimensional problem where curvature effect is absent, for which an interfacial curve is given by one point, say $x = \zeta(t)$, so that the time evolutional equation for $\zeta(t)$ is simply given by

$$\zeta'(t) = 0, \tag{34}$$

which says that the interfacial point does not move for any location in the interval $(-L, L)$. However, if ε is not zero, the correponding layer in (25) has to generically move. It is shown in [8] that its velocity is extremely slow.

In the following, we fix $a = 1/2$ (equi-stable case) and consider (30) within a rectangular domain $\Omega = \{(x,y) \in \mathbb{R}^2 \mid -L < x < L, -L < y < L\}$ with zero-flux boundary conditions.

Take an initial interfacial curve $x = \xi(y)$ specified by

$$
\begin{aligned}
u(0, x, y) = 1 & \quad -L < x < \xi(y), \quad -L < y < L, \\
u(0, x, y) = 0 & \quad \xi(y) < x < L, \quad\quad -L < y < L,
\end{aligned}
\tag{35}
$$

where $x = \xi(y)$ is given by the line $x = 0$ with a small deformation. As shown in Fig. 3, numerical simulation indicates that the deformation is instantly decaying and the resulting interfacial curve becomes almost straight and moves extremely slowly in the one-dimension fashion. We thus find that the final state is very simple, that is, either $u = 1$ or $u = 0$. This result is not surprising, because the system consists of diffusion and bistable reaction.

Fig. 3. Transversal stability of interfacial curve in (30)

Finally, in higher dimensions with $n \geq 3$, (31) can be extended to

$$
V(t) = c(a) - \varepsilon(n-1)\kappa, \quad t > 0,
\tag{36}
$$

where n is the spatial dimenionality and κ is the mean curvature of the interface. For the property of time evolution of interfaces, the readers should refer to [9, 10] and the references therein.

3.2 Two component systems of bistable RD equations

We consider the following two component RD system, with a small parameter $\varepsilon > 0$

$$
\begin{aligned}
\varepsilon \tau u_t &= \varepsilon^2 \Delta u + f(u) - v \\
v_t &= d \Delta u + u - \gamma v
\end{aligned}
, \quad t > 0, \ x \in \Omega \subset \mathbb{R}^2,
\tag{37}
$$

where τ, ε and γ are positive constants and $f(u) = u(1 - u)(u - a)$ with $0 < a < 1$. This is one of typical RD systems which Turing proposed to explain diffusion induced instability, where the unknowns u and v are respectively called the activator and its inhibitor. Suppose that γ is suitably large such that the nulclines $f(u) = v$ and $u = \gamma v$ intersect at three points, say, P, Q and R, as in Fig. 4. If γ tends to infinity, v becomes zero so that the first equation of (37) reduces to the scalar equation of the same type as (24). In this sense, (37) is a natural extension of (24).

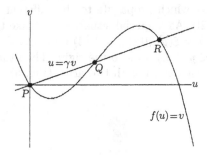

Fig. 4. Curves of $f(u) = v$ and $u = \gamma v$ in (u, v)-plane

First, we consider the diffusionless system of (37), that is,

$$\begin{aligned} \varepsilon\tau u_t &= f(u) - v \\ v_t &= u - \gamma v \end{aligned} \quad , \quad t > 0. \tag{38}$$

One finds that the critical points P and R are asymptotically stable, while Q is unstable. That is, (37) is a bistable RD system. Let $\gamma = \gamma^*$ be the value such that P and R are odd symmetric with respect to Q. This situation implies that P and R are equi-stable (which corresponds to $a = 1/2$ in (24)). Consider (37) in a rectangluar domain $\Omega = \{(x,y) \in \mathbb{R}^2 \mid -L < x < L, -L < y < L\}$ with zero-flux boundary conditions. We first note that P and R are stable spatially homogeneous equilibrium solutions of (37). Therefore, the following question arises: "Are there any other asymptotic states of solutions in addition to the spatial constant equilibria P and R?". In order to answer this question, we fix $\gamma = \gamma^*$ and also fix ε sufficiently small and a such that $0 < a < 1/2$, respectively. We take t as a free parameter. Specify the following initial conditions, which are similar to the scalar case (35), as

$$\begin{aligned} (u(0,x,y), v(0,x,y)) &= P & -L < x < \xi(y), \ -L < y < L, \\ (u(0,x,y), v(0,x,y)) &= R & \xi(y) < x < L, \ -L < y < L. \end{aligned} \tag{39}$$

Since ε is sufficiently small, singular perturbation arguments say that there appear transition sharp layers in the u-component, while the other v is smooth

[11]. We are now interested in time evolution of layer in the u-component only. For large τ, the initial deformation instantly decays and the resulting layer becomes uniform with respect to the y-direction and it slowly moves to the center line $x = 0$, where the solution tends to an odd symmetric equilibrium. This can be expected, because the nonlinearity of f and g takes odd-symmetry. On the other hand, for small τ, the situation is drastically changed. Some parts of the curve $x = \xi(y)$ move to the left direction and some parts move to the right direction so that the resulting curve becomes very complicated. Numerical calculation strongly suggests the coexistence of two travelling fronts which propagate to the different directions (for the rigorous proof, see [12]). As a special situation, we take the initial interface such that one part of $x = \xi(y)$ $(0 < y < L/2)$ moves to the left direction and the other part $(L/2 < y < L)$ does to the right. Then the resulting pattern exhibits a steadily rotating spiral with two arms (Fig. 5).

Fig. 5. Spiral wave with two arms arising in (37)

For intermediate values of τ, the resulting pattern is totally different from the aboves. We take the same initial interface as the above. Initial pattern begins to envolve into spiral but it gradually breaks down and changes to a very complicated labyrinthine pattern after large time (Fig. 6). The occurrence of such spiral and labyrinthine patterns can not be expected from the corresponding ODEs (38).

This numerical simulations show that, although the interaction of activator and its inhibitor possesses simple bistable mechanism, the interplay of this interaction with diffusion possibly generates unexpected complex spatiotemporal patterns, even if the system looks so simple. It is now clear that the study of qualitative properties of solutions of RD systems is very challenging from not only applied sciences but also purely mathematical viewpoints.

Fig. 6. From break up of spiral to labyrinthine pattern in (31)

4 Resource-consumer RD systems

In this section, we will discuss a class of RD systems which possess consumer-resource interaction, which arise in fields of biology and chemistry.

4.1 Bacterial colony model

The diversity and complexity of spatio-temporal patterns arising in biological systems often take place through the intertwinning of complex biological factors and environmental conditions. Intensive investigations have shown that the growth of bacterial colonies produces various complex patterns, depending on the species and environmental conditions. For instance, the bacterial species called Bacillus subtilis is known to exhibit qualitatively different two-dimensional colony patterns as a result of growth and cell division on the surface of thin agar plates by feeding on nutrients [13], [16]. Colony patterns change drastically when the concentrations of agar and nutrient, say C_a and C_n, are globally varied. They are qualitatively classified into five types, each of which is observed in the regions labeled $A - E$ in the (C_a, C_n^{-1})-plane, as in Fig. 7. In the region A (hard agar medium containing poor nutrient), colony patterns exhibit tip-splitting growth with characteristically branched structures (Fig. 8). These patterns are similar to those observed in diffusion limited processes in solidification from a supersaturated solution, solidification from an undercooled liquid, and electro-chemical deposition [14]. Also, it was reported that averages over about 25 samples of colony patterns in experiments yield a fractal dimension of 1.72 ± 0.02. This is indeed in good agreement with the results from two-dimensional DLA particle model. Increasing C_n with high fixed C_a values (corresponding to the change from the region A to B), the branch thickness of the colony increases gradually and colony patterns eventually become Eden-like. For high C_n and low C_a (soft agar medium with rich nutrient) in the region D, colony patterns drastically change to be spread homogeneously, which look macroscopically like a perfect disk. It is likely that the movement of bacterial cells inside a colony can be described in terms of diffusion, and there is no microscopic branching at all. In the region E, between the regions A and D, there emerge colony patterns clearly reminiscent of the so-called dense-branching morphology (DBM). Though the branching is very dense, the advancing envelope looks characteristically smooth compared with DLA-like colonies in the region A. Finally, in the region C, between the regions B and D, colonies spread and rest alternately, leaving stationary concentric ring-like patterns.

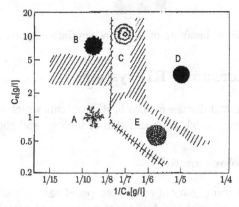

Fig. 7. Phase diagram of colony patterns

Fig. 8. Branched pattern in Region A

The above experimental observations lead to the following naive question

Is the diversity of colony patterns observed in experiments caused by different effects or governed by the same underlying mechanisms?

To theoretically understand the mechanism generating branched colony patterns, a cell-walker discrete model coupled with a diffusion equation for the nutrient concentration was proposed [15]. The numerical results confirm that the model generates patterns similar to the ones observed in the regions A and E.

On the other hand, a continuous model of RD equation for growth patterns in the region D was proposed [16]. Let $b(t, x)$ be the population density of bacterial cells at time t and position x. The equation describing $b(t, x)$ is

$$b_t = d_b \Delta b + (b_0 - \mu b), \tag{40}$$

where d_b is the diffusion coefficient of bacterial cells, b_0 is the initial concentration of nutrients, and μ is the intraspecific competition rate of cells. The parameters are all positive constants. The type of (40) is called the Fisher equation in mathematical genetics. If the initial function $b(0, x)$ is a point-like distribution, the solution $b(t, x)$ spreads like an expanding disk. It is confirmed that colony patterns in the region D are consistent with the behavior of solutions of (40) [16]. As a natural extension of (40), there is a model for the nutrient-limited growth of bacteria. It is given by

$$b_t = d_b \Delta b + \omega g(n)b,$$
$$n_t = d_n \Delta n - g(n)b, \tag{41}$$

where n is the concentration of nutrients, d_n is the diffusion coefficient of the nutrients, $g(n)$ is the growth rate. A simple form is the Malthusian rate $g(n) = n$. The positive constant ω is the conversion rate of nutrients from intake to growth. The initial conditions for (41) are

$$b(0, x) = b_0(x),$$
$$n(0, x) = n_0, \tag{42}$$

where $b_0(x)$ is a delta-like distribution of the initial density of the bacteria and n_0 is the initial concentration of nutrients which is distributed uniformly in space. We note that the parameters (n_0, d_b) in the model (41), (42) correspond to (C_n, C_a^{-1}) in the experiments. However, even if two parameters are globally varied, the resulting patterns generate disk-like patterns only. In order to improve this model so as to generate branched patterns, Kawasaki et al. [17] proposed a nonlinear diffusion model which is described by

$$b_t = \Delta(\sigma(x, y)nb\Delta b) + nb,$$
$$n_t = d_n \Delta n - nb, \tag{43}$$

where b and n are the same as the ones in (41) and $\sigma(x, y)$ is a diffusion coefficient with small spatial fluctuations. Furthermore, along the same line as (41), Kitsunezaki [18] proposed a density-dependent diffusion model including the death term

$$b_t = \Delta(d_b b \Delta b) + nb - \alpha b,$$
$$n_t = d_n \Delta n - nb,$$
(44)

where d_b and α are positive constants. Bacterial corpse, say $w(t,x)$, is given by $w_t = \alpha b$. Numerical simulations of these two models reproduce branched patterns ranging from DBM-like patterns to disk-like patterns.

From the modelling perspective, we still have the following problem

Is there any unified model which generates all of growth patterns observed in experiments when two parameters corresponding to C_n and C_a are varied globally?

For this problem, we propose a continuous model for the density of bacterial cells and the concentration of nutrient at the macroscopic level [19]. The essential ansatz in our modelling is the introduction of the internal state in each cell, so that, if this state variable is increasing, then the bacteria actively move, grow, and perform cell-division; while if it is decreasing, then it does nothing at all. In order to model this situation in a simple way, we assume that the bacterial cells consist of two types: active cells and inactive ones. In fact, a group of actively moving cells is clearly observed in the tip of each growing finger in the region E, as if they were a finger nail (Fig. 9).

Fig. 9. Caricature of finger-nail pattern

They seem to drive the growth of the finger tip, leaving inactive cells behind. Let $b(t,x)$ and $s(t,x)$ be the concentrations of the active and inactive cells, respectively, (hence the sum $b(t,x) + s(t,x)$ is the concentration of all bacterial cells) and $n(t,x)$ is the concentration of nutrients. The model for b, s and n is given by

$$b_t = \Delta(d_b \Delta b) + \omega g(n)b - a(b,n)b,$$
$$n_t = d_n \Delta n - g(n)b,$$
(45)
$$s_t = a(b,n)b,$$

where $d_b = d(k)$ and d_n are the diffusion rates of the bacterial cells and the nutrient, respectively. Here k is the parameter corresponding to agar concentrations. A plausible assumption is that $d(k)$ is monotone decreasing with respect to k. A simple form of $d(k)$ is $d(k) = d_0/(1+k)$, with some positive constant d_0. The growth rate of active cells is $\omega g(n)$. We assume $g(n)$ to be an exponential (or Malthusian) growth rate of cells. The rate of change from the active cells to the inactive ones is $a(b, n)$, which is specified by the mechanism of the internal state. We neglect the rate of change from the inactive cells to the active ones, because it is observed that once active cells become inactive, they never become active again unless food is added artificially. It is plausible that $a(\cdot, n)$ decreases with an increase in nutrient concentration n. However, the dependency of $a(b, \cdot)$ on the active cell population b is still unclear. Since it is observed that, if the cell population b becomes quite small, each cell is not very active, we assume that $a(b, \cdot)$ is a decreasing function of b. Hereafter, for concreteness and simplicity, we adopt the functional form $a(b, n) = a_0(1 + a_1 b)^{-1}(1 + a_2 n)^{-1}$ with positive coefficients a_i $(i = 0, 1, 2)$. It should be stressed, however, that this particular form is not essential to our conclusion below, and other functions which decrease with b and n will work as well. We rewrite the first two equations of (45) as the following nondimensionalized form

$$
\begin{aligned}
b_t &= d\Delta b + nb - a(b, n)b, \\
n_t &= \Delta n - nb,
\end{aligned}
\tag{46}
$$

where d is the ratio of the diffusion rates d_b and d_n. It is noted that these two equations (46) are closed for b and n. The initial conditions for (46) are

$$
\begin{aligned}
b(0, x) &= b_0(x), \\
n(0, x) &= n_0,
\end{aligned}
\tag{47}
$$

which turn out to be the same as (42). Consequently, we can find that the RD model (46) is a slight modification of (41). The only difference is the presence of the conversion term $a(b, n)b$ in the first equation. The third variable s can be obtained in terms of b and n by solving

$$
s_t = a(b(t, x), n(t, x))b(t, x), \tag{48}
$$
$$
s(0, x) = 0. \tag{49}
$$

We start with showing the following theorem (cf. [20])

Theorem 4.1. *Consider (46), (47) in a bounded domain Ω under zero flux boundary conditions. Then there is some constant $n_\infty > 0$ such that*

$$
\lim_{t \to \infty} (b(t, x), n(t, x)) = (0, n_\infty).
$$

This theorem says that the active cells density $b(t, x)$ does not generate any pattern, even simple disk-like patterns never occur. This conclusion seems to give negative information to us. However, what we want to know is how the total bacterial density $b(t, x) + s(t, x)$ forms spatial patterns. The above theorem also tells that there is $s_\infty(x) \geq 0$ such that

$$
\lim_{t \to \infty} (s(t, x)) = s_\infty(x).
$$

Therefore, we find

$$\lim_{t \to \infty} (b(t,x) + s(t,x)) = s_\infty(x),$$

which gives colony patterns we want to know. Unfortunately, there is no analytic methods to do it at present, so that we rely on numerical methods. Let us solve (46)-(49) in a two-dimensional rectangular domain, taking n_0 and d as free parameters. For quite small n_0 and d (hard and poor environment), one point distribution of $b(x)$ expands and breaks up into several spots, and then each spot splits into two smaller ones repeatedly as time goes on so as to generate many smaller spots which appear but eventually fade out. It is noted that splitting process of a single spot of b corresponds to tip-splitting process of branched patterns of s. Through this process, $b(t,x) + s(t,x)$ exhibits very complex branched patterns, which resemble colony patterns observed in the region A (Fig. 10).

Fig. 10. Numerical branched pattern in (46)-(49)

The fractal dimension is about 1.67. These ehibit characteristic of DLA growth. Increasing d with the same value of n_0, the pattern changes to a DBM-like pattern. On the other hand, for large d and n_0 (soft and rich environment), $b(t,x)$ shows an expanding ring pattern where no unstable mechanism occurs, so that the pattern of $b(t,x) + s(t,x)$ simply forms an expanding disk which is similar to the ones in region D. If d is decreasing with fixed large n_0, a pattern of growth and no growth occurs alternately and the ever-expanding disk pattern changes to an immobilized concentric-ring pattern. In spite of the simplicity of the system, it is surprising that the model (46)-(49) reproduces four qualitatively different patterns corresponding to those in the regions A, C, D, and E, depending on values of d and n_0. Thus, summarizing the above results, our reaction-diffusion model exhibits the phase diagram of colony patterns, as in Fig. 11.

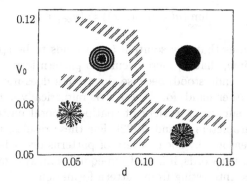

Fig. 11. Phase diagram of numerical colony patterns in (46)-(49)

For Eden-like patterns in region B (hard agar medium with rich nutrient), experimental observations indicate that bacterial cells hardly move, but cells grow moderately because of an adequate supply of nutrient, so that the mechanism of expansion of colony is different from the others. We will not touch with this (see [20]).

4.2 Grey-Scott model

Theoretical understanding of spatio-temporal patterns arising in far from equilibrium states have been investigated by using RD systems describing autocatalytic chemical reactions with feed processes. For example, there is the following cubic autocatalytic chemical reaction processes

$$2V + U \to 3V,$$
$$V \to W.$$

Letting u, v and w be the concentrations of the chemical species U, V and W, the model describing the above processes is

$$
\begin{aligned}
u_t &= d_u \Delta u - uv^2 + F(1 - u), \\
v_t &= d_v \Delta v + uv^2 - (F + k)v,
\end{aligned}
\tag{50}
$$

where d_u and d_v are diffusion coefficients of U and V, respectively, and parameters $F \geq 0$ and $k > 0$ are constants. (50) is called the Grey-Scott model [21]. When $F > 0$, (50) provides feeding processes so that it is called an open system of the Grey-Scott model, while, when $F = 0$, (50) is

$$
\begin{aligned}
u_t &= d_u \Delta uv^2 - uv^2, \\
v_t &= d_v \Delta v + uv^2 - kv,
\end{aligned}
\tag{51}
$$

which is called a closed system. For (51) in a bounded domain with zero-flux boundary conditions, it is known that there is some constant $u_\infty > 0$ such that

$$\lim_{t\to\infty} (u(t,x), v(t,x)) = (u_\infty, 0).$$

This result indicates that any solution (u,v) tends to be spatially homogeneous for large time, that is, there occurs no pattern formation. This result can be intuitively understood, because the system does not provide feeding process. On the other hand, for $F > 0$, it is numerically demonstrated that (50) generates a surprising variety of spatio-temporal patterns, sensitively depending on parameters F and k [22]. For these results, we had long believed that the reason why such variety of patterns occur is that (50) is an open system, in other words, it is far from equilibrium system, and the closed system (51) is less interesting from pattern formation.

However, we should notice that (n, b) in the bacterial colony model (46) possesses similar interaction to (u, v) in (51). Assume that the product W is immobile so that the resulting equation for w is

$$w_t = kv. \tag{52}$$

This indicates that w corresponds to s in (48), which suggests that w exhibits complex branched patterns. In order to confirm it, we write (51) in two-dimension rectangular domain Ω with zero flux boundary conditions

$$\partial u/\partial \nu = \partial v/\partial \nu, \qquad t > 0, \ x \in \partial\Omega, \tag{53}$$

where ν is the outward normal unit vector on the boundary $\partial\Omega$. The initial conditions are

$$u(0,x) = u_0(x) \geq 0, \quad v(0,x) = v_0(x) \geq 0, \quad w(0,x) = 0 \quad x \in \Omega. \tag{54}$$

Here we simply assume that $u_0(x) = u_0$ is constant in Ω. In a way similar to that of the theorem for the bacterial colony model, we can prove for (51)-(54) that there is some function $w_\infty(x) \geq 0$, depending on both u_0 and v_0, such that

$$\lim_{t\to\infty} w(t,x) = w_\infty(x). \tag{55}$$

Spatial pattern of $w_\infty(x)$ is numerically studied, by taking d and u_0 as free parameters. For small d and u_0, as is expected, $v(t,x)$ generates several spots and each small spot irregularly moves outward, splitting into two smaller spots. Of course, after large time, all the spots fade out (Fig. 12-(a)). On the other hand, $w(t,x)$, which is given by

$$w(t,x) = k \int_t v(\tau, x) d\tau, \qquad x \in \Omega,$$

records the history of v and exhibits a complex branched pattern, as shown in Fig. 12-(b). One can thus understand that the dynamics of v and w quitely resembles the ones of n and b in the bacteria colony model (46).

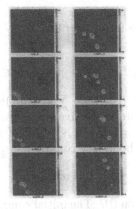 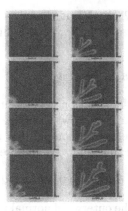

Fig. 12. (a) Spot-splitting of $v(t, x)$ (b) Tip-splitting of $w(t, x)$

Generally speaking, these systems fall in the framework of limited resource-comsumer RD systems. Quite recently, it has been reported that this kind of complex pattern also appears in experiments on combustion in micro-gravity [23]. Furthermore, we find that this models can be described by a limited resource-comsumer RD system.

We now conclude that, though any solution of a class of RD systems with limited resource-consumer interaction tends to be spatially homogeneous, the history of the consumer such as active bacteria s in (48) and the final product w in (52) forms very complex branched patterns. It turns out that the study of this class of systems is also an interesting problem from pattern formation viewpoint, even if there are closed systems.

5 Competition-diffusion systems and singular limit procedures

Understanding of spatial and/or temporal behaviors of ecologically interacting species is a central problem in population ecology. As for competitive interaction of ecological species, problems of coexistence or exclusion have been theoretically investigated by using different types of mathematical models [24]. Especially, a variety of RD equations have been proposed to study spatial segregation of competing species [25]. Quite recently, in order to understand the evolutional behavior of spatially segregating regions of competing species, the methods called spatial segregation limits have been successfully developed. These enable us to derive evolutional equations describing the boundaries of spatially segregating regions of competing species. In some situations, the derived equations are described by new types of free boundary problems.

As one of the well-known models, we consider a RD system of Gause-Lotka-Volterra type. Let $u_i(t, x)$ be the population density of the i-th species

U_i $(i = 1, 2, \ldots, N)$ at time $t > 0$ and the position $x \in \Omega$, where Ω is a bounded domain in $\mathbb{R}^2$. The resulting system for u_i $(i = 1, 2, \ldots, N)$ is given by

$$u_{it} = d_i \Delta u_i + f_i(u_1, u_2, \ldots, u_n) \qquad (i = 1, 2, \ldots, N), \quad t > 0, \ x \in \Omega. \quad (56)$$

Here d_i is the diffusion coefficient of u_i and $f_i = (r_i - \sum_{i,j} a_{ij} u_j) u_i$, where r_i is the intrinsic growth rate, a_{ii} and a_{ij} are respectively the intraspecific and the interspecfic competition rates $(i, j = 1, 2, \ldots, N)$. All of the rates are positive constants. We impose zero-flux boundary conditions to (56)

$$\partial u_i / \partial \nu = 0 \quad (i = 1, 2, \ldots, N), \qquad t > 0, \quad x \in \partial \Omega, \qquad (57)$$

where ν is the outerward normal unit vector on $\partial \Omega$. The initial conditions are

$$u_i(0, x) = u_0 i(x) \qquad (i = 1, 2, \ldots, N), \quad x \in \Omega. \qquad (58)$$

The simplest system of (56) is the case with $N = 2$, that is,

$$\begin{aligned} u_{1t} &= d_1 \Delta u_1 + (r_1 - a_1 u_1 - b_1 u_2) u_1 \\ u_{2t} &= d_2 \Delta u_2 + (r_2 - b_2 u_1 - a_2 u_2) u_2 \end{aligned}, \qquad t > 0, \ x \in \Omega. \qquad (59)$$

With the same boundary and initial conditions as (57), (58), qualitative properties of non-negative solutions of (59) have been intensively studied in mathematical cummunities. The first remark is that the stable attractor of (59) consists of equilibrium solutions only [1]. By this information, one finds that existence and stability of non-negative equilibrium solutions play an important role on the study of asymptotic behavior of solutions.

We ecologically assume the situation where two species are strongly competing, that is,

$$a_1/b_2 < r_1/r_2 < b_1/a_2, \qquad (60)$$

for which, stable spatially constant equilibrium solutions are $(u_1, u_2) = (r_1/a_1, 0)$ and $(0, r_2/a_2)$ only. These solutions mean that one of the competing species survives and the other is extinct. When the domain Ω is convex, all spatially non-constant equilibrium solution is unstable, even if they exist [26]. In other words, stable equilibria are only $(r_1/a_1, 0)$ and $(0, r_2/a_2)$, which ecologically indicates that two strongly competing species can never coexist in convex habitats. That is, if the domain is convex, the asymptotic behavior is extremely simple. However, if Ω is not convex, the structure of equilibrium solutions is not so simple but depends on the shape of Ω. If Ω takes suitable dumb-bell shape, for instance, there exist stable non-constant equilibrium solutions, which exhibit spatial segregation of the two competing species in a sense that one region is nearly occupied by the species U_1, while the other does by U_2, that is, two competing species possibly coexist, if the habitat is suitably nonconvex [27].

For multi-competing species case, the situation is drastically different from the case $N = 2$. Even if we restrict ourselves to the diffusionless system of (56), the solution-structures are rather complicated. There is possible coexistence of a variety of periodically and aperiodically solutions, in addition to equilibria, depending on parameters r_i and a_{ij} (for instance, [28]). There has been little work for the RD system (56) with $N \geq 3$, except for some results which are discussed by using singular perturbation methods [29].

5.1 Spatial segregating limits of two competing species model

The asymptotic behavior of solution $(u_1(t, x), u_2(t, x))$ to (59) is already known. If the domain is convex, the solution tends to either $(r_1/a_1, 0)$ and $(0, r_2/a_2)$. Even if we know this result, an ecologically important problem so as to know the transient behavior of solutions is still unsolved. For this purpose, the methods which are called spatial segregation limits have been developed. Assume that either the diffusion coefficients d_1 and d_2 are sufficiently small or all of the other rates r_i, a_i, and b_i ($i = 1, 2$) are sufficiently large. Then (59) is rewritten as

$$
\begin{aligned}
u_{1t} &= \varepsilon^2 \Delta u_1 + (r_1 - a_1 u_1 - b_1 u_2)u_1 \\
u_{2t} &= d\varepsilon^2 \Delta u_2 + (r_2 - b_2 u_1 - a_2 u_2)u_2
\end{aligned}
\quad , \qquad t > 0, \ x \in \Omega, \qquad (61)
$$

in which ε is a small parameter. Under the assumption (60), it is expected that there occur transition sharp layer regions which separate two subregions $\Omega_1(t) = \{x \in \Omega \mid (u_1(t, x), u_2(t, x)) \approx (r_1/a_1, 0)\}$ and $\Omega_2(t) = \{x \in \Omega \mid (u_1(t, x), u_2(t, x)) \approx (0, r_2/a_2)\}$, which ecologically mean segregating regions of competing species. To study the time evolution of $\Omega_i(t)$ ($i = 1, 2$), we take the limit as $\varepsilon \downarrow 0$ in (61), so that the layers become interfaces, say $\Gamma(t)$, which represent the boundary between two different regions $\Omega_1(t)$ and $\Omega_2(t)$. Using the singular limit analysis, the following evolution equation to describe the motion of the interface $\Gamma(t)$ can be derived as follows [30]

$$
\nabla(t) = c - \varepsilon L(d)\kappa, \qquad (62)
$$

where $V = \varepsilon \nabla$ is the normal velocity of the interface. $L(d)$ is a positive constant depending on d such that $L(1) = 1$ and c is the velocity of the one dimensional travelling wave solution (u_1, u_2) of

$$
\begin{aligned}
u_{1t} &= u_{1xx} + (r_1 - a_1 u_1 - b_1 u_2)u_1 \\
u_{2t} &= d u_{2xx} + (r_2 - b_2 u_1 - a_2 u_2)u_2
\end{aligned}
\quad , \qquad t > 0 \quad -\infty < x < \infty, \qquad (63)
$$

with the boundary conditions

$$
(u_1, u_2)(t, -\infty) = (r_1/a_1, 0) \quad \text{and} \quad (u_1, u_2)(t, \infty) = (0, r_2/a_2). \qquad (64)
$$

For (63), (64), the velocity c of the travelling wave solution $(u_1(z), u_2(z))$ ($z = x - ct$) is unique for fixed values of r_i, a_i and b_i ($i = 1, 2$) [31]. It is very interesting that the interface equation (62) is the same as (31) for the scalar bistable RD equation (30). This result shows that qualitative properties of solutions of (62) and the scalar bistable RD equation (30) are quite similar in the limit $\varepsilon \downarrow 0$.

There is another spatial segregating limit. It is the situation when the interspecific competition rates b_1 and b_2 are very large [32]. To study this situation, it is convenient to rewrite (59) as

$$\begin{aligned} u_{1t} &= d_1 \Delta u_1 + r_1(1 - u_1)u_1 - bu_1 u_2 \\ u_{2t} &= d_2 \Delta u_2 + r_2(1 - u_2)u_2 - \alpha bu_1 u_2 \end{aligned} \quad , \quad t > 0, \ x \in \Omega, \quad (65)$$

where b and α are positive constants. We assume that b is the only parameter which is large and that all other parameters are of order $O(1)$. The coefficient α is the competition rate between two species u_1 and u_2. If $\alpha > 1$, then u_1 has a competitive advantage over u_2, while if $\alpha < 1$, the situation is reversed. When b is sufficiently large, in other words, the interspecific competiton is very strong, one can expect that u_1 and u_2 have disjoint supports (habitats) with only curves which separate the habitats of the competing species. We now derive the time evolution equation of the supports of u_1 and u_2, taking the limit as $b \uparrow \infty$. The limiting system can be described by a free boundary problem which is a two phase Stefan-like problem with reaction terms [33]. Let $\Gamma(t)$ be the interface which separates the two subregions $\Omega_1(t) = \{x \in \Omega \mid u_1(t,x) > 0, u_2(t,x) = 0\}$ and $\Omega_2(t) = \{x \in \Omega \mid u_1(t,x) = 0, u_2(t,x) > 0\}$ in Ω. Then u_1 and u_2 satisfy equations

$$\begin{aligned} u_{1t} &= d_1 \Delta u_1 + r_1(1 - u_1)u_1 \ , & t > 0, \ x \in \Omega_1(t), \\ u_{2t} &= d_2 \Delta u_2 + r_2(1 - u_2)u_2 \ , & t > 0, \ x \in \Omega_2(t) \end{aligned} \quad (66)$$

and boundary conditions

$$\partial u_i/\partial \nu = 0 \quad (i = 1, 2), \quad t > 0, \ x \in \partial \Omega. \quad (67)$$

On the interface $\Gamma(t)$, we have

$$u_1(t, x) = 0, \quad u_2(t, x) = 0, \quad t > 0, \ x \in \Gamma(t) \quad (68)$$

and

$$0 = -\alpha d_1 \partial u_1/\partial \zeta - d_2 \partial u_2/\partial \zeta \ , \quad t > 0, \ x \in \Gamma(t), \quad (69)$$

where ζ is a unit vector normal to $\Gamma(t)$. The initial conditions are given by

$$u_i(0, x) = u_{i0}(x) \quad (i = 1, 2), \quad x \in \Omega, \quad (70)$$

and are such that their support is separated by

$$\Gamma(0) = \Gamma_0. \quad (71)$$

The problem is to find $(u_1(t, x), u_2(t, x))$ and $\Gamma(t)$ which satisfy (66)-(71). If this problem can be solved, the interface $\Gamma(t)$ determines segregating patterns between the two strongly competing species.

5.2 Dynamics of triple junctions arising in three competing species model

We consider (22) with $N = 3$ in two dimensions, that is,

$$u_{it} = d_i \Delta u_i + f_i(u_1, u_2, u_3), \qquad t > 0, \ x \in \Omega \subset \mathbb{R}^2, \tag{72}$$

where $f_i(u_1, u_2, u_3) = (r_i - a_{i1}u_1 - a_{i2}u_2 - a_{i3}u_3)u_i$ $(i = 1, 2, 3)$, or simply

$$U_t = D\Delta U + F(U), \qquad t > 0, \ x \in \Omega, \tag{73}$$

where $U = (u_1, u_2, u_3)$, D is the 3×3 diagonal matrix with positive elements $\{d_i\}$ $(i = 1, 2, 3)$, and $F = (f_1, f_2, f_3)$. We first consider the diffusionless system corresponding to (73)

$$U_t = F(U), \qquad t > 0. \tag{74}$$

Assume that a_{ij} $(i \neq j)$ are large comparing with others a_{ii}, in order to require that $P_1 = (r_1/a_{11}, 0, 0)$, $P_2 = (0, r_2/a_{22}, 0)$ and $P_3 = (0, 0, r_3/a_{33})$ are stable and other critical points are all unstable, that is, three species are in strong competition. It is thus shown that almost all non-negative solutions $U(t)$ of (74) tend to one of P_i $(i = 1, 2, 3)$ [34]. This ecologically implies that one of the competing species can survive and the other two are extinct. This result indicates that the dynamics of $U(t)$ of (74) is extremely simple. Under the non-coexistence situation stated above, we consider what kind of patterns RD system (74) generates by the interplay of diffusion and reaction. In order to study this problem, we assume that the diffusion rates D are sufficiently small, that is, $D = \varepsilon \bar{D}$ with a sufficiently small parameter ε, where $\bar{D}$ is a diagonal matrix with elements $\{\bar{d}_i\}$ $(i = 1, 2, 3)$. The resulting system from (73) is written as

$$U_t = \varepsilon^2 D\Delta U + F(U), \qquad t > 0, \ x \in \Omega, \tag{75}$$

where we wrote $\bar{D}$ as the original D. The initial and boundary conditions are

$$U(0, x) = U_0(x), \qquad x \in \Omega \tag{76}$$

and

$$\partial U/\partial \nu = 0, \qquad t > 0, \ x \in \partial\Omega, \tag{77}$$

where ν is the outward normal unit vector on $\partial\Omega$. Since ε is sufficiently small, one can expect that the behavior of solutions of (75)-(77) essentially consists of two stages. The first stage is the occurrence of transition layer regions (when ε tends to zero, they become interfacial curves), which generally divide the domain Ω into three subdomains Ω_1, Ω_2, and Ω_3, where the solution $U(t, x)$ is close to one of P_i $(i = 1, 2, 3)$. This indicates the appearance of spatial segregation of three competing species with triple junctions, where three interfacial curves meet. The second stage is the motion of interfacial curves with triple junctions, that is, the dynamics of segregating patterns.

We numerically consider (75)-(77) in a rectangular domain Ω. As was stated above, the first stage is the occurrence of segregating patterns of (u_1, u_2, u_3) so that the domain Ω is clearly divided into three subdomains Ω_1, Ω_2 and Ω_3, which are separated by interfacial curves with triple junctions (Fig. 13).

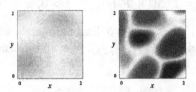

Fig. 13. Occurrence of transition layers among three segregating regions

Next consider the second stage. The first we will discuss is the completely symmetric case where $d_i = d$, $a_{ii} = a$, $a_{ij} = b$ $(i = 1, 2, 3)$. The dynamics of segregating pattern changes slowly (Fig. 14).

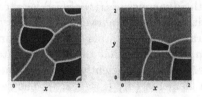

Fig. 14. Segregation regions with triple junctions

Here we note two points

(i) if interfacial curves are almost straight, they move very slowly;
(ii) angles between any two neighboring interfacial curves are almost equal.

This angle condition of interfaces can be intuitively understood by the fact that three competing species possess completely symmetric property.

The next cases are non-symmetric. The first is a semi-symmetric case where $d_i = d$ $(i = 1, 2, 3)$, $a_{12} = a_{21} = b_1, a_{23} = a_{32} = b_2, a_{31} = a_{13} = b_3$, but b_1, b_2 and b_3 are not necessarily equal. The dynamics of segregating pattern is qualitatively similar to the above symmetric case except for the angles of interfaces at triple junction areas, which seem to be not necessarily equal.

The second is the case with ordering property where $a_{12} < a_{21}$, $a_{13} < a_{31}$ and $a_{23} < a_{32}$, that is, the species U_1 is the strongest of the three. The resulting dynamics of pattern is much faster than the previous two cases and, as is easily expected, the domain is gradually occupied by the strongest species U_1.

The third is a symmetrically cyclic case where $a_{12} = a_{23} = a_{31} = b$ and $a_{21} = a_{32} = a_{13} = b'$ with $b \approx b'$. As the cyclic property suggests, one can expect that there appear steadily rotating spiral patterns with three arms (Fig. 15).

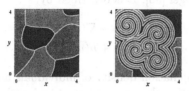

Fig. 15. Perfect spiral waves with three arms

Finally, the fourth is a general case with cyclic property where $a_{12} < a_{21}$, $a_{23} < a_{32}$ and $a_{31} < a_{13}$. There is no longer any stationary rotating spiral but complex spatio-temporal pattern with several clustering spirals, where each spiral seems to be steadily rotating in a vicinity of triple junction areas (Fig. 16). The precise discussion is stated in [35].

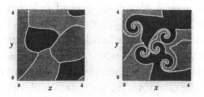

Fig. 16. Clusters of spiral waves

We have observed that segregating patterns of three competing species are drastically different from the ones of two competing species, that is, even if Ω is convex domain in $\mathbb{R}^2$, it is possible for the three species to coexist but the resulting pattern is not stationary but very dynamic.

6 Chemotactic patterns - aggregation and colonies

In the movement of biological individuals, it is often observed that they have a tendency to aggregate by moving preferentially toward higher concentrations of chemotactic substances. As an example, *Dictyostelium discoideum* (Dd) is a kind of microorganisms which show aggregation of single cells to form cellular aggregates, as if it were amoebae, at some phase in their life cycles. It is observed that multicellular development is induced by lack of nutrients and leads to the formation of a motile slug-like organism, eventually transforming into a fruiting body. Following starvation, they acquire the ability to respond to extracellular substance, which is called cAMP with intracellular synthesis and secretion of cAMP and also with chemotactic movement towards higher gradient of cAMP concentration.

RD model which takes chemotactic effect into account was originally proposed by Keller and Segel [36]. It is described by

$$n_t = d_n \Delta n - \nabla(n \nabla_\chi(u)),$$
$$u_t = d_u \Delta u + h(u, n),$$

$$(78)$$

where $n(t, x)$ is the population density of individuals and $u(t, x)$ is the concentration of chemotactic substance at time t and position x. d_n and d_u are respectively the diffusion coefficients of n and u. $\chi(u)$ is a sensitivity function of chemotaxis and $\nabla_\chi(u)$ is the velocity of the direct movement of n due to chemotaxis, where $\chi(u)$ generally satisfies $\chi'(u) \geq 0$ for $u > 0$. Some plausible forms of $\chi(u)$ are proposed such as

$$\chi(u) = q \log u, \quad \chi(u) = qu, \quad \text{and} \quad \chi(u) = qu^2(s^2 + u^2)^{-1},$$

with positive constants q and s. $h(u, n)$ is the reaction term leading to production and degradation of u so that $\partial h / \partial n > 0$ and $\partial h / \partial u < 0$ are assumed for any $u > 0$ and $n > 0$. A simplest form of h is $h(u, n) = \beta n - \gamma u$ with positive coefficients β and γ. The system (78) implies that individuals move by diffusion and chemotaxis and the total number of individuals is preserved, because neither growth nor death occur in the process. Under the concept of chemotaxis-induced instability, there exist stable non-constant equilibrium solutions, which indicate chemotactic aggregation of individuals, in a bounded domain with zero flux boundary conditions [37]. Furthermore, the blow-up problem, which indicates chemotactic collapse, is discussed in higher dimensions [38].

6.1 Aggregation of Dictyostelium discoideum

Following starvation, Dd acquire the ability to respond to extracellular substance, which is called cAMP with intracellular synthesis and secretion of cAMP and also with chemotactic movement towards higher gradient of cAMP concentration. Concentric and spiral waves of cAMP are observed and cell movement towards the centers of the wave patterns takes place. However, cell movement subsequently ceases and the wave patterns break up and amoebae organize in a pattern of branching cell stream.

Several continuous models describing aggregation process of Dd have been proposed so far. They are basically described by the following three component system ([39], [40], for instance)

$$u_t = d\Delta u + \lambda a(n) f(u, v),$$
$$v_t = g(u, v), \tag{79}$$
$$n_t = D\Delta n - \nabla(\chi(v) n \nabla u),$$

where u, v and n denote the concentration of extracellular cAMP, the fraction of active cAMP cell membrane receptors per cell and the cell density (number of cells per area), respectively. $f(u, v)$ is the rate of cAMP synthesis and degradation per cell. The total production per unit area depends on the cell density with $a(n)$. $g(u, v)$ is the kinetics of desensitization and the recovery of the active form is the ratio of the characteristic rates for cAMP synthesis and receptor desensitization. When the cell density is constant in space and time, that is, n is constant, the first two equations for u and v are qualitatively similar to activator-inhibitor typed RD systems (20), which was introduced by Turing. The third equation for n describes the dynamics of cell movement with diffusivity D and the chemotactic movement with the velocity $\chi(v)\nabla u$, which is similar to (78). Numerical computation of (79) demostrates that pattern formation consists of two stages: the first stage is the occurrence of spiral waves. Its mechanism is similar to the one suggested by Hodgkin and Huxley and the second is broken up of spiral waves into streaming patterns, which is chemotaxis-induced instability. We thus find that the ideas by Turing, Hodgkin and Huxley are included in the model (79).

6.2 New pattern arising in a chemotaxis-diffusion-growth system

In this section, we consider the following two component system including diffusion, chemotaxis and growth

$$n_t = d_n \Delta n - \nabla(n \nabla \chi(u)) + f(n),$$
$$u_t = d_u \Delta u + \beta n - \gamma u, \tag{80}$$

where $f(n)$ is the growth term of the form $f(n) = (g(n) - \alpha)n$ with the growth rate $g(n)$ including cooperation and competition effects and with the degradation rate α due to exterior forces such as predation and/or intoxication. By suitable transformations, (80) can be written as

$$n_t = d_n \Delta n - \nabla(n \nabla \chi(u)) + f(n),$$
$$u_t = d_u \Delta u + n - \gamma u. \tag{81}$$

We assume that $f(n) = n(1 - n)(n - a)$ with small $a > 0$. We first that (81) has three constant equilibrium solutions $(n, u) = (0, 0), (1, 1/\gamma)$ and $(a, a/\gamma)$; the first two are stable, while the last is unstable, that is, (81) is a bistable system. The boundary condition is

$$\lim_{|x| \to \infty} (n(t, x), u(t.x)) = (0, 0), \qquad t > 0. \tag{82}$$

In the absence of chemotaxis, (81) reduces to (24) for which, if $n(0, x)$ is a relatively large delta-like function, it expands uniformly and forms a disk-like pattern. We now address the following question: if the chemotactic effect becomes stronger, are there any other patterns generated by (81)? In order to answer to this question, we assume the situation where (i) the movement of individuals is mainly due to chemotaxis, (ii) the chemotactic substance diffuses so fast compared with the migration of individuals. We thus introduce a small parameter ε into (81) and rewrite it as

$$
\begin{aligned}
n_t &= \varepsilon d_n \Delta n - k\nabla(n\nabla\chi(u)) + \varepsilon^{-1}f(n) \\
\varepsilon u_t &= d_u \Delta u + n - \gamma u
\end{aligned}
\qquad , \qquad t > 0, \ x \in \mathbb{R}^2, \qquad (83)
$$

where k is a positive constant such that $\chi(u)$ is normalized to satisfy $\max_{u>0} \chi'(u) = 1$. It is first noted that, when $k = 0$, the first of (83) reduces to (30) so that there appear transition layers between two states where u takes values nearly 1 and 0, and furthermore, in the limit $\varepsilon \downarrow 0$, the aggregating region is described by the resulting interface $\Gamma(t)$ and the time evolution of $\Gamma(t)$ is

$$
V(t) = c(a) - \varepsilon\kappa, \qquad t > 0,
$$

where $V(t)$ is the normal velocity of the interface $\Gamma(t)$ and κ is the curvature of the interfacial curve [41]. As a situation similar to (30), system (83) generates transional layers between two states: u takes values nearly 1 and 0. In ecological terms, the region where u is near 1 results an aggregating region of individuals. Moreover, in the limit as $\varepsilon \downarrow 0$, the resulting interface $\Gamma(t)$ is described by

$$
V(t) = c(a) - \varepsilon\kappa - k\chi'(u_I), \qquad t > 0,
$$

where u_I is the value of u on the interface $\Gamma(t)$ [41]. Noting that $c(a) > 0$, because a is small, and $\chi'(u) > 0$, one can expect the appearance of localized solutions under suitable balance between the growth and chemotaxis. In fact, singular perturbation analysis shows that there exist radially symmetric equilibrium solutions in a suitable range of k for sufficiently small but not zero ε. Moreover, the stability of this solution can be discussed. If $\chi''(u) \le 0$, it is stable, while if $\chi''(u) > 0$, the stability depends the values of d (for a precise discussion, see [40]). With the sensitive function $\chi(u) = qu^2(s^2 + u^2)^{-1}w$, we only show numerical simulations of solutions which are destabilized from radially symmetric equilibrium solutions, as shown in Fig. 17. Typical patterns are a sponge pattern (Fig. 17-(a)), network one (Fig. 17-(b)) and fingering one (Fig. 17-(c)), depending on values of k.

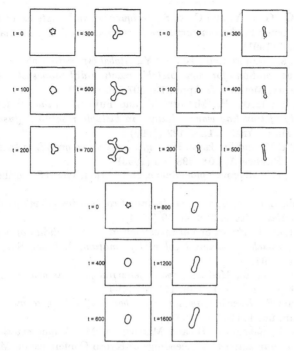

Fig. 17. (a) Sponge pattern   (b) Network pattern   (c) Fingering pattern

References

[1]   Hirsch, M.W., *Differential equations and convergence almost everywhere of strongly monotone flows*, Contemp. Math., **17**, Amer. Math. Soc., 267-285 (1983)

[2]   Casten, R.G. and Holland, C.J., *Instability results for reaction-diffusion equations with Neumann boundary conditions*, J. Differential Equations, **27**, 266-273 (1978)

[3]   Matano, H., *Asymptotic behavior and stability of solutions of semilinear parabolic equations*, Pub. R.I.M.S., Kyoto Univ., **15**, 401-454 (1979)

[4]   Turing, A., *The chemical basis of morphogenesis*, Phil. Trans. Roy. Soc. Lond. B, **237**, 23-72 (1952)

[5]   Hodgkin, A.L. and Huxley, A.F., *A quantitative description of membrane and its application to conduction and excitation in nerve*, J. Physiol., **117**, 500-544 (1952)

[6]   de Mottoni, P. and Schatzman, M., *Development of interfaces in* $\mathbf{R}^N$, Proc. Royal Soc. Edinburgh Sect. A, **116**, 207-220 (1990)

[7]   Grayson, M., *The heat equation shrinking convex plane curves*, J. Differential Geometry, **26**, 285-314 (1987)

[8]   Carr, J. and Pego, R.L., *Metastable patterns in solutions of* $u_t = \varepsilon^2 u_{xx} - f(u)$, Comm. Pure Appl. Math., **42**, 523-576 (1989)

[9]   Evans, L.C., and Spruck, J. , *Motion by level sets by mean curvature I*, J. Differential Geometry, **33**, 635-681 (1991)

[10] Chen, Y.G., Giga, Y. and Goto, S., *Uniqueness and existence of viscosity solutions of generalized mean curvature flow equations*, J. Differential Geometry, **33**, 749-786 (1991)

[11] Ikeda, H., Mimura M. and Nishiura, Y., *Global bifurcation phenomena of travelling wave solutions for some bistable reaction-diffusion systems*, Nonlinear Anal. Theory, Methods & Appl., **13**, 507-526 (1989)

[12] Nishiura, Y., Ikeda, H., Mimura, M. and Fujii, H., *Singular limit analysis of stability of traveling wave solutions in bistable reaction-diffusion systems*, SIAM J. Math. Anal., **21**, 85-122 (1990)

[13] Matsushita, M. and Fujikawa, H., *Diffusion-limited growth in bacterial colony formation*, Physica A, **168**, 498-506 (1990)

[14] Vicsek, T., *Fractal growth phenomena*, 2nd edition. World Scientific, Singapore (1992)

[15] Ben-Jacob, E. et al., *Generic modelling of cooperative growth patterns in bacterial colonies*, Nature, **368**, 46-49 (1994).

[16] Wakita, J. et al., *Experimental investigation on the validity of population dynamics approach to bacterial colonial formation*, J. Phys. Soc. Japan, **63**, 1205-1211 (1994).

[17] Kawasaki K., et al., *Modelling spatio-temporal patterns generated by Bacillus subtilis*, J. Theoret. Biol., **188**, 175-185 (1997)

[18] Kitsunezaki S., *Interface dynamics for bacterial colony formation*, J. Phys. Soc. Japan, **66**, 1544-1550 (1997)

[19] Mimura, M., Sakaguchi, H. and Matsushita, M., *A mathematical model of bacterial colony patterns*, Proceedings of Kyoto Conference on Mathematical Biology (1996)

[20] Mimura, M., Sakaguchi, H. and Matsushita, M., *Reaction-diffusion modelling of bacterial coloniy patterns*, Physica A, **282**, 283-303 (2000)

[21] Gray, P. and Scott, S.K., *Autocatalytic reactions in the isothermal, continuous stirred tank reactor: oscillations and itstabilities in the system $A + B \to 3B$, $B \to C$*, Chem. Eng. Sci., **39**, 1087-1097 (1984)

[22] Pearson, J.E., *Complex patterns in a simple system*, Science, **261**, 189-192 (1993)

[23] Lubkin, G.B., *Combustion in two dimensions yields fingering instability*, Physics Today, **Jan.**, 19-21 (1999)

[24] Gause, G.F., *The Struggle for Existence*, The Williams & Wilkins Company, Baltimore, 1934.

[25] Okubo A. et al., *On the spatial spread of the grey squirrel in Britain*, Proceedings of the Royal Society of London, Series B, **238**, 113-125 (1989)

[26] Kishimoto, K. and Weiberger, H., *The spatial homogeneity of stable equilibria of some reaction-diffusion systems on convex domains*, J. Differential Equations, **58**, 15-21 (1985)

[27] Matano, H. and Mimura, M., *Pattern formation in competition-diffusion systems in non-convex domains*, Pub. R.I.M.S., Kyoto Univ., **19**, 1049-1079 (1983)

[28] May. R and Leonard, W.J., *Nonlinear aspects of competition between species*, SIAM J. Appl. Math., **29**, 243-275 (1975)

[29] Mimura, M. and Fife, P., *A 3-component system of competition and diffusion*, Hiroshima Math. J., **16**, 189-207 (1986)

[30] Ei, S.-I. and Yanagida, E., *Dynamics of interfaces in competition-diffusion systems*, SIAM J. Appl. Math., **54**, 1355-1373 (1994)

[31] Kan-on, Y., *Parameter dependence of propagation speed of travelling waves for competition-diffusion equations*, SIAM J. Math. Anal., **26**, 340-363 (1995)

[32] Dancer, E.N., Hilhorst, D., Mimura, M. and Peletier, L.A., *Spatial segregation limit of a competition-diffusion system*, European J. Appl. Math., **10**, 97-115 (1999)

[33] Ockendon, J.R. and Hodgkins, W.R., eds., *Moving boundary problems in heat flow and diffusion*, Oxford Univ. Press, Oxford (1975)

[34] Van den Driessche, P. and Zeeman, M.L., *Three-dimensional competitive Lotka-Volterra systems with no periodic orbits*, SIAM J. Appl. Math., **58**, 227-234 (1998)

[35] Ei, S.-I., Ikota, R. and Mimura, M., *Segregating partition problem in competition-diffusion systems*, Interface and Free Boundaries, **1**, 57-80 (1999)

[36] Keller, E.F. and Segel, L.A., *Initiation of slime mold aggregation viewed as an instability*, J. Theor. Biol., **26**, 399-415 (1970)

[37] Schaaf, R., *Stationary solutions of chemotaxis systems*, Trans. Amer. Math. Soc., **292**, 531-556 (1985)

[38] Herrero, M.A. and Velazquez, J.J.L., *Chemotactic collapse for the Keller-Segel model*, J. Math. Biol., **35**, 177-194 (1996)

[39] Hoefer, T., Sherratt, J.A. and Maini, P.K. *Cellular pattern formation during Dictyosteleium aggregation*, Physica D, **85**, 425-444 (1995)

[40] Vasiev, B. NB., Hogeweg, P.H. and Panfilov, A.V., *Simulation of Dictyosteleium Discideum aggregation vis reaction-diffusion model*, Phys. Rev. Letters, **73**, 3173-3176 (1994)

[41] Mimura, M. and Tsujikawa, T., *Aggregating pattern dynamics in a chemotaxis model including growth*, Physica A, **230**, 499-543 (1996)

Lectures on Evolution Free Boundary Problems: Classical Solutions

Vsevolod A. Solonnikov

Saint-Petersburg Department of Steklov Mathematical Institute,
Fontanka, 27, 191011 S-Petersburg, Russia
solonnik@pdmi.ras.ru

Introduction

The aim of these notes is to present some methods of the proof of the solvability of free boundary problems for the parabolic and nonstationary Navier–Stokes equations. As "test problems" we have chosen the one-phase Stefan problem and the problem of evolution of an isolated liquid mass. These methods allow us to show that the above problems (and some others to which they are applicable) possess unique classical solutions defined on some finite time intervals. In addition, we prove that the second problem is solvable in an infinite time interval $t > 0$, when initial data are close to an equilibrium state. Of course, for one-phase Stefan problem, this result is far from being optimal, since the free boundary may be analytical for analytical data (see [18] and more references in the survey [28]). Additional references and results on the Stefan problem, including weak solutions, may be found, for instance, in the books [23], [15] or [45].

The proof consists in the transformation of the free boundary problems into nonlinear evolution problems in given domains and in a careful analysis of linearized problems. Sharp (coercive) estimates for solutions of linearized problems permit us to obtain local solutions of nonlinear problems using the contraction mapping principle. The solutions we obtain do not loose regularity with time and are at least as smooth as initial data.

The mapping of the unknown domain with a free boundary onto a given domain is made in a different way in the parabolic and in the Navier–Stokes case. In the first case the free boundary Γ_t is given by an equation $x = \xi + \rho(\xi, t) N(\xi)$, where ξ is an arbitrary point of the given surface Γ_0, $N(\xi)$ is a smooth non-tangential vector field on Γ_0 and $\rho(\xi, t)$ is a certain (unknown) function, whereas in the hydrodynamical problems this mapping is defined by relationship between the Eulerian and the Lagrangean coordinates.

A central role in our arguments is played by analysis of linearized problems.

In both cases they are non-standard, and the proof of coercive estimates relays on the study of corresponding non-standard model problems in the half-space. We are able to write solutions of these problems in the form of sums of potentials and prove the pointwise estimates of their kernels which makes it possible to obtain necessary estimates of solutions.

Major part of the results presented here is published in [7, 8, 35, 40].

These notes have been developed during a visit to the Centro de Matemática e Aplicações Fundamentais at the University of Lisbon, supported by a grant of the Portuguese Fundação para a Ciência e Tecnologia.

Contents

Part I – Parabolic free boundary problems

1 Introduction

Let $\Omega \subset \mathbb{R}^n$, $n \geq 2$, be a bounded domain with the boundary $S = \partial\Omega$ divided into two subdomains, Ω_1 and Ω_2, by a closed surface Γ which is bounded away from S, and let $\partial\Omega_2 = \Gamma$, $\partial\Omega_1 = S \cup \Gamma$. One phase Stefan problem consists in the determination of one parameter family of surfaces $\Gamma(t)$ such that $\Gamma(0) = \Gamma$ and of the function $u(x, t)$ given in the domain $\Omega_1(t)$ with the boundary $\partial\Omega_1(t) = \Gamma(t) \cup S$ and satisfying the equations

$$\frac{\partial u}{\partial t} - a\,\Delta u = 0, \quad x \in \Omega_1(t) \,,$$

$$u|_{t=0} = u_0(x), \quad x \in \Omega_1(0) = \Omega_1 \,,$$

$$u|_{x \in S} = b(x,t) \,,$$

$$u|_{x \in \Gamma(t)} = 0, \quad V_n = -c_0 \left.\frac{\partial u}{\partial n}\right|_{x \in \Gamma(t)} \,,$$

(1)

where c_0, a are positive constants and V_n is the velocity of the evolution of $\Gamma(t)$ in the direction of interior normal n. It is natural to assume that $b(x,t) > 0$ and $u_0(x) > 0$ inside Ω_1. We also formulate the closely related two-phase Stefan problem and the Muskat–Verigin problem considered in [2, 3, 6, 17, 22, 26, 27]. In the two-phase Stefan problem it is required to find $\Gamma(t)$ and the functions

$$u_i(x,t), \quad x \in \Omega_i(t), \quad i = 1,2, \quad (\Omega_2(t) = \Omega - \overline{\Omega_1(t)}) \,,$$

satisfying the equations

$$\frac{\partial u_i}{\partial t} - a_i\,\Delta u = 0, \quad x \in \Omega_i(t), \quad i = 1,2 \,,$$

$$u_i|_{t=0} = u_{0i}(x), \quad x \in \Omega_i(0) = \Omega_i \,,$$

$$u_1|_{x \in S} = b(x,t) \,,$$

$$u_i|_{x \in \Gamma(t)} = 0 \,,$$

(2)

$$V_n = -c_0\left(\lambda_1 \frac{\partial u_1}{\partial n} - \lambda_2 \frac{\partial u_2}{\partial n}\right)\Bigg|_{x \in \Gamma(t)} \equiv -c_0\left[\lambda \frac{\partial u}{\partial n}\right]\Bigg|_{x \in \Gamma(t)} \,,$$

where a_i, λ_i, c_0 are positive constants. Finally, the Verigin problem consists in the determination of $\Gamma(t)$, $u_1(x,t)$, $u_2(x,t)$, such that

$$\frac{\partial u_i}{\partial t} - a_i\,\Delta u = 0, \quad x \in \Omega_i(t), \quad i = 1,2 \,,$$

$$u_i|_{t=0} = u_{0i}(x), \quad x \in \Omega_i(0) \equiv \Omega_i, \quad u_1|_{x \in S} = b(x,t) \,,$$

(3)

$$[u]|_{x \in \Gamma(t)} = 0, \quad \left[\lambda \frac{\partial u}{\partial n}\right]\Bigg|_{x \in \Gamma(t)} = 0, \quad V_n = -c_0\lambda_1 \left.\frac{\partial u_1}{\partial n}\right|_{x \in \Gamma(t)} \,.$$

It will be shown that the problem (1) is uniquely solvable in a certain finite time interval and that the solution belongs to a certain Hölder space of functions. We recall the definition of these spaces. Let ℓ be an arbitrary positive non-integral number and let Ω be a domain in $\mathbb{R}^n$. By $C^\ell(\Omega)$ we mean the set of functions defined in Ω and having a finite norm

$$|u|_{C^\ell(\Omega)} \equiv |u|_\Omega^{(\ell)} = \sum_{0 \leq |j| < \ell} \sup_\Omega |\mathcal{D}^j u(x)| + [u]_\Omega^{(\ell)} \, ,$$

$$[u]_\Omega^{(\ell)} = \sum_{|j|=[\ell]} \sup_{x,y \in \Omega} |x-y|^{[\ell]-\ell} \left| \mathcal{D}_x^j u(x) - \mathcal{D}_y^j u(y) \right| \, ,$$

where $[\ell]$ is the integral part of ℓ, $j = (j_1, ..., j_n)$, j_k are non-negative integers, $|j| = j_1 + \cdots + j_n$, $\mathcal{D}^j u(x) = \frac{\partial^{|j|} u(x)}{\partial x_1^{j_1} \cdots \partial x_n^{j_n}}$. The expression $[u]_\Omega^{(\ell)}$ is referred to as a principal part of the Hölder norm $|u|_\Omega^{(\ell)}$, and separate terms in this expression are Hölder constants of the derivatives $\mathcal{D}^j u(x)$. The boundedness of these constants M_j means that all the derivatives $\mathcal{D}^j u(x)$, $|j| = [\ell]$, satisfy the Hölder condition

$$\left| \mathcal{D}^j u(x) - \mathcal{D}^j u(y) \right| \leq M_j |x-y|^{\ell - [\ell]} \, ,$$

where $M_j = \sup |x-y|^{-\ell+[\ell]} |\mathcal{D}^j u(x) - \mathcal{D}^j u(y)|$.

Further, by anisotropic Hölder space $C^{\ell,\ell/2}(Q_T)$, $Q_T = \Omega \times (0,T)$, we mean the set of functions $u(x,t)$, $x \in \Omega$, $t \in (0,T)$, with a finite norm

$$|u|_{Q_T}^{(\ell,\ell/2)} = \sup_{t \in (0,T)} |u(\cdot,t)|_\Omega^{(\ell)} + \sup_{x \in \Omega} |u(x,\cdot)|_{(0,T)}^{(\ell/2)} \, .$$

It is known that if $u \in C^{\ell,\ell/2}(Q_T)$ and $|j| + 2k < \ell$, then $\mathcal{D}_t^k \mathcal{D}_x^j u \in C^{\ell-2k-|j|, \frac{1}{2}(\ell-2k-|j|)}(Q_T)$. Hence, a widely used norm

$$\|u\|_{Q_T}^{(\ell,\ell/2)} = \sum_{0 \leq |j|+2k < \ell} |\mathcal{D}_x^j \mathcal{D}_t^k u|_{Q_T}^{(\ell-2k-|j|, \frac{1}{2}(\ell-2k-|j|))}$$

is equivalent to $|u|_{Q_T}^{(\ell,\ell/2)}$. By $[u]_{Q_T}^{(\ell,\ell/2)}$ we mean

$$[u]_{Q_T}^{(\ell,\ell/2)} = \sup_{t \in (0,T)} [u(\cdot,t)]_{\Omega_1}^{(\ell)} + \sup_{x \in \Omega} [u(x,\cdot)]_{(0,T)}^{(\ell/2)} \, .$$

The space $C^{\ell,\ell/2}(\Gamma \times (0,T))$ is defined in a standard way, i.e. with the help of local maps and of the partition of unity on Γ. Finally, one can introduce the space $C^{\ell,\ell/2}(Q_T')$ in a non-cylindrical domain $Q_T' = \{x \in \Omega(t), t \in (0,T)\}$ by mapping this domain onto a cylinder $\Omega \times (0,T)$ by an appropriate coordinate transformation. An example of such a transformation is given below.

The theorem on the solvability of one-phase Stefan problem may be stated as follows.

Theorem 1.1. *Assume that the surfaces S and Γ belong to the class $C^{2+\alpha}$, $\alpha \in (0,1)$, and that $b(x,t) > 0$, $b \in C^{2+\alpha,1+\alpha/2}(S \times (0,T))$. Further, let $u_0 \in C^{2+\alpha}(\Omega)$ satisfy the compatibility conditions*

$$u_0(x)|_{x \in \Gamma} = 0, \qquad u_0(x)|_{x \in S} = b(x, 0), \tag{4}$$

$$b_t(x, 0) - \Delta u_0(x)|_{x \in S} = 0, \tag{5}$$

$$\left(\Delta u_0 - c_0 \left(\frac{\partial u_0}{\partial n} \right)^2 \right) \Bigg|_{x \in \Gamma} = 0, \tag{6}$$

and the conditions $u_0(x) \geq 0$,

$$\frac{\partial u_0}{\partial n} \Bigg|_{\Gamma} \geq d_0 > 0.$$

Then the problem (1) has a unique solution $(\Gamma(t), u(x, t))$ defined for $t \in (0, T_0)$, $T_0 \in (0, T)$ and such that $\Gamma(t)$ belongs to the class $C^{2+\alpha}$, $u_t(\cdot, t) \in C^\alpha(\Omega(t))$, $u(\cdot, t) \in C^{2+\alpha}(\Omega_1(t))$ for all $t \in (0, T_0)$, and

$$\sup_{t < T_0} |u_t(\cdot, t)|_{\Omega_1(t)}^{(\alpha)} + \sup_{t < T_0} |u(\cdot, t)|_{\Omega_1(t)}^{(2+\alpha)} \leq c \left(|u_0|_{\Omega_1}^{(2+\alpha)} + |b|_{S \times (0, T_0)}^{(2+\alpha, 1+\alpha/2)} \right).$$

In fact, the solution possesses some additional properties (see Theorem 3.1 below).

Similar theorems hold for the two-phase Stefan problem, the Verigin problem and some other problems which may be reduced to the pseudodifferential equation $A v = h$ on $\Gamma \times (0, T)$ where A is a pseudodifferential operator with the principal symbol $s + \phi(\xi, s)$ (see remark in Section 2). Our methods may be also applicable to a larger class of problems, for instance, problems containing mean curvature in the boundary conditions or quasistationary problems (see [10, 11, 13, 14, 21, 27, 29, 47] and the bibliography in these papers) but we do not discuss it here.

The Dirichlet condition $u|_s = b$ in (1) can be replaced by another boundary condition, for instance, $\frac{\partial u}{\partial n} + \beta u|_s = b(x, t)$, $b \in C^{1+\alpha, \frac{1+\alpha}{2}}(S \times (0, T))$. In this case, condition (5) has no sense.

The compatibility conditions (4)–(6) are necessary for the existence of the solution with the derivatives u_t and $\frac{\partial^2 u}{\partial x_i \partial x_j}$ continuous in the closed domain $\overline{Q}_T$. In particular, (6) is a consequence of the equation $\Delta u_0 = u_t|_{t=0}, \ x \in \Gamma$ and of the boundary conditions on Γ. Indeed, differentiation of $u(x, t)|_{x \in \Gamma(t)} = 0$ with respect to t leads to

$$u_t = -\nabla u \cdot x_t = -\frac{\partial u}{\partial n} V_n = c_0 \left(\frac{\partial u}{\partial n} \right)^2,$$

hence,

$$u_t|_{t=0,\, x\in\Gamma} = \Delta u_0 = c_0 \left(\frac{\partial u_0}{\partial n}\right)^2.$$

The restrictions (5), (6) on $u_0(x)$ have no physical meaning and they can be avoided, if the solution is sought in larger spaces, for instance in weighted Hölder spaces $C_s^{\ell,\ell/2}(Q_T)$, $s \in [0,\ell]$, or in the Sobolev spaces. The norm in $C_s^{\ell,\ell/2}(Q_T)$ is given by

$$|u|_{s,Q_T}^{(\ell,\ell/2)} = |u|_{Q_T}^{(s,s/2)} + \sum_{s<2k+|j|<\ell}\sup_{t<T} t^{\frac{2k+|j|-s}{2}}|\mathcal{D}_t^k\mathcal{D}_x^j u(x,t)| + \sup_{t<T} t^{\frac{\ell-s}{2}}[u]_{Q_t'}^{(\ell,\ell/2)},$$

$$Q_t' = \Omega \times (t/2, t).$$

It is clear that $C_\ell^{\ell,\ell/2}(Q_T) = C^{\ell,\ell/2}(Q_T)$ but if $s < \ell$, then the derivatives $\mathcal{D}_t^k\mathcal{D}_x^j u(x,t)$ of the elements of $C_s^{\ell,\ell/2}(Q_T)$ with $s < 2k + |j| < \ell$ may have singularities for $t = 0$. Therefore the equation $u_t - \Delta u|_{t=0} = 0$ for the elements of $C_s^{2+\alpha,1+\alpha/2}(Q_T)$ with $s \in (1,2)$ has no sense, and the first order compatibility conditions (5), (6) are not necessary.

Theorems on the solvability of the two-phase Stefan problem and of the Verigin problem in usual and weighted Hölder spaces may be found in [6, 8]. Different variants of L^p-theory for parabolic free boundary problems are presented in [14, 43]. Further results on the classical solutions approach can also be found in [29], where the combustion free boundary problem can be obtained as a limit case of the Stefan problem as well as an application to a superconductivity model (see also [46]).

2 The model problem

An important role in the proof of Theorem 1.1 is played by the estimates of the solution of the following model problem in the half-space $\mathbb{R}_+^n = \{x_n > 0\}$

$$v_t - a\,\Delta v = f(x,t), \qquad x \in \mathbb{R}_+^n, \quad t > 0,$$

$$v|_{t=0} = 0, \qquad\qquad\qquad\qquad\qquad\qquad (7)$$

$$v_t + b\cdot\nabla v|_{x_n=0} = g(x',t), \qquad x' = (x_1, ..., x_{n-1}) \in \mathbb{R}^{n-1}.$$

Here $a = \text{const} > 0$ and $b = (b_1, ..., b_n)$ is a constant vector. This problem does not belong to the class of parabolic initial-boundary value problems, if $n > 2$ (because the operator on the boundary $\frac{\partial}{\partial t} + b\cdot\nabla$ does not satisfy a standard "complementing" condition which guarantees the solvability). Nevertheless, if $b_n < 0$, the problem (7) is well posed.

Theorem 2.1. *Assume that*

$$b_n < 0, \quad f \in C^{\alpha,\,\alpha/2}(\mathbb{R}^n_+ \times (0,T)), \quad g \in C^{1+\alpha,\,\frac{1+\alpha}{2}}(\mathbb{R}^{n-1} \times (0,T))$$

and

$$f(x,0) = 0, \quad g(x',0) = 0 .$$

Then the problem (7) has a unique solution $v \in C^{2+\alpha,\,1+\alpha/2}(\mathbb{R}^n_+ \times (0,T))$ *with* $v_t \in C^{1+\alpha,\,\frac{1+\alpha}{2}}(\mathbb{R}^n \times (0,T))$, *and*

$$[v]^{(2+\alpha,\,1+\alpha/2)}_{\mathbb{R}^n_+ \times (0,T)} + [v_t|_{x_n=0}]^{(1+\alpha,\,\frac{1+\alpha}{2})}_{\mathbb{R}^{n-1} \times (0,T)} \le c\Big([f]^{(\alpha,\alpha/2)}_{\mathbb{R}^n_+ \times (0,T)} + [g]^{(1+\alpha,\,\frac{1+\alpha}{2})}_{\mathbb{R}^{n-1} \times (0,T)}\Big), \quad (8)$$

with the constant independent of T.

Proof. We construct the solution in the form

$$v(x,t) = v_1(x,t) + u(x,t),$$

where $v_1(x,t)$ and $u(x,t)$ are solutions of the problems

$$v_{1t} - a\,\Delta v_1 = f(x,t), \quad x \in \mathbb{R}^n_+ ,$$
$$v_1|_{t=0} = 0, \quad v_1|_{x_n=0} = 0$$

and

$$u_t - a\,\Delta u = 0, \quad x \in \mathbb{R}^n_+ , \quad u|_{t=0} = 0 ,$$
$$u_t + b \cdot \nabla u|_{x_n=0} = g - b \cdot \nabla v_1|_{x_n=0} \equiv h(x',t) , \qquad (9)$$

respectively. Both v_1 and u can be written explicitly in the form of potentials, namely,

$$v_1(x,t) = \int_0^t \int_{\mathbb{R}^n_+} \Big(\Gamma(x-y,t-\tau) - \Gamma(x-y^*,t-\tau)\Big) f(y,\tau)\, dy\, d\tau ,$$

where $y^* = (y', -y_n)$ and $\Gamma(x,t)$ is a fundamental solution of the heat equation

$$\Gamma(x,t) = \frac{1}{(4\pi a t)^{n/2}}\, e^{-\frac{|x|^2}{4at}}, \quad t > 0 ,$$

$$\Gamma(x,0) = 0, \quad t < 0 ,$$

and

$$u(x,t) = \int_0^t \int_{\mathbb{R}^n} G(x'-y', x_n, t-\tau)\, h(y',\tau)\, dy'\, d\tau, \qquad (10)$$

where

$$G(x,t) = -2a \int_0^t \frac{\partial \Gamma(x - b\lambda, t - \lambda)}{\partial x_n} \, d\lambda$$

(see [7]). It is well known that

$$[v_1]_{\mathbb{R}_+^n \times (0,T)}^{(2+\alpha, 1+\frac{\alpha}{2})} \leq c[f]_{\mathbb{R}_+^n \times (0,T)}^{(\alpha, \alpha/2)} , \tag{11}$$

hence,

$$[h]_{\mathbb{R}^{n-1} \times (0,T)}^{(1+\alpha, \frac{1+\alpha}{2})} \leq [g]_{\mathbb{R}^{n-1} \times (0,T)}^{(1+\alpha, \frac{1+\alpha}{2})} + c[f]_{\mathbb{R}_+^n \times (0,T)}^{(\alpha, \alpha/2)} .$$

Let us consider the potential (10). First of all, we estimate the derivatives of the kernel $G(x,t)$. We have

$$|\mathcal{D}_x^j G(x,t)| \leq c \int_0^t (t - \lambda)^{-\frac{n+1+|j|}{2}} e^{-c\frac{|x - b\lambda|^2}{t - \lambda}} \, d\lambda$$

$$\leq c \left(\left(\frac{t}{2} \right)^{-\frac{n+1+|j|}{2}} \int_0^{t/2} e^{-c\frac{|x|^2 + \lambda^2}{t - \lambda}} \, d\lambda \right.$$

$$+ \left. \int_{t/2}^t (t - \lambda)^{-\frac{n+1+|j|}{2}} e^{-c\frac{|x|^2 + t^2/4}{t - \lambda}} \, d\lambda \right)$$

$$\leq c \left(t^{-\frac{n+|j|}{2}} e^{-\frac{c|x|^2}{t}} + (x^2 + t^2)^{-\frac{n+|j|-1}{2}} \right),$$

which implies

$$\int_0^t |\mathcal{D}_x^j G(x', x_n, t - \tau)| \, d\tau \leq \frac{c}{|x|^{n+|j|-2}} \tag{12}$$

and, as a consequence,

$$\int_0^t d\tau \left| \int_{|x' - y'| \leq a} \frac{\partial}{\partial x_j} G(x' - y', x_n, t - \tau) \, dy' \right| \leq c \tag{13}$$

$$j = 1, ..., n-1, \qquad \forall a > 0 .$$

In an analogous way it can be proved (see [8]) that

$$\int_{\mathbb{R}^{n-1}} |\mathcal{D}_t^k G(x', x_n, t)| \, dx' \leq ct^{-1/2-k}, \qquad k = 0, 1, 2 . \tag{14}$$

Let us pass to the estimate of $[u]_{\mathbb{R}_+^n \times (0,T)}^{(2+\alpha, 1+\alpha/2)}$. Consider the difference

$$\frac{\partial^2 u(x,t)}{\partial x_j \, \partial x_k} - \frac{\partial^2 u(z,t)}{\partial z_j \, \partial z_k}, \qquad j, k = 1, ..., n-1, \quad x, z \in R_+^n .$$

It can be represented in the form

$$\frac{\partial^2 u(x,t)}{\partial x_j\,\partial x_k} - \frac{\partial^2 u(z,t)}{\partial z_j\,\partial z_k}$$

$$= \int_0^t d\tau \int_{|x'-y'|\le 2\rho} \frac{\partial G(x'-y',x_n,t-\tau)}{\partial x_j}\left[\frac{\partial h(y',\tau)}{\partial y_k} - \frac{\partial h(x',\tau)}{\partial x_k}\right] dy'$$

$$- \int_0^t d\tau \int_{|x'-y'|\le 2\rho} \frac{\partial G(z'-y',z_n,t-\tau)}{\partial z_j}\left[\frac{\partial h(y',\tau)}{\partial y_k} - \frac{\partial h(z',\tau)}{\partial z_k}\right] dy'$$

$$+ \int_0^t \left[\frac{\partial h(x',\tau)}{\partial x_j} - \frac{\partial h(z',\tau)}{\partial z_j}\right] d\tau \int_{|x'-y'|\le 2\rho} \frac{\partial G(x'-y',x_n,t-\tau)}{\partial x_j} dy'$$

$$+ \int_0^t d\tau \int_{|x'-y'|\ge 2\rho}\left[\frac{\partial G(x'-y',x_n,t-\tau)}{\partial x_j} - \frac{\partial G(z'-y',z_n,t-\tau)}{\partial z_j}\right]$$

$$\times \left[\frac{\partial h(y',\tau)}{\partial y_k} - \frac{\partial h(z',\tau)}{\partial z_k}\right] dy' \ ,$$

where $\rho = 2\,|x-z|$. Using inequalities (12), (13), we obtain

$$\left|\frac{\partial^2 u(x,t)}{\partial x_j\,\partial x_k} - \frac{\partial^2 u(z,t)}{\partial z_j\,\partial z_k}\right|$$

$$\le c \sup_{\tau<t}[h_{y_k}(\cdot,\tau)]_{\mathbb{R}^2}^{(\alpha)}\left(\int_{|x'-y'|\le 2\rho}\frac{|x'-y'|^\alpha}{|x'-y'|^{n-1}}\,dy'\right.$$

$$+ \int_{|x'-y'|\le 2\rho}\frac{|z'-y'|^\alpha}{|z'-y'|^{n-1}}\,dy'\bigg) + \rho^\alpha + \rho\int_{|x'-y'|\ge 2\rho}\frac{|z'-y'|^\alpha}{|z'-y'|^n}\,dy'\bigg)$$

$$\le c\,\rho^\alpha \sup_{\tau<t}[h_{y_k}(\cdot,\tau)]_{\mathbb{R}^2}^{(\alpha)}.$$

$$(15)$$

Similarly, the representation formula

$$\frac{\partial u(x,t+r)}{\partial t} - \frac{\partial u(x,t)}{\partial t}$$

$$= \int_{-\infty}^{t+r} d\tau \int_{\mathbb{R}^{n-1}} \frac{\partial G(x'-y',x_n,t+r-\tau)}{\partial t}\left[h(y',\tau) - h(y',t+r)\right] dy'$$

$$- \int_{-\infty}^{t} d\tau \int_{\mathbb{R}^{n-1}} \frac{\partial G(x'-y',x_n,t-\tau)}{\partial t}\left[h(y',\tau) - h(y',t)\right] dy'$$

$$= \int_{t-r}^{t+r} d\tau \int_{\mathbb{R}^{n-1}} \frac{\partial G(x'-y', x_n, t+r-\tau)}{\partial t} \left[h(y', \tau) - h(y', t+r) \right] dy'$$

$$- \int_{t-r}^{t} d\tau \int_{\mathbb{R}^{n-1}} \frac{\partial G(x'-y', x_n, t-\tau)}{\partial t} \left[h(y', \tau) - h(y', t) \right] dy'$$

$$+ \int_{\mathbb{R}^{n-1}} \left[h(y', t+r) - h(y', t) \right] G(x'-y', x_n, r) \, dy'$$

$$+ \int_{-\infty}^{t-r} d\tau \int_{\mathbb{R}^{n-1}} \left[\frac{\partial G(x'-y', x_n, t+r-\tau)}{\partial t} - \frac{\partial G(x'-y', x_n, t-\tau)}{\partial t} \right]$$

$$\times \left[h(y', \tau) - h(y', t+r) \right] dy' ,$$

where $h(y', \tau) = 0$ for $\tau < 0$, implies

$$\left| \frac{\partial u(x, t+r)}{\partial t} - \frac{\partial u(x, t)}{\partial t} \right|$$

$$\le c \sup_{\mathbb{R}^{n-1}} [h(y', \cdot)]_{(0, t+r)}^{(\frac{1+\alpha}{2})} \cdot \left(\int_{t-r}^{t+r} \frac{d\tau}{(t+r-\tau)^{1-\alpha/2}} \right.$$

$$\left. + \int_{t-r}^{t} \frac{d\tau}{(t-\tau)^{1-\alpha/2}} + r^{\alpha/2} + r \int_{-\infty}^{t-r} \frac{d\tau}{(t+r-\tau)^{2-\alpha/2}} \right)$$

$$\le c h^{\alpha/2} \sup_{\mathbb{R}^{n-1}} [h(y', \cdot)]_{(0, t+r)}^{(\frac{1+\alpha}{2})} ,$$

i.e.,

$$\sup_{\mathbb{R}_+^n} [u_t(\cdot, t)]_{(0,T)}^{(\alpha/2)} \le c \sup_{\mathbb{R}^{n-1}} [h(\cdot, t)]_{(0,T)}^{(\frac{1+\alpha}{2})} . \tag{16}$$

Finally, we estimate the Hölder constant

$$M_n \left[\frac{\partial^2 u}{\partial x_n^2} \right] \equiv \sup_{r>0} \sup_{\mathbb{R}_+^n \times (0,T)} r^{-\alpha} \left| \frac{\partial^2 u(x', x_n+r, t)}{\partial x_n^2} - \frac{\partial^2 u(x', x_n, t)}{\partial x_n^2} \right| .$$

Since $u_{x_n x_n} = a^{-1} u_t - \sum_{j=1}^{n-1} u_{x_j x_j}$, we have

$$M_n[u_{x_n x_n}] \le a^{-1} M_n[u_t] + \sum_{j=1}^{n-1} M_n[u_{x_j x_j}] .$$

The last term is already estimated, and the first term we estimate by interpolation inequality

$$M_n[u_t] \leq \varepsilon M_n[u_{x_n x_n}] + c(\varepsilon) \sup_{\mathbb{R}^n_+} [u_t(x, \cdot)]^{(\alpha)}_{(0,T)},$$

(see [30]) with $\varepsilon \ll 1$. This yields, by virtue of (15) and (16),

$$M_n[u_{x_n x_n}] \leq c\,[h]^{(1+\alpha, \frac{1+\alpha}{2})}_{\mathbb{R}^{n-1} \times (0,T)}.$$

This estimate, together with (15) and (16), implies

$$[u]^{(2+\alpha, 1+\alpha/2)}_{\mathbb{R}^n_+ \times (0,T)} \leq c\,[h]^{(1+\alpha, \frac{1+\alpha}{2})}_{\mathbb{R}^{n-1} \times (0,T)}. \tag{17}$$

Inequality (8) follows from (11), (17) and from the equation $v_t + b \cdot \nabla v|_{x_n=0} = g$.

Now, it can be easily shown that the function $v(x,t)$ is bounded; indeed,

$$|v(x,t)| \leq \int_0^t |v_\tau(x,\tau)| d\tau \leq \int_0^t \left| v_\tau(x,\tau) - v_\tau(x,0) \right| d\tau \leq \frac{T^{1+\alpha}}{1+\alpha}\, [v]^{(2+\alpha, 1+\alpha/2)}_{\mathbb{R}^n_+ \times (0,T)}.$$

Hence, the derivatives $\frac{\partial v}{\partial x_i}$ and $\frac{\partial^2 v}{\partial x_i\, \partial x_m}$ are also bounded for arbitrary $t \in (0,T)$, $T < \infty$, and $v \in C^{2+\alpha, 1+\alpha/2}(\mathbb{R}^n_+ \times (0,T))$.

Let us show that the solution of the problem (7) constructed above is unique in the class $C^{2+\alpha, 1+\alpha/2}(\mathbb{R}^n_+ \times (0,T))$, i.e., that arbitrary solution $u_0 \in C^{2+\alpha, 1+\alpha/2}(\mathbb{R}^n_+ \times (0,T))$ of a homogeneous problem vanishes. At first we assume additionally that u_0 and its derivatives belong to $L_2(\mathbb{R}^n_+ \times (0,T))$. Then there holds the energy equality

$$
\begin{aligned}
0 &= \int_{\mathbb{R}^n_+} u_0\,(u_{0t} - a\,\Delta u_0)\, dx\, d\tau \\
&= \frac{1}{2}\frac{d}{dt}\|u_0(\cdot,t)\|^2_{L_2(\mathbb{R}^n_+)} + a\,\|\nabla u_0(\cdot,t)\|^2_{L_2(\mathbb{R}^n_+)} + a \int_{\mathbb{R}^{n-1}} u_0 \left.\frac{\partial u_0}{\partial x_3}\right|_{x_3=0} dx' \\
&= \frac{1}{2}\frac{d}{dt}\left(\|u_0(\cdot,t)\|^2_{L_2(\mathbb{R}^n_+)} - \frac{a}{b_n} \|u_0|_{x_3=0}\|^2_{L_2(\mathbb{R}^{n-1})} \right) + a\,\|\nabla u_0(\cdot,t)\|^2_{L_2(\mathbb{R}^n_+)},
\end{aligned}
$$

from which it follows that $u_0 = 0$. Now, we assume the regularity $C^{2+\alpha, 1+\alpha/2}(\mathbb{R}^n_+ \times (0,T))$ for u_0 and we introduce the family of cut-off functions $\zeta_R(x) = \zeta(\frac{x}{R})$, $R \gg 1$, where $\zeta \in C_0^\infty(\mathbb{R}^n_+)$, $\zeta(z) = 1$ for $|z| < \frac{1}{2}$, $\zeta(z) = 0$ for $|z| > 1$. It is clear that the function $u_0^R = u_0\,\zeta_R$ is a solution of the problem (7) with $f = -2a\,\nabla\zeta_R \cdot \nabla u_0 - a\,u_0\,\nabla^2\zeta_R$, $g = u_0\,b \cdot \nabla\zeta_R$. Since, $u_0^R \in L_2(\mathbb{R}^n_+)$, $\forall t \in (0,T)$, it is expressed in terms of f and g as indicated above (because it can be easily verified that the solution constructed above belongs to $L_2(\mathbb{R}^n_+)$, if f and g have a compact support). Hence

$$|u_0^R|^{(2+\alpha, 1+\alpha/2)}_{\mathbb{R}^n_+ \times (0,T)} \leq |f|^{(\alpha, \alpha/2)}_{\mathbb{R}^n_+ \times (0,T)} + |g|^{(1+\alpha, \frac{1+\alpha}{2})}_{\mathbb{R}^n_+ \times (0,T)} \leq c\,R^{-1}$$

and, as a consequence, $u_0 = u_0^R(x,t) \to 0$ as $R \to \infty$ for arbitrary $x \in \mathbb{R}^n_+$, $t \in (0,T)$, and the theorem is proved. ∎

Remark 2.1. Let us come back to the problem (9). Making the Fourier–Laplace transformation with respect to (x', t), we can write the solution in the form

$$\tilde{u}(\xi, x_n, t) = \frac{\tilde{h}(\xi, s)}{s + \sqrt{\frac{s}{\alpha} + \xi^2}} \, e^{-\sqrt{\frac{s}{\alpha} + \xi^2} \, x_n},$$

where $\tilde{u}, \tilde{h}$ are transformed functions.

Similar formulas hold for the solutions of model problems arising in the study of free boundary problems (2), (3), but the function $\sqrt{\frac{s}{a} + \xi^2}$ is replaced with some other homogeneous functions analytical in the domain $\operatorname{Re} s > -\delta \xi^2$, $\delta \ll 1$.

3 Transformation of the problem (1)

It is convenient to write the problem (1) as a nonlinear initial-boundary value problem in a given domain Ω. It can be achieved by a special mapping of the domain $\Omega_1(t)$ onto Ω.

Let $N(\xi), \xi \in \Gamma$, be a regular (at least of class $C^{2+\alpha}(\Gamma)$) vector field such that

$$N(\xi) \cdot n_0(\xi) \geq \nu_0 > 0,$$

where $n_0(\xi)$ is a unit interior (with respect to Ω_1) normal to Γ. For instance, we can set

$$N(\xi) = \frac{\sum_{k=1}^{M} \varphi_k(\xi) \, n_0(\xi_k)}{\left| \sum_{k=1}^{M} \varphi_k(\xi) \, n_0(\xi_k) \right|},$$

where $\{\varphi_k(\xi)\}_{k=1,\ldots,M}$ is a partition of unity on Γ (and in a certain neighbourhood of Γ) subordinate to the covering of Γ by the balls $|\xi - \xi_k| < \delta \ll 1$, $\xi_k \in \Gamma$. It is easily seen that the set

$$\mathcal{N} = \left\{ \xi + N(\xi)\lambda, \; 0 < \lambda < \lambda_0, \; \xi \in \Gamma \right\} \cup \Omega$$

contains a certain neighbourhood Ω' of Γ and that the equation

$$x = \xi + N(\xi)\lambda$$

defines the functions $\xi(x), \lambda(x)$ of the class $C^{2+\alpha}(\Omega')$ (we observe that if $N(\xi) = n_0(\xi)$, then $\xi(x)$ and $\lambda(x)$ belong to $C^{1+\alpha}(\Omega')$). Hence, for small t the surface $\Gamma(t)$ can be defined by the equation $\lambda = \rho(\xi, t), \xi \in \Gamma$, i.e.,

$$\Gamma(t) = \left\{ x \in \mathbb{R}^n \colon \; x = \xi + N(\xi)\rho(\xi, t), \; \xi \in \Gamma \right\},$$

where ρ is a certain (unknown) function of the class $C^{2+\alpha, 1+\alpha/2}(\Gamma \times (0, T))$. To every such function we put into correspondence its extension

$$\rho^*(\xi, t) = \mathcal{P}\rho(\xi, t),$$

where $\mathcal{P}\colon C^{2+\alpha,\,1+\alpha/2}(\Gamma \times (0,T)) \to C^{2+\alpha,\,1+\alpha/2}(\Omega_1 \times (0,T))$ is a continuous linear operator such that

$$\rho^*(\xi,t)|_{\xi \in \Gamma} = \rho(\xi,t), \qquad \left.\frac{\partial \rho^*(\xi,t)}{\partial n_0}\right|_{\xi \in \Gamma} = 0, \qquad \rho^*(\xi,t)|_{\xi \in S} = 0,$$

and

$$|\rho^*|_{\Omega_1 \times (0,t)}^{(2+\alpha,\,1+\alpha/2)} \le c|\rho|_{\Gamma \times (0,t)}^{(2+\alpha,\,1+\alpha/2)}.$$

We also extend $N(\xi)$ from Γ into Ω_1: first into Ω' by the formula

$$N(x) = N(\xi(x))$$

and then into Ω_1 with the preservation of class $C^{2+\alpha}$ and in such a way that $N(x) = 0$ near S. Consider the mapping

$$x = y + N(y)\,\rho^*(y,t) \equiv e_\rho(y,t), \qquad y \in \Omega_1 .$$

For small ρ, this is an invertible mapping of Ω_1 onto $\Omega_1(t)$, whose Jacobi matrix $J = (J_{km})_{k,m=1,\dots,n}$ has the elements

$$J_{km} = \delta_{km} + \frac{\partial}{\partial y_m} N_k \rho^* = \delta_{km} + N_k \frac{\partial \rho^*}{\partial y_m} + \frac{\partial N_k}{\partial y_m} \rho^* .$$

By J^{km} we mean the elements of J^{-1}. They are algebraic functions of ρ^* and $\frac{\partial \rho^*}{\partial y_m}$. The normal $n_0(y)$ to Γ and the normal $n(e_\rho(y,t))$ to $\Gamma(t)$ are related to each other as follows

$$n = \frac{J^{-T} n_0}{|J^{-T} n_0|},$$

$J^{-T} = (J^{-1})^T$. It is easily seen that (1) is equivalent to the problem of determination of $\rho(y,t)$, $y \in \Gamma$, and of $U(y,t) = u(e_\rho(y,t),t)$, $y \in \Omega_1$, such that

$$\mathcal{L}_\rho U = 0, \qquad y \in \Omega, \ t \in (0,T),$$

$$U|_{t=0} = u_0(y), \qquad \rho|_{t=0} = 0, \tag{18}$$

$$U|_{y \in S} = b(y,t), \qquad U|_{y \in \Gamma} = 0,$$

$$\rho_t(N \cdot J^{-T} n_0) + c_0\, n_0 \cdot \mathcal{A}\nabla U|_{y \in \Gamma} = 0, \tag{19}$$

where $\mathcal{A} = J^{-1} J^{-T}$ is the matrix with the elements

$$A_{mp} = \sum_{k=1}^{n} J^{mk} J^{pk}, \qquad \nabla = \left(\frac{\partial}{\partial y_1}, \dots, \frac{\partial}{\partial y_n}\right).$$

and

$$\mathcal{L}_\rho U = \frac{\partial U}{\partial t} - (N \cdot J^{-T} \nabla U)\, \rho_t^* - a\left(\sum_{m,p=1}^n A_{mp} \frac{\partial^2 U}{\partial y_m \partial y_p} + \sum_{k,m,p=1} J^{mk} \frac{\partial J^{pk}}{\partial y_m}\frac{\partial U}{\partial y_p} \right)$$

(we have used the equation $V_n = \frac{\partial x}{\partial t}\cdot n|_{\Gamma(t)} = \rho_t \frac{N\cdot J^{-T} n_0}{|J^{-T} n_0|}\big|_\Gamma$). Since $U|_\Gamma = 0$, the condition (19) is equivalent to

$$\rho_t + c_0\, S_\rho \left.\frac{\partial U}{\partial n_0}\right|_\Gamma = 0 , \tag{20}$$

where

$$S_\rho = (N \cdot J^{-T} n_0)^{-1}\, (n_0 \cdot A\, n_0) .$$

Further, we introduce auxiliary functions $\rho_0(\xi,t)$, $\xi \in \Gamma$, and $U_0(y,t)$, $y \in \Omega$, satisfying the same initial conditions as ρ and U, namely,

$$\rho_0|_{t=0} = 0 , \qquad \rho_{0_t}|_{t=0} = -c_0\, (N\cdot n_0)^{-1} \frac{\partial u_0}{\partial n_0} ,$$

$$U_0|_{t=0} = u_0 , \qquad U_{0t}|_{t=0} = (N\cdot \nabla u_0)\,\rho_{0_t}^* + a\, \Delta u_0 .$$

We construct ρ_0 making use of the following proposition (cf. [8])

Proposition 3.1. *For arbitrary $f_0 \in C^{1+\alpha}(\Gamma)$ there exists*

$$\rho^{(0)} \equiv R\, f_0 \in C^{2+\alpha,\,1+\alpha/2}(\Gamma \times d(0,T))$$

$$\text{such that } \rho_t^{(0)} \in C^{1+\alpha,\,\frac{1+\alpha}{2}}(\Gamma \times d(0,T)) ,$$

$$\rho^{(0)}|_{t=0} = 0 , \qquad \rho_t^{(0)}|_{t=0} = f_0 ,$$

and

$$|\rho^{(0)}|_{\Gamma \times (0,T)}^{(2+\alpha,\,1+\alpha/2)} + |\rho_t^{(0)}|_{\Gamma \times (0,T)}^{(1+\alpha,\,\frac{1+\alpha}{2})} \le c\, |f_0|_\Gamma^{(1+\alpha)}$$

with the constant independent of T. The operator R is linear.

We set $\rho_0 = R(-c_0\, (N\cdot n_0)^{-1}\frac{\partial u_0}{\partial n_0})$ and define U_0 as a solution of the problem

$$\mathcal{L}_{\rho_0} U_0 = 0 , \qquad y \in \Omega_1 ,$$

$$U_0|_{t=0} = u_0(y) ,$$

$$U_0|_{y\in S} = b , \qquad U_0|_{y\in\Gamma} = 0 .$$

In virtue of (4)–(6), necessary compatibility conditions are satisfied, and this problem has a unique solution $U_0 \in C^{2+\alpha,\,1+\alpha/2}(\Omega_1 \times (0,T))$ satisfying the inequality

$$|U_0|^{(2+\alpha,\,1+\alpha/2)}_{\Omega_1\times(0,T)} \le c\left(|b|^{(2+\alpha,\,1+\alpha)}_{S\times(0,T)} + |u_0|^{(2+\alpha)}_\Omega\right).$$

Further, we set $r = \rho - \rho_0$, $V = U - U_0$,

$$\delta\,S_{\rho_0} = \frac{d}{d\lambda}S_{\rho_0+\lambda r}|_{\lambda=0}\,, \qquad \delta\,\mathcal{L}_{\rho_0} = \mathcal{L}_{\rho_0+\lambda r}|_{\lambda=0}\,,$$

and we write (18), (20) as the problem of determination of r and V such that

$$\mathcal{L}_{\rho_0}V + \delta\,\mathcal{L}_{\rho_0}U_0 = \mathcal{F}(r,V)\,, \qquad y \in \Omega\,,$$
$$V|_{t=0} = 0\,, \qquad r|_{t=0} = 0\,,$$
$$V|_{y\in S} = 0\,, \qquad V|_{y\in\Gamma} = 0\,, \tag{21}$$
$$r_t + c_0\left(S_{\rho_0}\frac{\partial V}{\partial n} + \delta\,S_{\rho_0}\frac{\partial U_0}{\partial n_0}\right) = \mathcal{G}(r,V) + \varphi(y,t)\,, \qquad y \in \Gamma\,,$$

where

$$\mathcal{F}(r,V) = -(\mathcal{L}_\rho - \mathcal{L}_{\rho_0} - \delta\,\mathcal{L}_{\rho_0})\,U_0 - (\mathcal{L}_{\rho_0} - \mathcal{L}_{\rho_0})\,V\,,$$
$$\mathcal{G}(r,V) = -c_0\,(S_\rho - S_{\rho_0} - \delta\,S_{\rho_0})\frac{\partial U_0}{\partial n_0} - c_0\,(S_\rho - S_{\rho_0})\frac{\partial V}{\partial n_0}\,, \tag{22}$$
$$\varphi = -\rho_{0_t} - c_0\,S_{\rho_0}\frac{\partial U_0}{\partial n_0}\,.$$

It is clear that $\varphi(y,0) = 0$.

The expressions $\delta\,S_{\rho_0}$ and $\delta\,\mathcal{L}_{\rho_0}$ depend on

$$\delta\,J_0^{-1} = -J_0^{-1}\,(\delta\,J_0)\,J_0^{-1} = -J_0^{-1}\,(\nabla N r^*)^T\,J_0^{-1}$$

and can be easily calculated. This leads to the following form of equations (21)

$$L_{\rho_0}V - (N\cdot J_0^{-T}\,\nabla U_0)\,L_{\rho_0}r^* + m(r^*,V) = \mathcal{F}(r,V)\,, \qquad y \in \Omega_1\,,$$
$$V|_{t=0} = 0\,, \qquad r|_{t=0} = 0\,,$$
$$V|_{y\in S} = 0\,, \qquad V|_{y\in\Gamma} = 0\,, \tag{23}$$
$$r_t + c_0\,S_{\rho_0}\frac{\partial V}{\partial n_0} + h\cdot\nabla_\Gamma r + h_0\,r|_{y\in\Gamma} = \mathcal{G}(r,V) + \varphi(y,t)\,.$$

Here $h = (h_1,...,h_m) \in C^{1+\alpha,\frac{1+\alpha}{2}}(\Gamma\times(0,T))$, $h_0 \in C^{1+\alpha,\frac{1+\alpha}{2}}(\Gamma\times(0,T))$, ∇_Γ is a gradient on Γ, $r^* = \mathcal{P}r$,

$$L_{\rho_0}w = w_t - a\sum_{i,j=1}^n A_{ij}^{(0)}\frac{\partial^2 w}{\partial y_i\,\partial y_j}\,, \qquad A_{ij}^{(0)} = A_{ij}|_{\rho=\rho_0}\,,$$

and $m(r^*, V)$ is a linear first order operator of the type

$$m(r^*, V) = \eta' \cdot \nabla r^* + \eta'' \cdot \nabla V + \eta_0' r^* + \eta_0'' V ,$$

with the coefficients belonging to $C^{\alpha, \alpha/2}(\Omega_1 \times (0, T))$. Theorem on the solvability of the problem (23) can be stated as follows.

Theorem 3.1. *Under the hypotheses of Theorem 1.1, the problem (23) has a unique solution (V, r) such that $V \in C^{2+\alpha, 1+\alpha/2}(\Omega_1 \times (0, T_0))$, $r \in C^{2+\alpha, 1+\alpha/2}(\Gamma \times (0, T_0))$, $r_t \in C^{1+\alpha, \frac{1+\alpha}{2}}(\Gamma \times (0, T_0))$, defined on a certain time interval $(0, T_0)$ and satisfying the inequality*

$$|V|_{\Omega_1 \times (0, T_0)}^{(2+\alpha, 1+\alpha/2)} + |r|_{\Gamma \times (0, T_0)}^{(2+\alpha, 1+\alpha/2)} + |r_t|_{\Gamma \times (0, T_0)}^{(1+\alpha, \frac{1+\alpha}{2})} \le c|\varphi|_{\Gamma \times (0, T_0)}^{(1+\alpha, \frac{1+\alpha}{2})} . \quad (24)$$

Theorem 1.1 follows from Theorem 3.1.

4 Proof of Theorem 3.1

Assume that $\rho_0(y, t)$, $y \in \Gamma$, $t \in (0, T)$, is so small that the transformation $X = e_{\rho_0}(y, t)$ is invertible, $N \cdot J_0^{-1} n_0$ and $|J_0^{-1} n_0|$ are strictly positive, and consider a linear problem

$$L_{\rho_0} V - q(y, t) L_{\rho_0} r^* + m(r^*, V) = f(y, t), \quad y \in \Omega_1, \ t \in (0, T) ,$$

$$V|_{t=0} = 0, \quad V|_{y \in S} = 0, \quad V|_{y \in \Gamma} = 0 , \quad (25)$$

$$r_t + b \frac{\partial V}{\partial n} + h \cdot \nabla_\Gamma r + h_0 r|_{y \in \Gamma} = g(y, t) ,$$

where $q \in C^{\alpha, \alpha/2}(\Omega_1 \times (0, T))$, $b \in C^{1+\alpha, \frac{1+\alpha}{2}}(\Gamma \times (0, T))$.

Theorem 4.1. *If $b \ge b_0 > 0$, $q \ge q_0 > 0$ then for arbitrary $f \in C^{\alpha, \alpha/2}(\Omega_1 \times (0, T))$, $g \in C^{1+\alpha, \frac{1+\alpha}{2}}(\Gamma \times (0, T))$ satisfying the conditions*

$$f(y, 0) = 0, \quad g(y, 0) = 0 ,$$

the problem (25) has a unique solution (V, r) such that $V \in C^{2+\alpha, 1+\alpha/2}(\Omega_1 \times (0, T))$, $r \in C^{2+\alpha, 1+\alpha/2}(\Gamma \times (0, T))$, $r_t \in C^{1+\alpha, \frac{1+\alpha}{2}}(\Gamma \times (0, T))$ and

$$\mathcal{Y}(T) \equiv |V|_{\Omega_1 \times (0, T)}^{(2+\alpha, 1+\alpha/2)} + |r|_{\Gamma \times (0, T)}^{(2+\alpha, 1+\alpha/2)} + |r_t|_{\Gamma \times (0, T)}^{(1+\alpha, \frac{1+\alpha}{2})}$$
$$\le c\left(|f|_{\Omega_1 \times (0, T)}^{(\alpha, \alpha/2)} + |g|_{\Gamma \times (0, T)}^{(1+\alpha, \frac{1+\alpha}{2})}\right) \equiv c\mathcal{N}(T) . \quad (26)$$

This theorem is proved in the same way as Theorem 4.2 in [8], and we restrict ourselves with sketching main ideas of the proof of the estimate (26).

We use classical Schauder's method and estimate at first the function $r(y, t)$, $y \in \Gamma$. Let us consider a neighbourhood of arbitrary point $y_0 \in \Gamma$. Without restriction of generality we can assume that $y_0 = 0$ and the y_n-axis is directed along the interior normal $n(y_0)$. Let

$$y_n = F(y'), \quad y' = (y_1, ..., y_{n-1}) \in K_d \equiv \{|y'| < d\}$$

be equation of Γ near the origin; clearly, $F \in C^{2+\alpha}(K_d)$ and

$$F(0) = 0, \quad \nabla' F(0) = (F_{y_1}(0), ..., F_{y_{n-1}}(0)) = 0 .$$

We make the change of variables

$$z' = y', \quad z_n = y_n - F(y'),$$

which "flattens" Γ near the origin. In the new variables (z, t) (25) takes the form

$$\widetilde{L}_{\rho_0} V - q(z, t) \widetilde{L}_{\rho_0}(r^*) + \widetilde{m}(r^*, V) = f(z, t) ,$$

$$V|_{t=0} = 0, \quad V|_{z_n=0} = 0 , \tag{27}$$

$$r_t + \widetilde{b}\, \frac{\partial V}{\partial z_n} + \widetilde{h} \cdot \nabla' r + h_0\, r|_{z_n=0} = g(z', t) ,$$

where

$$\widetilde{L}_{\rho_0} = \frac{\partial}{\partial t} - a \sum_{i,j=1}^{n} \widetilde{A}_{ij} \frac{\partial^2}{\partial z_i\, \partial z_j} ,$$

$$\widetilde{A}_{ij}(0, t) = A_{ij}(0, t) , \quad \widetilde{A}_{ij}(0, 0) = A_{ij}(0, 0) = \delta_{ij} ,$$

and $\widetilde{m}$ is an operator of the same type as m. We introduce a new unknown function

$$W(z, t) = V(z, t) - q(0, 0)\, r^*(z, t)$$

and obtain, instead of (27),

$$\widetilde{L}_{\rho_0} W = \Big(q(z, t) - q(0, 0)\Big) \widetilde{L}_{\rho_0} r^* - \widetilde{m}\Big(r^*, W + q(0, 0)\, r^*\Big) + f(z, t) ,$$

$$W|_{t=0} = 0 ,$$

$$W_t - q(0, 0)\, \widetilde{b}(z, t)\, W_{z_n} + \Big(\widetilde{h} - \widetilde{b}(z, t)\, q(0, 0)\, \nabla' F\Big) \cdot \nabla' W + h_0\, W|_{z_n=0}$$

$$= -q(0, 0)\, g'(z', t) . \tag{28}$$

Finally, we multiply (28) by the cut-off function $\zeta_\lambda(z)$, $\lambda < d$ (ζ_λ is defined above) and we set

$$w(z,t) = W(z,t)\,\zeta_\lambda(z) \quad \text{if } z_n > 0, \ |z| < \lambda ,$$

$$w(z,t) = 0 \qquad\qquad \text{if } z_n > 0, \ |z| \geq \lambda .$$

This function can be considered as a solution of the problem

$$\frac{\partial w}{\partial t} - a\,\Delta w = f_1(z,t), \quad w|_{t=0} = 0 ,$$

$$w_t - \beta(0,0)\cdot \nabla w|_{z_n=0} = g_1(z,t) ,$$

where

$$f_1(z,t) = a\sum_{i,j=1}^{n}\left(\widetilde{A}_{ij}(z,t) - \widetilde{A}_{ij}(0,0)\right) w_{z_i z_j} - \left(L_{\rho 0}\zeta_\lambda - \zeta_\lambda L_{\rho 0}\right) W$$

$$+ \left(q(z,t) - q(0,0)\right)\zeta_\lambda L_{\rho 0} r^* - \zeta_\lambda \widetilde{m}\left(r^*, W + q(0,0)r^*\right) + f\zeta_\lambda ,$$

$$g_1(z',t) = \left(\beta(z,t) - \beta(0,0)\right)\cdot \nabla w + W\,\beta(z,t)\cdot \nabla \zeta_\lambda$$

$$- h_0\,w - \zeta_\lambda\,q(0,0)\,g(z,t) \quad \text{if } |z| < \lambda ,$$

$$f_1(z,t) = 0, \quad g(z',t) = 0 \quad \text{if } |z| \geq \lambda ,$$

$$\beta(z,t) = \left(\widetilde{b}(z,t)\,q(0,0)\,\nabla' F - \widetilde{h}, \ q(0,0)\,\widetilde{b}(z,t)\right) .$$

In virtue of Theorem 2.1, we have

$$[w]_{\mathbb{R}^n_+\times(0,\tau)}^{(2+\alpha,\,1+\alpha/2)} + [w_t]_{\mathbb{R}^{n-1}\times(0,\tau)}^{(1+\alpha,\,1+\alpha/2)} \leq c\left([f_1]_{\mathbb{R}^n_+\times(0,\tau)}^{(\alpha,\alpha/2)} + [g_1]_{\mathbb{R}^{n-1}\times(0,\tau)}^{(1+\alpha,\,\frac{1+\alpha}{2})}\right) ,$$

for arbitrary $\tau < T$. Hence,

$$X(\tau) \equiv |w|_{\mathbb{R}^n_+\times(0,\tau)}^{(2+\alpha,\,1+\alpha/2)} + |w_t|_{\mathbb{R}^{n-1}\times(0,\tau)}^{(1+\alpha,\,\frac{1+\alpha}{2})}$$

$$\leq c\left(\left(\tau^{\frac{\alpha}{2}} + \lambda^\alpha\right)X(\tau) + \left(\frac{\tau}{\lambda^2}\right)^{\alpha/2}\mathcal{Y}(\tau)\right) + c(\lambda)\,\mathcal{N}(\tau) ,$$

where $\mathcal{Y}(\tau)$ and $\mathcal{N}(\tau)$ are defined in (26). Since the point $y_0 \in \Gamma$ is arbitrary, this inequality implies, for small τ and λ,

$$|r|_{\Gamma\times(0,\tau)}^{(2+\alpha,\,1+\alpha/2)} + |r_t|_{\Gamma\times(0,\tau)}^{(1+\alpha,\,\frac{1+\alpha}{2})} \leq c\left(\frac{\tau}{\lambda^2}\right)^{\alpha/2}\mathcal{Y}(\tau) + c(\lambda)\,\mathcal{N}(\tau) .$$

Now, we consider V as a solution of a parabolic problem

$$L_{\rho 0}V + m(r^*, V) = f - q\,L_{\rho 0}r^* ,$$

$$V|_{t=0} = 0, \quad V|_{y\in S} = 0, \quad V|_{y\in\Gamma} = 0 ,$$

with a given $r(y,t)$ and obtain

$$\mathcal{Y}(\tau) \leq c\left(\frac{\tau}{\lambda^2}\right)^{\alpha/2} \mathcal{Y}(\tau) + c(\lambda)\,\mathcal{N}(\tau)\ .$$

If τ/λ^2 is small, this inequality yields (26). Hence, (26) is proved in a small time interval $(0,\tau)$. As shown in [8], this result can be easily generalized to the case of arbitrary $T > 0$.

Theorem 3.1 can be now deduced from the contraction mapping principle. We write the problem (23) in the form

$$(V,r) = \mathcal{M}(\mathcal{F}, \mathcal{G} + \varphi) \equiv \mathcal{M}(\mathcal{F}, \mathcal{G}) + (V_0, r_0)\ , \tag{29}$$

where $(V_0, r_0) = \mathcal{M}(0, \varphi)$, and $\mathcal{M}$ is a linear operator which makes correspond to (f, g) the solution of a linear problem

$$L_{\rho_0} V - (N \cdot J_0^{-T} \nabla U_0)\, L_{\rho_0} r^* + m(r^*, V) = f(y,t)\ ,$$

$$V|_{t=0} = 0\,, \quad r|_{t=0} = 0\,, \quad V|_{y \in S} = 0\,, \quad V|_{y \in \Gamma} = 0\ ,$$

$$r_t + c_0\, S_{\rho_0}\, \frac{\partial V}{\partial n} + h \cdot \nabla_\Gamma r + h_0\, r|_{y \in \Gamma} = g(y,t)\ .$$

We consider (29) in the space $\mathcal{X}_\alpha$ of couples of functions (V, r) with the norm

$$\|(V,r)\|_{\mathcal{X}_T} = |V|_{\Omega_1 \times (0,T)}^{(2+\alpha,\,1+\alpha/2)} + |r|_{\Gamma \times (0,T)}^{(2+\alpha,\,1+\alpha/2)} + |r_t|_{\Gamma \times (0,T)}^{(1+\alpha,\,\frac{1+\alpha}{2})}\ .$$

It can be easily verified that the functions $\mathcal{F}(r^*, V)$ and $\mathcal{G}(r^*, V)$ defined in (22) satisfy the inequalities

$$|\mathcal{F}(r^*, V)|_{\Omega_1 \times (0,\tau)}^{(\alpha,\alpha/2)} + |\mathcal{G}(r^*, V)|_{\Gamma \times (0,\tau)}^{(1+\alpha,\,\frac{1+\alpha}{2})} \leq c\tau^{\frac{1+\alpha}{2}}\, \|(V,r)\|_{\mathcal{X}_\alpha(\tau)}^2\ ,$$

$$\left|\mathcal{F}(r^*, V_1) - \mathcal{F}(r_2^*, V_2)\right|_{\Omega_1 \times (0,\tau)}^{(\alpha,\alpha/2)} + \left|\mathcal{G}(r^*, V_1) - \mathcal{G}(r_2^*, V_2)\right|_{\Gamma \times (0,\tau)}^{(1+\alpha,\,\frac{1+\alpha}{2})}$$

$$\leq c\tau^{\frac{1+\alpha}{2}}\left(|r_1 - r_2|_{\Gamma \times (0,\tau)}^{(2+\alpha,\,1+\alpha/2)} + |V_1 - V_2|_{\Gamma \times (0,\tau)}^{(1+\alpha,\,\frac{1+\alpha}{2})}\right)\left(\|(V_1, r_1)\|_{\mathcal{X}_\tau} + \|(V_2, r_2)\|_{\mathcal{X}_\tau}\right), \tag{30}$$

if ρ_0, r, r_1, r_2 are sufficiently small. Therefore, for small $T = T_0$, the contraction mapping principle guarantees the existence of a unique solution of the equation (29) in the ball $\|(V, r)\|_{\mathcal{X}_{T_0}} \leq 2\,\|(V_0, r_0)\|_{\mathcal{X}_{T_0}}$. There is no other solution with a finite norm $\|(V, r)\|_{\mathcal{X}_{T_0}}$ which can be easily deduced from the estimates (26) and (30). Thus, Theorem 3.1 is proved. ∎

Part II – Evolution free boundary problems for the Navier–Stokes equation

5 Introduction

A typical example of evolving interface is provided by a free surface of a moving liquid or an interface between two different moving liquids. We mention here two evolution free boundary problems for viscous flow studied in the most detailed way. The first of them concerns the motion of a viscous incompressible liquid over a rigid bottom and can be stated as follows. It is necessary to find the domain $\Omega_t = \{-b < x_3 < \eta(x', t), \ x' = (x_1, x_2) \in \mathbb{R}^2\}$, i.e., the function $\eta(x', t)$, the velocity vector field $\mathbf{v}(x, t) = (v_1(x, t), v_2(x, t), v_3(x, t))$ and a scalar pressure $p(x, t)$, $x \in \Omega_t$, satisfying the Navier–Stokes equations

$$\mathbf{v}_t + (\mathbf{v} \cdot \nabla)\mathbf{v} - \nu \nabla^2 \mathbf{v} + \nabla p = 0,$$

$$\nabla \cdot \mathbf{v} = 0, \qquad x \in \Omega_t, \ t > 0,$$

and the following initial and boundary conditions

$$\mathbf{v}(x, 0) = \mathbf{v}_0(x), \qquad x \in \Omega_0 = \{-b < x_3 < \eta(x', 0)\}$$
$$\eta(x', 0) = \eta_0(x'), \qquad x' \in \mathbb{R}^2, \tag{31}$$

$$\eta_t = v_3 - \sum_{j=1}^{2} \eta_{x_j} v_j, \qquad x \in \Gamma_t = \{x_3 = \eta(x', t)\}, \tag{32}$$

$$T(\mathbf{v}, p)\mathbf{n} = \sigma H \mathbf{n} - g \eta(x', t)\mathbf{n}, \qquad x \in \Gamma_t, \tag{33}$$
$$\mathbf{v}|_{x_3 = b} = 0. \tag{34}$$

Here $g, \nu > 0$ and $\sigma \geq 0$ are given constants, namely, gravitational constant, coefficient of viscosity and of the surface tension, respectively, $\mathbf{n}$ is an exterior normal to Γ_t, $T(\mathbf{v}, p) = -pI + \nu S(\mathbf{v})$ and $S(\mathbf{v}) = (\frac{\partial v_j}{\partial x_k} + \frac{\partial v_k}{\partial x_j})_{j,k=1,2,3} \equiv (S_{jk})_{j,k=1,2,3}$ are the stress tensor and the doubled rate-of-strain tensor, respectively,

$$H = \sum_{j=1}^{2} \frac{\partial}{\partial x_j} \frac{\eta_{x_j}}{\sqrt{1 + \eta_{x_1}^2 + \eta_{x_2}^2}}$$

is the doubled mean curvature of Γ_t,

$$\nabla = \left(\frac{\partial}{\partial x_1}, \frac{\partial}{\partial x_2}, \frac{\partial}{\partial x_3} \right), \qquad \nabla p = \operatorname{grad} p, \quad \nabla \cdot \mathbf{v} = \sum_{j=1}^{3} \frac{\partial v_j}{\partial x_j} = \operatorname{div} \mathbf{v}.$$

The kinematic boundary condition (32) tells that the velocity of the evolution of the free surface Γ_t in the normal direction

$$V_n = \frac{\eta_t}{\sqrt{1 + \eta_{x_1}^2 + \eta_{x_2}^2}}$$

coincides with $v_n = \mathbf{v} \cdot \mathbf{n}$. The dynamic boundary condition (33) expresses the balance of forces at the free boundary Γ_t. Finally, (34) is a standard no-slip condition at the bottom $x_3 = -b$.

The problem governing the evolution of an isolated liquid mass consists in the determination of a bounded domain $\Omega_t \subset \mathbb{R}^3$, and of the functions $\mathbf{v}(x,t) = (v_1, v_2, v_3)$ and $p(x,t)$ satisfying the relations

$$\mathbf{v}_t + (\mathbf{v} \cdot \nabla)\mathbf{v} - \nu\nabla^2\mathbf{v} + \nabla p = 0, \quad \nabla \cdot \mathbf{v} = 0, \quad x \in \Omega_t, \ t > 0,$$

$$\mathbf{v}|_{t=0} = \mathbf{v}_0(x), \quad x \in \Omega_0 \ (\Omega_0 \equiv \Omega \text{ is given}), \tag{35}$$

$$T(\mathbf{v}, \rho)\mathbf{n} - \sigma H\mathbf{n}|_{\Gamma_t} = 0, \quad V_n = \mathbf{v} \cdot \mathbf{n}|_{\Gamma_t}.$$

Problems (31)–(34) and (35) have been studied in the papers [1, 4, 5, 33–37]. It has been shown that both of them are uniquely solvable in a certain finite time interval, but if $\sigma > 0$ and the initial data are close to the rest state, then the solution exists for all $t > 0$. The same is true for the solution of the problem (35) with $\sigma = 0$, if $\mathbf{v}_0(x)$ is small and satisfies additional condition expressing the fact that the total momentum of the liquid equals zero. The solutions were found in anisotropic Sobolev spaces. There have been studied other related problems, such as evolution of isolated mass of viscous compressible liquid or non-steady motion of two liquids (see for instance [12, 44]) but they are not treated here.

In what follows we give the main ideas of the proof of the solvability of the problem (35) in the Hölder spaces of functions (more elementary than in [24]).

Our main results may be stated as follows.

Theorem 5.1. *Let $\Gamma \equiv \partial\Omega \in C^{2+\alpha}$, $\alpha \in (0,1)$. For arbitrary $\mathbf{v}_0 \in C^{2+\alpha}(\Omega)$ satisfying the compatibility conditions*

$$\nabla \cdot \mathbf{v}_0 = 0, \quad S(\mathbf{v}_0)\mathbf{n}_0 - \mathbf{n}_0(\mathbf{n}_0 \cdot S(\mathbf{v}_0)\mathbf{n}_0)|_\Gamma = 0, \tag{36}$$

where $\mathbf{n}_0$ is an exterior normal to Γ, problem (35) with $\sigma = 0$ has a unique solution defined in a certain finite time interval $(0, T_0)$ and possessing the following properties: $\Gamma_t \in C^{2+\alpha}$, $\mathbf{v}(\cdot,t) \in C^{2+\alpha}(\Omega_t)$, $\mathbf{v}_t(\cdot,t) \in C^\alpha(\Omega_t)$, $p(\cdot,t) \in C^{1+\alpha}(\Omega_t)$ for all $t \in [0, T_0]$. The solution satisfies the inequality

$$\sup_{t<T_0} |\mathbf{v}_t(\cdot,t)|_{\Omega_t}^{(\alpha)} + \sup_{t<T_0} |\mathbf{v}(\cdot,t)|_{\Omega_t}^{(2+\alpha)} + \sup_{t<T_0} |p(\cdot,t)|_{\Omega_t}^{(1+\alpha)}$$
$$\leq c|\mathbf{v}_0|_\Omega^{(2+\alpha)}(1+|\mathbf{v}_0|_\Omega^{(2+\alpha)}). \tag{37}$$

The number T_0 depends on $|\mathbf{v}_0|_\Omega^{(2+\alpha)}$ and it tends to infinity as $|\mathbf{v}_0|_\Omega^{(2+\alpha)}$ tends to 0.

Theorem 5.2. *Let* $\sigma > 0$, $\Gamma \in C^{3+\alpha}$, $\alpha \in (0,1)$. *For arbitrary* $\mathbf{v}_0 \in C^{3+\alpha}(\Omega)$ *satisfying the conditions (36) the problem (35) has a unique solution* $\Gamma_t \in C^{2+\alpha}$, $\mathbf{v}(\cdot,t) \in C^{2+\alpha}(\Omega_t)$, $\mathbf{v}_t(\cdot,t) \in C^{\alpha}(\Omega_t)$, $p(\cdot,t) \in C^{1+\alpha}(\Omega_t)$ $\forall t \in [0,T_0]$, *and*

$$\sup_{t<T_0} |\mathbf{v}_t(\cdot,t)|_{\Omega_t}^{(\alpha)} + \sup_{t<T_0} |\mathbf{v}(\cdot,t)|_{\Omega_t}^{(2+\alpha)} + \sup_{t<T_0} |p(\cdot,t)|_{\Omega_t}^{(1+\alpha)}$$

$$\leq c\left(|\mathbf{v}_0|_{\Omega}^{(2+\alpha)}\left(1 + |\mathbf{v}_0|_{\Omega}^{(2+\alpha)}\right) + \sigma |H_0|_{\Gamma}^{(1+\alpha)}\right), \tag{38}$$

where H_0 *is the doubled mean curvature of* $\Gamma = \partial\Omega$.

Theorem 5.3. *Assume that* Γ *and* $\mathbf{v}_0(x)$ *satisfy the hypotheses of Theorem 5.1 and that*

$$|\mathbf{v}_0|_{\Omega}^{(2+\alpha)} \leq \varepsilon \ll 1, \tag{39}$$

$$\int_{\Omega_0} \mathbf{v}_0(x)\,dx = 0, \qquad \int_{\Omega_0} [\mathbf{v}_0 \times x]\,dx = 0. \tag{40}$$

Then the solution of the problem (35) with $\sigma = 0$ *is defined for all* $t > 0$, $\Gamma_t \in C^{2+\alpha}$ *for* $t > 0$ *and (37) holds for* $T = \infty$, *moreover,*

$$|\mathbf{v}_t(\cdot,t)|_{\Omega_t}^{(\alpha)} + |\mathbf{v}(\cdot,t)|_{\Omega_t}^{(2+\alpha)} + |p(\cdot,t)|_{\Omega_t}^{(1+\alpha)} \leq c\,e^{-bt} |\mathbf{v}_0|_{\Omega}^{(2+\alpha)}. \tag{41}$$

The formulation of the global existence theorem in the case $\sigma > 0$ needs some preparations. We observe that the problem (35) possesses some integrals of motion (both for $\sigma > 0$ and for $\sigma = 0$), namely,

$$\int_{\Omega_t} \mathbf{v}(x,t)\,dx = \int_{\Omega} \mathbf{v}_0(x)\,dx \quad \text{(conservation of momentum)},$$

$$\int_{\Omega_t} [\mathbf{v}(x,t) \times x]\,dx = \int_{\Omega} [\mathbf{v}_0(x) \times x]\,dx \quad \text{(conservation of an angular momentum)},$$

$$|\Omega_t| = |\Omega| \quad \text{(conservation of mass)}.$$

Without restriction of generality, it can be assumed that

$$\int_{\Omega_t} \mathbf{v}(x,t)\,dx = 0, \qquad \int_{\Omega_t} [\mathbf{v} \times x]\,dx = \beta\,\mathbf{e}_3, \quad \mathbf{e}_3 = (0,0,1),$$

because to this case one can always arrive by a passage to a new inertial coordinate system.

Moreover, problem (35) has a stationary solution corresponding to the rotation of the liquid as a rigid body about the x_3-axis with the angular momentum β. We denote this solution by $\mathbf{v}_\infty, p_\infty, \Omega_\infty$. It is defined by

$$\mathbf{v}_\infty = -\frac{\beta}{I_\infty}\mathbf{h}(x)\,, \quad p_\infty(x) = \frac{\beta}{2\,I_\infty^2}|\mathbf{h}(x)|^2 + p_1\,,$$

$$\mathbf{h}(x) = (-x_2, x_1, 0)\,, \quad \Omega_\infty = \left\{|x| < R_\infty\left(\frac{x}{|x|}\right)\right\}\,,$$

where $p_1 = \text{const}$, $I_\infty = \int_{\Omega_\infty}|\mathbf{h}(x)|^2\,dx$ and R_∞ is a solution of the following differential equation on the unit sphere $S_1 \subset \mathbb{R}^3$

$$H[R_\infty] + \frac{\beta^2}{2\,I_\infty^2}\,R_\infty^2(y)\,(y_1^2 + y_2^2) + p_1 = 0\,, \quad |y| = 1\,. \tag{42}$$

Here $H[R]$ is the doubled mean curvature of the surface $|x| = R(\frac{x}{|x|})$, i.e.,

$$H[R] = \frac{1}{R}\left(\frac{\Delta_{S_1}R}{\sqrt{g}} + \nabla_{S_1}\frac{1}{\sqrt{g}}\cdot\nabla_{S_1}R - \frac{2R}{\sqrt{g}}\right)\,, \tag{43}$$

$g = R^2 + |\nabla_{S_1}R|^2$, ∇_{S_1} and Δ_{S_1} are the gradient and the Laplacean on S_1, respectively.

It is shown in [37] that for small β equation (42) has a unique solution $R_\infty \in C^\infty(S_1)$ such that

$$|\Omega_\infty| = |\Omega| \quad \text{and} \quad \sup_{y\in S_1}|\nabla_{S_1}^j R_\infty(y)| \le c(|j|)\,|\beta|\,.$$

The condition $|\Omega_\infty| = |\Omega|$ defines the constant p_1 in a unique way.

Theorem 5.4. *Assume that the surface $\Gamma = \partial\Omega$ is given by the equation*

$$|x| = R\left(\frac{x}{|x|}, 0\right)$$

on S_1. If the constant β is small and, in addition,

$$|\mathbf{v}_0 - \mathbf{v}_\infty|_\Omega^{(2+\alpha)} + \left|R(\cdot, 0) - R_\infty\right|_{S_1}^{(3+\alpha)} \le \varepsilon_1 \ll 1\,,$$

then the solution of the problem (35) is defined for $t > 0$, the surface Γ_t is given by the equation

$$|x| = R\left(\frac{x}{|x|}, t\right) \tag{44}$$

and

$$|\mathbf{v}_t(\cdot, t)|_{\Omega_t}^{(\alpha)} + \left|\mathbf{v}(\cdot, t) - \mathbf{v}_\infty\right|_{\Omega_t}^{(2+\alpha)} + \left|p(\cdot, t) - p_\infty(\cdot)\right|_{\Omega_t}^{(1+\alpha)} + \left|R(\cdot, t) - R_\infty(\cdot)\right|_{S_1}^{(3+\alpha)}$$

$$\le c\,e^{-bt}\left(|\mathbf{v}_0 - \mathbf{v}_\infty|_\Omega^{(2+\alpha)} + \left|R(\cdot, 0) - R_\infty\right|_{S_1}^{(3+\alpha)}\right)\,.$$

$$\tag{45}$$

6 Linear and model problems

The proof of Theorems 5.1–5.4 is based on the analysis of two linear problems

$$\mathbf{v}_t - \nu \nabla^2 \mathbf{v} + \nabla p = \mathbf{f}(x,t), \quad \nabla \cdot \mathbf{v} = g(x,t), \quad x \in \Omega, \ t > 0 \,,$$

$$\mathbf{v}|_{t=0} = \mathbf{v}_0(x) \,, \tag{46}$$

$$T(\mathbf{v},p)\,\mathbf{n}_0|_\Gamma = \mathbf{a}(x,t),$$

and

$$\mathbf{v}_t - \nu \nabla^2 \mathbf{v} + \nabla p = \mathbf{f}(x,t), \quad \nabla \cdot \mathbf{v} = g(x,t), \quad x \in \Omega, \ t > 0 \,,$$

$$\mathbf{v}|_{t=0} = \mathbf{v}_0(x) \,,$$

$$T(\mathbf{v},p)\,\mathbf{n}_0 - \sigma\,\mathbf{n}_0 \left(\mathbf{n}_0 \cdot \Delta_\Gamma \int_0^t \mathbf{v}(x,\tau)\,d\tau \right)\bigg|_\Gamma = \mathbf{b} + \mathbf{n}_0 \int_0^t B\,d\tau \,, \tag{47}$$

where Ω is a given bounded domain in $\mathbb{R}^3$, $\Gamma \equiv \partial\Omega$, $\mathbf{n}_0$ is an exterior normal to Γ and Δ_Γ is the Laplace–Beltrami operator on Γ.

   Problem (46) is the second (Neumann) initial-boundary value problem for the Stokes equations studied in the papers [31, 32, 39]. The following result is obtained in [39].

Theorem 6.1. *Let* $\Gamma \in C^{2+\alpha}$, $\alpha \in (0,1)$, $\mathbf{v}_0 \in C^{2+\alpha}(\Omega)$, *and let* $\mathbf{f}$, g, $\mathbf{a}$ *be continuous functions such that* $\mathbf{f} \in C^\alpha(\Omega)$, $g \in C^{1+\alpha}(\Omega)$, $\mathbf{a} \in C^{1+\alpha}(\Gamma)$ *for all* $t \in [0,T]$, *and*

$$\Pi_0 \mathbf{a} \equiv \mathbf{a} - \mathbf{n}_0\,(\mathbf{n}_0 \cdot \mathbf{a}) \in C^{1+\alpha, \frac{1+\alpha}{2}}(\Gamma \times (0,T)) \,.$$

Assume also that the compatibility conditions

$$\nabla \cdot \mathbf{v}_0 = g(x,0) \,,$$

$$\Pi_0(S(\mathbf{v}_0)\,\mathbf{n}_0)|_\Gamma = \Pi_0\,\mathbf{a}(x,0) \,, \tag{48}$$

are satisfied and that

$$g(x,t) = \nabla \cdot \mathbf{h}(x,t) \,, \tag{49}$$

with $\mathbf{h}_t \in C^\alpha(\Omega)$, $\forall t \in [0,T]$. *Then the problem (46) has a unique solution such that* $\mathbf{v} \in C^{2+\alpha}(\Omega)$, $\mathbf{v}_t \in C^\alpha(\Omega)$, $p \in C^{1+\alpha}(\Omega)$ *for all* $t \in [0,T]$, *and*

$$\sup_{t<T} |\mathbf{v}_t(\cdot,t)|_\Omega^{(\alpha)} + \sup_{t<T} |\mathbf{v}(\cdot,t)|_\Omega^{(2+\alpha)} + \sup_{t<T} |p(\cdot,t)|_\Omega^{(1+\alpha)}$$

$$\leq c \bigg(\sup_{t<T} |\mathbf{f}(\cdot,t)|_\Omega^{(\alpha)} + \sup_{t<T} |g(\cdot,t)|_\Omega^{(1+\alpha)} + \sup_{t<T} |\mathbf{h}_t(\cdot,t)|_\Omega^{(\alpha)}$$

$$+ |\mathbf{v}_0|_\Omega^{(2+\alpha)} + \sup_{t<T} |\mathbf{a} \cdot \mathbf{n}_0|_\Gamma^{(1+\alpha)} + [\Pi_0\,\mathbf{a}]_{\Gamma \times (0,T)}^{(1+\alpha, \frac{1+\alpha}{2})} \bigg) \,. \tag{50}$$

Similar theorem holds for the problem (47).

Theorem 6.2 ([40]). *Let $\Gamma \in C^{2+\alpha}$, $\alpha \in (0,1)$. For arbitrary $\mathbf{f}$, g, $\mathbf{v}_0$, $\mathbf{b}$, B, such that $\mathbf{v}_0 \in C^{2+\alpha}(\Omega)$, $\mathbf{f} \in C^{\alpha}(\Omega)$, $g \in C^{1+\alpha}(\Omega)$, $\mathbf{b} \in C^{1+\alpha}(\Gamma)$, $B \in C^{\alpha}(\Gamma)$ for $\forall t \in [0,T]$, and $\Pi_0 \mathbf{b} \in C^{1+\alpha, \frac{1+\alpha}{2}}(\Gamma \times (0,T))$, satisfying the compatibility conditions (48) and the condition (49) with $\mathbf{h}_t \in C^{\alpha}(\Omega)$, $t \in [0,T]$, the problem (47) has a unique solution $\mathbf{v} \in C^{2+\alpha}(\Omega)$, $p \in C^{1+\alpha}(\Omega)$ with $\mathbf{v}_t \in C^{\alpha}(\Omega)$, $\forall t \in [0,T]$, and the solution satisfies the inequality*

$$\sup_{t<T} |\mathbf{v}_t(\cdot,t)|_{\Omega}^{(\alpha)} + \sup_{t<T} |\mathbf{v}(\cdot,t)|_{\Omega}^{(2+\alpha)} + \sup_{t<T} |p(\cdot,t)|_{\Omega}^{(1+\alpha)}$$

$$\le c \left(\sup_{t<T} |\mathbf{f}(\cdot,t)|_{\Omega}^{(\alpha)} + \sup_{t<T} |g(\cdot,t)|_{\Omega}^{(1+\alpha)} + \sup_{t<T} |\mathbf{h}_t(\cdot,t)|_{\Omega}^{(\alpha)} \right. \tag{51}$$

$$\left. + |\mathbf{v}_0|_{\Omega}^{(2+\alpha)} + \sup_{t<T} |\mathbf{b}\cdot\mathbf{n}|_{\Gamma}^{(1+\alpha)} + [\Pi_0 \mathbf{b}]_{\Gamma \times (0,T)}^{(1+\alpha, \frac{1+\alpha}{2})} + \sup_{t<T} |B|_{\Gamma}^{(\alpha)} \right) .$$

We give here the idea of the proof of the estimates (50) and (51). It is based on the analysis of the corresponding model problems in the half-space $\mathbb{R}_+^3 = \{x_3 > 0\}$, namely,

$$\mathbf{v}_t - \nu \nabla^2 \mathbf{v} + \nabla p = \mathbf{f}(x,t), \quad \nabla \cdot \mathbf{v} = g(x,t), \quad x \in \mathbb{R}_+^3, \quad t > 0$$

$$\mathbf{v}|_{t=0} = \mathbf{v}_0(x) ,$$

$$\nu \left(\frac{\partial v_j}{\partial x_3} + \frac{\partial v_3}{\partial x_j} \right)\bigg|_{x_3=0} = a_j(x',t), \quad x' = (x_1, x_2) \in \mathbb{R}^2, \quad j = 1, 2 , \tag{52}$$

$$-p + 2\nu \frac{\partial v_3}{\partial x_3}\bigg|_{x_3=0} = a_3(x',t)$$

and

$$\mathbf{v}_t - \nu \nabla^2 \mathbf{v} + \nabla p = \mathbf{f}(x,t), \quad \nabla \cdot \mathbf{v} = g(x,t), \quad x \in \mathbb{R}_+^3, \quad t > 0 ,$$

$$\mathbf{v}|_{t=0} = \mathbf{v}_0(x) ,$$

$$\nu \left(\frac{\partial v_j}{\partial x_3} + \frac{\partial v_j}{\partial x_3} \right)\bigg|_{x_3=0} = b_j(x',t), \quad x' \in \mathbb{R}^2, \quad j = 1, 2 , \tag{53}$$

$$-p + 2\nu \frac{\partial v_3}{\partial x_3} + \sigma \nabla'^2 \int_0^t v_3(x',\tau)\, d\tau\bigg|_{x_3=0} = b_3(x',t) + \int_0^t B(x',\tau)\, d\tau ,$$

where $\nabla' = (\frac{\partial}{\partial x_1}, \frac{\partial}{\partial x_2})$. We assume at first that $\mathbf{f} = 0$, $g = 0$, $\mathbf{v}_0 = 0$ and that $\mathbf{a}$, $\mathbf{b}$, B are given for $t > 0$, decay at infinity sufficiently rapidly and satisfy the compatibility conditions

$$a_1(x',0) = a_2(x',0) = 0, \quad b_1(x',0) = b_2(x',0) = 0 . \tag{54}$$

Then we can reduce these problems to the boundary value problems for the system of ordinary differential equations on the half-line $x_3 > 0$ with the aid of the Fourier–Laplace transformation

$$\widehat{u}(\xi,s,x_3) = \int_{\mathbb{R}^2} e^{-ix'\cdot\xi'} \, dx' \int_0^\infty e^{-st} \, u(x,t) \, dt .$$

Making this transformation we obtain

$$-\nu \frac{d^2\widehat{v}}{dx_3^2} + \nu r^2 \, \widehat{v} + \widehat{\nabla}\widehat{p} = 0 ,$$

$$\widehat{\nabla}\cdot\widehat{v} = i\,\xi_1\,\widehat{v}_1 + i\,\xi_2\,\widehat{v}_2 + \frac{d\widehat{v}_3}{dx_3} = 0 , \quad x_3 > 0 ,$$

$$\widehat{v}(\xi,s,x_3), \ \widehat{p}(\xi,s,x_3) \to 0 \quad (x_3 \to +\infty) ,$$

$$\nu\left(\frac{d\widehat{v}_j}{dx_3} + i\,\xi_j\,\widehat{v}_3\right)\bigg|_{x_3=0} = \widehat{a}_j, \quad j = 1,2 , \tag{55}$$

$$-\widehat{p} + 2\nu\frac{d\widehat{v}_3}{dx_3}\bigg|_{x_3=0} = \widehat{a}_3 ,$$

where $\widehat{\nabla} = (i\,\xi_1, i\,\xi_2, \frac{d}{dx_3})$, $r = \sqrt{s\nu^{-1} + |\xi|^2}$, $\arg r \in (-\frac{\pi}{2}, \frac{\pi}{2}]$ and

$$-\nu \frac{d^2\widehat{v}}{dx_3^2} + \nu r^2 \, \widehat{v} + \widehat{\nabla}p = 0 , \quad \widehat{\nabla}\cdot\widehat{v} = 0 ,$$

$$\widehat{v}(\xi,s,x_3), \ \widehat{p}(\xi,s,x_3) \to 0 \quad (x_3 \to +\infty) ,$$

$$\nu\left(\frac{d\widehat{v}_j}{dx_3} + i\,\xi_j\,\widehat{v}_3\right)\bigg|_{x_3=0} = \widehat{b}_j, \quad j = 1,2 , \tag{56}$$

$$-\widehat{p} + 2\nu\frac{d\widehat{v}_3}{dx_3} - \frac{\sigma|\xi|^2}{s}\widehat{v}_3\big|_{x_3=0} = \widehat{b}_3 + \frac{\widehat{B}}{s} .$$

Both problems (55) and (56) can be solved explicitly (see [31, 38]). The solution of (55) has the form

$$\widehat{v}_j = -\frac{(1-\delta_{j3})\,\widehat{a}_j}{\nu\,r}\,e^{-rx_3} + \sum_{k=1}^3 \frac{R_{jk}\,\widehat{a}_k}{\nu\,r\,Q}\,e^{-rx_3} + \sum_{k=1}^3 \frac{S_{jk}\,\widehat{a}_k}{\nu\,Q}\,\frac{e^{-rx_3} - e^{-|\xi|x_3}}{r - |\xi|} ,$$

$$\widehat{p} = -\widehat{a}_3\,e^{-|\xi|x_3} + \sum_{k=1}^3 P_k\,\widehat{a}_k\,e^{-|\xi|x_3} ,$$

where

$$P_j(s,\xi) = i\xi_j \frac{2r\,(r+|\xi|)}{Q(s,\xi)}, \qquad j=1,2\,,$$

$$P_3(s,\xi) = 2\,|\xi|^2\,\frac{r-|\xi|}{Q(s,\xi)}\,,$$

$$Q(s,\xi) = r^2 + |\xi|\,r^2 + 3\,r\,\xi^2 - |\xi|^3 = (r+|\xi|)\,M(s,\xi), \quad M(s,\xi) = \frac{s}{\nu} + 4|\xi|^2 \frac{r}{r+|\xi|}\,,$$

and R_{jk}, S_{jk} are elements of the matrices

$$\mathcal{R} = \begin{pmatrix} \xi_1^2(3r-|\xi|) & \xi_1\xi_2\,(3r-|\xi|) & i\xi_1 r\,(r-|\xi|) \\ \xi_1\xi_2\,(3r-|\xi|) & \xi_2^2\,(3r-|\xi|) & i\xi_2 r\,(r-|\xi|) \\ -i\xi_1 r\,(r-|\xi|) & -i\xi_2\,(r-|\xi|) & -|\xi|\,r\,(r+|\xi|) \end{pmatrix}$$

$$S = \begin{pmatrix} -2r\,\xi_1^2 & -2r\,\xi_1\xi_2 & -i\xi_1\,(r^2+\xi^2) \\ -2r\,\xi_1\xi_2 & -2r\,\xi_2^2 & -i\xi_2\,(r^2+\xi^2) \\ -2\,i\,\xi_1\,|\xi|^r & -2\,i\,\xi_2\,|\xi|\,r & |\xi|\,(r^2+\xi^2) \end{pmatrix}$$

and it can be considered as a solution of the Neumann problem for the system of 3 heat equations

$$\mathbf{v}_t - \nu\nabla^2\mathbf{v} = -\nabla p, \qquad x\in\mathbb{R}_+^3,\ t>0\,,$$

$$\mathbf{v}|_{t=0} = 0\,,$$

$$\left.\frac{\partial\mathbf{v}}{\partial x_3}\right|_{x_3=0} = \mathbf{d}\,. \tag{57}$$

Here,

$$\widehat{d}_j = \frac{\widehat{a}_j}{\nu} - i\xi_j\,\widehat{v}_3|_{x_3=0} = -\frac{i\xi_j}{\nu r}\sum_{k=1}^3 \frac{R_{3k}}{Q}\,\widehat{a}_k + \frac{\widehat{a}_j}{\nu}\,, \qquad j=1,2\,,$$

$$\widehat{d}_3 = -\sum_{j=1}^2 i\xi_j\,\widehat{v}_j|_{x_3=0} = \sum_{j=1}^2 \frac{i\xi_j}{\nu r}\,\widehat{a}_j - \sum_{j=1}^2\sum_{k=1}^3 \frac{i\xi_j}{\nu r}\frac{R_{jk}}{Q}\,\widehat{a}_k\,, \tag{58}$$

and $p(x,t)$ is a solution of the Dirichlet problem

$$\nabla^2 p(x) = 0\,, \qquad x_3 > 0\,,$$

$$p(x)\to 0\ \ (|x|\to\infty)\,, \qquad p|_{x_3=0} = -a_3 + 2\nu\,d_3(x',t)\,. \tag{59}$$

Moreover, the solution of the problem (56) can be regarded as a solution of the problem (55) with $a_k(x', t)$ related to $b_m(x', t)$ as follows

$$a_j(x', t) = b_j(x', t), \quad j = 1, 2,$$

$$\widehat{a}_3(\xi, s) = \widehat{b}_3 \left(1 - \frac{\sigma |\xi|^3}{P(s, \xi)} \right) + \widehat{B} \frac{\nu M(s, \xi)}{P(s, \xi)}$$

$$+ \sigma \left(\frac{|\xi|^2}{P(s, \xi)} - \frac{2 |\xi|^3}{(r + |\xi|) P(s, \xi)} \right) \sum_{j=1}^{2} i \xi_j \widehat{b}_j,$$

$$P(s, \xi) = \nu s M(s, \xi) + \sigma |\xi|^3.$$

(60)

Equations (58) and (60) are equivalent to

$$d_j(x', t) = \frac{1}{\nu} a_j(x', t) - \sum_{m=1}^{3} \frac{\partial}{\partial x_j} \int_0^t \int_{\mathbb{R}^2} K_{3m}(x' - y', t - \tau) a_m(y', \tau) \, dy' \, d\tau,$$

$$d_3(x', t) = \sum_{j=1}^{2} \frac{\partial}{\partial x_j} \left(2 \int_0^t \int_{\mathbb{R}^2} \Gamma(x' - y', 0, t - \tau) a_j(y', \tau) \, dy' \, d\tau \right.$$

$$\left. - \sum_{m=1}^{3} \int_0^t \int_{\mathbb{R}^2} K_{jm}(x' - y', t - \tau) a_m(y', \tau) \, dy' \, d\tau \right),$$

$$a_3(x', t) = b_3(x', t) + \sum_{j=1}^{2} \frac{\partial}{\partial x_j} \int_0^t \int_{\mathbb{R}^2} V_j(x' - y', t - \tau) b_3(y', \tau) \, dy' \, d\tau$$

$$+ \int_0^t \int_{\mathbb{R}^2} V(x' - y', t - \tau) B(y', \tau) \, dy' \, d\tau$$

$$+ \sum_{j=1}^{2} \frac{\partial}{\partial x_j} \int_0^t \int_{\mathbb{R}^2} W(x' - y', t - \tau) b_j(y', \tau) \, dy' \, d\tau,$$

(61)

where $\Gamma(x, t) = \frac{1}{(4 \pi \nu t)^{3/2}} e^{-\frac{|x|^2}{4 \nu t}}$ and K_{jm}, V_j, V, W are defined by

$$\widehat{K}_{im} = \frac{R_{im}}{\nu r Q}$$

(62)

$$\widehat{V}_j = \frac{\sigma i \xi_j |\xi|}{P(s, \xi)}, \quad \widehat{V} = \frac{\nu M}{P}, \quad \widehat{W} = -\sigma \left(\frac{2 |\xi|^3}{(r + |\xi|) P} - \frac{|\xi|^2}{P} \right). \quad (63)$$

The following propositions provide pointwise estimates of these kernels.

Proposition 6.1. *Assume that $\widehat{K}(\zeta, s) = \widehat{K}(\zeta_1, ..., \zeta_n, s)$ is defined for all complex-valued $\zeta_j = \xi_j + i\eta_j$, $j = 1, ..., n$, and s, such that*

$$\operatorname{Re} s + \ae \left| \operatorname{Im} s \right| \geq -\delta_1 |\xi|^2, \quad |\eta| \leq \delta_2 |\xi|, \quad \xi \neq 0 , \tag{64}$$

where $\ae$, δ_1, δ_2 are positive constants, is analytic in the domain (64), homogeneous, i.e.

$$\widehat{K}(\lambda \zeta, \lambda^2 s) = \lambda^{m-q}\, \widehat{K}(\zeta, s)$$

and

$$|\widehat{K}(\zeta, s)| \leq \frac{c\,|\xi|^m}{(|s| + |\xi|^2)^{q/2}} ,$$

where q and m are integral numbers, $q > 0$, $m > -n$. Then the Fourier–Laplace pre-image of $\widehat{K}$,

$$K(x, t) = \frac{1}{(2\pi)^{n+1} i} \int_{\mathbb{R}^n} e^{ix \cdot \xi}\, d\xi \int_{\operatorname{Re} s = a > 0} e^{st} \widehat{K}(\xi, s)\, ds, \quad x \in \mathbb{R}^n ,$$

satisfies the inequalities

$$|\mathcal{D}_t^k \mathcal{D}_x^j K(x, t)| \leq \frac{c\, t^{q/2 - 1 - k}}{(|x|^2 + t)^{\frac{m+n+|j|}{2}}}, \quad |j| \geq 0, \ k \geq 0 ,$$

and $K(x, t) = 0$ for $t < 0$.

Corollary 6.1. *The kernels $K_{jm}(x', t)$ (62) satisfy the inequalities*

$$|\mathcal{D}_t^k \mathcal{D}_{x'}^j K_{jm}(x, t)| \leq \frac{c(k, j)}{t^{\frac{1}{2}+k}(|x'|^2 + t)^{\frac{2+|j|}{2}}}, \quad k \geq 0, \ |j| \geq 0 , \tag{65}$$

and they vanish for $t < 0$.

Proposition 6.2. *The kernels V_j, V, W_j satisfy the inequalities*

$$|\mathcal{D}_{x'}^j V_j(x, t)| + |\mathcal{D}_{x'}^j V(x, t)| + |\mathcal{D}_{x'}^j W_j(x', t)| \leq \frac{c(j, t)}{(|x| + t)^{2+|j|}} , \tag{66}$$

where $c(j, t)$ are increasing functions of t, and

$$V_j(x', t) = V(x', t) = W_j(x', t) = 0 \quad \text{for } t < 0 .$$

Proposition 6.1 is proved exactly as Lemma 3.2 in [39] (see also [41], Proposition 2.1). Proposition 6.2 is proved in [40].

Using estimates (65)–(66), it can be shown by standard methods of the potential theory (see [39, 40] and the proof of (17) that

$$\sum_{k=1}^{2} [d_k]_{\mathbb{R}^2 \times (0,T)}^{(1+\alpha, \frac{1+\alpha}{2})} \leq c \left(\sum_{j=1}^{2} [a_j]_{\mathbb{R}^2 \times (0,T)}^{(1+\alpha, \frac{1+\alpha}{2})} + \sup_{t<T} [a_3(\cdot, t)]_{\mathbb{R}^2}^{(1+\alpha)} \right) , \tag{67}$$

and that for the function $a_3(x',t)$ in $(61)_3$ the inequality

$$\sup_{t<T}[a_3(\cdot,t)]_{\mathbb{R}^2}^{(1+\alpha)} \le c(T)\left(\sum_{k=1}^{3}\sup_{t<T}[b_k(\cdot,t)]_{\mathbb{R}^2}^{(1+\alpha)} + \sup[B(\cdot,t)]_{\mathbb{R}^2}^{(\alpha)}\right) \quad (68)$$

holds.

Now, we can estimate the solutions of the problems (52) and (53).

Theorem 6.3. *If* $\mathbf{v}_0 \in C^{2+\alpha}(\mathbb{R}_+^3)$, $\mathbf{f} \in C^{\alpha}(\mathbb{R}_+^3)$, $g(\cdot,t) \in C^{1+\alpha}(\mathbb{R}_+^3)$, $a_3(\cdot,t) \in C^{1+\alpha}(\mathbb{R}^2)$ *for all* $t \in [0,T]$, $a_j \in C^{1+\alpha,\frac{1+\alpha}{2}}(\mathbb{R}^2 \times (0,T))$, $j = 1,2$, *the compatibility conditions*

$$\nabla \cdot \mathbf{v}_0 = g(x,0), \quad \nu\left(\frac{\partial v_{0j}}{\partial x_3} + \frac{\partial v_{03}}{\partial x_j}\right)\bigg|_{x_3=0} = a_j(x',0), \quad j = 1,2, \quad (69)$$

hold and

$$g(x,t) = \nabla \cdot \mathbf{h}(x,t), \quad \mathbf{h}_t \in C^{\alpha}(\mathbb{R}_+^3), \quad \forall t \in [0,T], \quad (70)$$

then the problem (52) has a unique solution $\mathbf{v} \in C^{2+\alpha}(\mathbb{R}_+^3)$, $p \in C^{1+\alpha}(\mathbb{R}_+^3)$, *such that* $\mathbf{v}_t \in C^{\alpha}(\mathbb{R}_+^3)$, $\forall t \in [0,T]$, *and this solution satisfies the inequality*

$$\sup_{t<T}[\mathbf{v}_t(\cdot,t)]_{\mathbb{R}_+^3}^{(\alpha)} + \sup_{t<T}[\mathbf{v}(\cdot,t)]_{\mathbb{R}_+^3}^{(2+\alpha)} + \sup_{t<T}[p(\cdot,t)]_{\mathbb{R}_+^3}^{(1+\alpha)}$$

$$\le c\left(\sup_{t<T}[\mathbf{f}(\cdot,t)]_{\mathbb{R}_+^3}^{(\alpha)} + \sup_{t<T}[g(\cdot,t)]_{\mathbb{R}_+^3}^{(1+\alpha)} + \sup_{t<T}[\mathbf{h}_t(\cdot,t)]_{\mathbb{R}_+^3}^{(\alpha)}\right.$$

$$\left. + \sup_{t<T}[a_3(\cdot,t)]_{\mathbb{R}^2}^{(1+\alpha)} + \sum_{j=1}^{2}[a_j]_{\mathbb{R}^2\times(0,T)}^{(1+\alpha,\frac{1+\alpha}{2})}\right), \quad (71)$$

with the constant independent of T.

Theorem 6.4. *If* $\mathbf{v}_0$, $\mathbf{f}$, g, b_k *satisfy the hypotheses of Theorem 6.3 and* $B(\cdot,t) \in C^{\alpha}(\mathbb{R}^2)$, $\forall t \in [0,T]$, *then the problem (53) has a unique solution* $\mathbf{v}(\cdot,t) \in C^{2+\alpha}(\mathbb{R}_+^3)$, $p(\cdot,t) \in C^{1+\alpha}(\mathbb{R}_+^3)$ *with* $\mathbf{v}_t(\cdot,t) \in C^{\alpha}(\mathbb{R}_+^3)$ *for all* $t \in [0,T]$, *and this solution satisfies the inequality*

$$\sup_{t<T}[\mathbf{v}_t(\cdot,t)]_{\mathbb{R}_+^3}^{(\alpha)} + \sup_{t<T}[\mathbf{v}(\cdot,t)]_{\mathbb{R}_+^3}^{(2+\alpha)} + \sup_{t<T}[p(\cdot,t)]_{\mathbb{R}_+^3}^{(1+\alpha)}$$

$$\le c(T)\left(\sup_{t<T}[\mathbf{f}(\cdot,t)]_{\mathbb{R}_+^3}^{(\alpha)} + \sup_{t<T}[g(\cdot,t)]_{\mathbb{R}_+^3}^{(1+\alpha)} + \sup_{t<T}[\mathbf{h}_t(\cdot,t)]_{\mathbb{R}_+^3}^{(\alpha)}\right.$$

$$\left. + \sup_{t<T}[b_3(\cdot,t)]_{\mathbb{R}^2}^{(1+\alpha)} + \sum_{j=1}^{2}[b_j]_{\mathbb{R}^2\times(0,T)}^{(1+\alpha,\frac{1+\alpha}{2})} + \sup_{t<T}[B(\cdot,t)]_{\mathbb{R}^2}^{(\alpha)}\right). \quad (72)$$

We restrict ourselves with the proof of the estimates (71), (72). We assume at first that $\mathbf{f} = \mathbf{0}$, $g = 0$ and consider the problem (57). Well known estimates for the parabolic Neumann problem and for harmonic functions imply

$$\sup_{t<T}[\mathbf{v}_t(\cdot,t)]_{\mathbb{R}^3_+}^{(\alpha)} + \sup_{t<T}[\mathbf{v}(\cdot,t)]_{\mathbb{R}^3_+}^{(2+\alpha)} \leq c\left(\sup_{t<T}[\nabla p(\cdot,t)]_{\mathbb{R}^3_+}^{(\alpha)} + [\mathbf{d}]_{\mathbb{R}^2\times(0,T)}^{(1+\alpha,\frac{1+\alpha}{2})}\right),$$

$$\sup_{t<T}[\nabla p(\cdot,t)]_{\mathbb{R}^3_+}^{(\alpha)} \leq c\left(\sup_{t<T}[a_3(\cdot,t)]_{\mathbb{R}^2}^{(1+\alpha)} + \sup_{t<T}[d_3(\cdot,t)]_{\mathbb{R}^2}^{(1+\alpha)}\right).$$

Now, we take account of (67) and arrive at the estimate (71) with $\mathbf{f} = \mathbf{0}$, $g = 0$, $\mathbf{h} = \mathbf{0}$, from which, due to (68), inequality (72) follows (also with $\mathbf{f} = \mathbf{0}$, $g = 0$, $\mathbf{h} = \mathbf{0}$).

The general case is treated by the construction of auxiliary functions satisfying non-homogeneous Stokes system. Let $\mathbf{f}^*$ and $\mathbf{v}_0^*$ be extensions of $\mathbf{f}(x,t)$ and $\mathbf{v}_0(x)$ into the whole space $\mathbb{R}^3$ such that

$$[\mathbf{f}^*(\cdot,t)]_{\mathbb{R}^3}^{(\alpha)} \leq c\,[\mathbf{f}(\cdot,t)]_{\mathbb{R}^3_+}^{(\alpha)}, \quad [\mathbf{v}_0^*]_{\mathbb{R}^3}^{(2+\alpha)} \leq c\,[\mathbf{v}_0]_{\mathbb{R}^3_+}^{(2+\alpha)}.$$

They can be defined, for instance, by the formulas

$$\mathbf{f}^*(x,t) = \mathbf{f}(x',-x_3,t), \qquad x_3 < 0,$$

$$\mathbf{v}_0^*(x) = \sum_{k=1}^{3}\lambda_k\,\mathbf{v}_0\left(x',-\frac{x_3}{k}\right),$$

where $(\lambda_1,\lambda_2,\lambda_3)$ is a solution of the system

$$\sum_{k=1}^{3}\lambda_k\left(-\frac{1}{k}\right)^s = 1, \quad s = 0,1,2.$$

Consider the functions

$$\mathbf{v}^{(1)}(x,t) = \int_0^t d\tau \int_{\mathbb{R}^3}\Gamma(x-y,t-\tau)\,\mathbf{f}^*(y,t)\,dy + \int_{\mathbb{R}^3}\Gamma(x-y,t)\,\mathbf{v}_0^*(y)\,dy,$$

$$p^{(1)}(x,t) = 0,$$

$$\mathbf{v}^{(2)}(x,t) = \nabla\int_{\mathbb{R}^3}\Big(E(x-y)-E(x-y^*)\Big)\Big(g-\nabla\cdot\mathbf{v}^{(1)}(y,t)\Big)\,dy,$$

$$p^{(2)}(x,t) = \nu\big(g-\nabla\cdot\mathbf{v}^{(1)}\big) + \int_{\mathbb{R}^3_+}\nabla_y\Big[E(x-y)-E(x-y^*)\Big]\Big(\mathbf{h}(y,t)-\mathbf{v}_t^{(1)}(y,t)\Big)\,dy,$$

where $y = (y',-y_3)$, and $E(x) = -\frac{1}{4\pi|x|}$ is a fundamental solution of the Laplace equation. It is easily seen that

$$\frac{\partial}{\partial t}(\mathbf{v}^{(1)}+\mathbf{v}^{(2)}) - \nu\nabla^2(\mathbf{v}^{(1)}+\mathbf{v}^{(2)}) + \nabla(p^{(1)}+p^{(2)}) = \mathbf{f},$$

$$\nabla\cdot(\mathbf{v}^{(1)}+\mathbf{v}^{(2)}) = g, \quad \mathbf{v}^{(1)}+\mathbf{v}^{(2)}|_{t=0} = \mathbf{v}_0(x),$$

so the functions

$$\mathbf{u}(x,t) = \mathbf{v} - \mathbf{v}^{(1)} - \mathbf{v}^{(2)}, \qquad q = p - p^{(1)} - p^{(2)}$$

satisfy (52) or (53) with $\mathbf{f} = \mathbf{0}$, $g = 0$. Hence, these functions can be estimated by inequalities (71) or (72) with $\mathbf{f} = \mathbf{0}$, $g = 0$, obtained above. Using this inequalities and classical estimates of Newtonian potential we easily prove (71), (72) in the general case.

We have considered model problems (52), (53). As for the estimates (50), (51) they are obtained by a standard Schauder's method and the solvability of the problems (46), (47) can be proved by the construction of the regularizer (see [31, 32, 39, 40]).

In addition to (71), (72), the solutions of the problems (46), (47) (with $g = 0$) satisfy coercive estimates (see [24, 31]), namely,

$$|\mathbf{v}|_{\Omega \times (0,T)}^{(2+\alpha,1+\alpha/2)} + |\nabla p|_{\Omega \times (0,T)}^{(\alpha,\alpha/2)} + \|p\|_{\Omega \times (0,T)}^{(\gamma,1+\alpha)}$$
$$\leq c\left(|\mathbf{f}|_{\Omega \times (0,T)}^{(\alpha,\alpha/2)} + |\mathbf{v}_0|_{\Omega}^{(2+\alpha)} + |\Pi_0\,\mathbf{a}|_{\Gamma \times (0,T)}^{(1+\alpha,\frac{1+\alpha}{2})} + \|\mathbf{a}\cdot\mathbf{n}\|_{\Gamma \times (0,T)}^{(\gamma,1+\alpha)} \right) \quad (73)$$

and

$$|\mathbf{v}|_{\Omega \times (0,T)}^{(2+\alpha,1+\alpha/2)} + |\nabla p|_{\Omega \times (0,T)}^{(\alpha,\alpha/2)} + \|p\|_{\Omega \times (0,T)}^{(\gamma,1+\alpha)}$$
$$\leq c\left(|\mathbf{f}|_{\Omega \times (0,T)}^{(\alpha,\alpha/2)} + |\mathbf{v}_0|_{\Omega}^{(2+\alpha)} + |\Pi_0\,\mathbf{b}|_{\Gamma \times (0,T)}^{(1+\alpha,\frac{1+\alpha}{2})} + \|\mathbf{b}\cdot\mathbf{n}\|_{\Gamma \times (0,T)}^{(\gamma,1+\alpha)} + |B|_{\Gamma \times (0,T)}^{(\alpha,\alpha/2)} \right),$$
$$(74)$$

where $\gamma \in (0,1)$,

$$\|u\|_{\Omega \times (0,T)}^{(\gamma,1+\alpha)} = \sup_{t,\tau \in (0,T)} |t - t'|^{-\frac{1+\alpha-\gamma}{2}} |u(\cdot,t) - u(\cdot,t')|_{\Omega}^{(\gamma)} \quad (75)$$

and $\|u\|_{\Gamma \times (0,T)}^{(\gamma,1+\alpha)}$ is defined in a similar way.

These estimates hold under the conditions $\nabla \cdot \mathbf{f} = 0$ and $B|_{t=0} = 0$.

7 Lagrangean coordinates and local existence theorems

In the problems of fluid mechanics, there exists a natural mapping of the domain Ω_t with a free boundary onto a given domain $\Omega_0 \equiv \Omega$. This mapping is given by a relationship between the Eulerian and the Lagrangean coordinates, i.e.,

$$\mathbf{x} = \xi + \int_0^t \mathbf{u}(\xi,\tau)\,d\tau \equiv X(\xi,t), \qquad \xi \in \Omega,$$

where

$$\mathbf{u}(\xi,t) = \mathbf{v}\Big(X(\xi,t),t\Big),$$

$$\mathbf{u}_t - \nu \nabla_u^2 \mathbf{u} + \nabla_u q = 0, \qquad \nabla_u \cdot \mathbf{u} = 0, \ \ \xi \in \Omega, \ \ t > 0 \ ,$$

$$\mathbf{u}|_{t=0} = \mathbf{v}_0(\xi) \ , \tag{76}$$

$$T_u(\mathbf{u}, q)\,\mathbf{n} - \sigma \, \Delta(t)\, X(\xi, t)|_\Gamma = 0 \ , \tag{77}$$

where $q(\xi, t) = p(X(\xi, t), t)$, $\nabla_u = J^{-T}\nabla$ is a transformed gradient, $J = (\frac{\partial x_k}{\partial x_m})_{k,m=1,2,3}$, $T_u = -q\,I + \nu\,S_u(\mathbf{u})$ and $S_u = (\nabla_u \mathbf{u}) + (\nabla_u \mathbf{u})^T$ are transformed stress and rate-of-strain tensor, respectively, $\Delta(t)$ is the Laplace–Beltrami operator on the surface

$$\Gamma_t = X(\Gamma, t) \ . \tag{78}$$

We have used the well-known formula $H\,\mathbf{n} = \Delta(t)\,X$.

By virtue of the equation $\nabla \cdot \mathbf{v} = 0$, the matrix J^{-T} coincides with the matrix of cofactors $\mathcal{A} = (A_{km})_{k,m=1,2,3}$ of the elements

$$a_{km}(\xi, t) = \frac{\partial x_k}{\partial \xi_m} = \delta_{km} + \int_0^t \frac{\partial u_k(\xi, \tau)}{\partial \xi_m}\, d\tau \tag{79}$$

of the matrix J, hence, $\nabla_u = \mathcal{A}\nabla$. Moreover, since $\sum_{k=1}^3 \frac{\partial}{\partial \xi_k} A_{km}(\xi, t) = 0$, we have

$$\nabla_u \cdot \mathbf{u} = \nabla \cdot \mathcal{A}\,\mathbf{u} \ . \tag{80}$$

We observe finally that the normal $\mathbf{n}$ to Γ_t at the point $X(\xi, t)$ and the normal $\mathbf{n}_0$ to Γ at the point ξ are related to each other by

$$\mathbf{n} = \frac{\mathcal{A}\,\mathbf{n}_0}{|\mathcal{A}\,\mathbf{n}_0|} \ . \tag{81}$$

It is clear that theorems 5.1 and 5.2 reduce to the proof of the solvability of the problem (76), (77) with $\sigma = 0$ or $\sigma > 0$.

For subsequent computations it is convenient to separable tangential and normal components in the boundary condition (77). Let $\Pi_0\,\mathbf{f} = \mathbf{f} - \mathbf{n}_0\,(\mathbf{n}_0 \cdot \mathbf{f})$ and $\Pi\mathbf{f} = \mathbf{f} - \mathbf{n}\,(\mathbf{n} \cdot \mathbf{f})$ be projections of the vector field $\mathbf{f}(\xi)$, $\xi \in \Gamma$, onto tangent planes to Γ and Γ_t, respectively. If $\mathbf{n} \cdot \mathbf{n}_0 > 0$ (which is certainly the case for small t), then (77) is equivalent to two equations

$$\Pi_0 \Pi\, S_u(\mathbf{u})\,\mathbf{n} = 0 \ ,$$

$$\mathbf{n}_0 \cdot T_u(\mathbf{u}, q)\,\mathbf{n} - \sigma\,\mathbf{n}_0 \cdot \Delta(t)\,X = 0, \qquad \xi \in \Gamma \ .$$

Making use of the identities

$$\mathbf{n}_0 \cdot \Delta(t)\,X = \mathbf{n}_0 \cdot \Delta(t)\,\xi + \mathbf{n}_0 \cdot \Delta(t) \int_0^t \mathbf{u}(\xi, \tau)\, d\tau$$

$$= \mathbf{n}_0 \cdot \Delta(0) \int_0^t \mathbf{u}(\xi, \tau)\, d\tau + H_0(\xi) + \mathbf{n}_0 \cdot \int_0^t \dot{\Delta}(\tau)\,\xi\, d\tau$$

$$+ \mathbf{n}_0 \cdot \int_0^t \left(\dot{\Delta}(\tau) \int_0^t \mathbf{u}(\xi, \tau')d\tau' + \Big(\Delta(t) - \Delta(0)\Big)\mathbf{u}(\xi, \tau) \right) d\tau \ ,$$

where $H_0(\xi) = n_0(\xi) \cdot \Delta(0)\,\xi$ is the doubled mean curvature of Γ_0 and $\dot{\Delta}(t) = \frac{d}{dt}\Delta(t)$ we write the problem (76), (77) in the form

$$\mathbf{u}_t - \nu\,\nabla^2\mathbf{u} + \nabla q = F_1(\mathbf{u}, q)\,,$$

$$\nabla \cdot \mathbf{u} = F_2(\mathbf{u})\,,$$

$$\mathbf{u}|_{t=0} = \mathbf{v}_0(\xi)\,,$$

$$\Pi_0\,S(\mathbf{u})\,\mathbf{n}_0|_\Gamma = F_3(\mathbf{u})\,, \tag{82}$$

$$\mathbf{n}_0 \cdot T(\mathbf{u}, q)\,\mathbf{n}_0 - \sigma\,\mathbf{n}_0 \cdot \Delta(0)\int_0^t \mathbf{u}(\xi, \tau)\,d\tau\big|_\Gamma$$

$$= \sigma\,H_0(\xi) + F_4(\mathbf{u}, q) + \sigma\,F_5(\mathbf{u}) + \sigma\int_0^t F_6(\mathbf{u})\,d\tau\,,$$

where

$$F_1(\mathbf{u}, q) = \nu\,(\nabla_u^2 - \nabla^2)\,\mathbf{u} + (\nabla - \nabla_u)\,q\,,$$

$$F_2(\mathbf{u}) = (\nabla - \nabla_u) \cdot \mathbf{u} = \nabla \cdot \mathbf{h}(\mathbf{u})\,, \quad \mathbf{h}(\mathbf{u}) = (I - \mathcal{A}^T)\,\mathbf{u}\,,$$

$$F_3(\mathbf{u}) = \Pi_0\Big(\Pi_0\,S(\mathbf{u})\,\mathbf{n}_0 - \Pi\,S_u(\mathbf{u})\,\mathbf{n}\Big)\Big|_\Gamma\,,$$

$$F_4(\mathbf{u}, q) = q\,\mathbf{n}_0 \cdot (\mathbf{n} - \mathbf{n}_0) + \nu\,\mathbf{n}_0 \cdot \Big(S(\mathbf{u})\,\mathbf{n}_0 - S_u(\mathbf{u})\,\mathbf{n}\Big)\Big|_\Gamma\,, \tag{83}$$

$$F_5(\mathbf{u}) = \mathbf{n}_0 \cdot \int_0^t \dot{\Delta}(\tau)\,\xi\,d\tau$$

$$F_6(\mathbf{u}) = \mathbf{n}_0 \cdot \Big(\dot{\Delta}(t)\int_0^t \mathbf{u}(\xi, \tau)\,d\tau + \Big(\Delta(t) - \Delta(0)\Big)\mathbf{u}(\xi, t)\Big)\,.$$

We shall need the following simple estimates of the elements $A_{ik}(\xi, t)$ of the matrix $\mathcal{A}$.

Proposition 7.1. Let $\mathbf{u}(\xi, t)$ and $\mathbf{u}'(\xi, t)$ be two vector fields such that

$$\int_0^t |\mathbf{u}(\cdot, \tau)|_\Omega^{(2+\alpha)}\,d\tau \le \mu\,, \quad \int_0^t |\mathbf{u}'(\cdot, \tau)|_\Omega^{(2+\alpha)}\,d\tau \le \mu \tag{84}$$

and let A_{ik} and A'_{ik} be corresponding cofactors of a_{ik}. There hold the estimates

$$\sup_{\tau < t}\Big|A_{ik}(\cdot, \tau) - A'_{ik}(\cdot, \tau)\Big|_\Omega^{(1+\alpha)} \le c(1 + \mu)\int_0^t |\mathbf{u} - \mathbf{u}'|_\Omega^{(2+\alpha)}\,d\tau\,,$$

$$\sup_\Omega\Big[A_{ik}(\xi, \cdot) - A'_{ik}(\xi, \cdot)\Big]_{(0,\tau)}^{(\frac{1+\alpha}{2})} \le c(1 + \mu)\,t^{\frac{1-\alpha}{2}}\sup_{\Omega \times (0,t)}\Big|\nabla\mathbf{u}(\xi, \tau) - \nabla\mathbf{u}'(\xi, \tau)\Big|$$

$$\tag{85}$$

and, in particular,

$$\sup_{t<\tau} |A_{ik}(\cdot,\tau) - \delta_{ik}| \leq c(1+\mu) \int_0^t |\mathbf{u}|_\Omega^{(2+\alpha)} \, d\tau ,$$

$$\sup_\Omega [A_{ik}(\xi,\cdot)]_{(0,t)}^{(\frac{1+\alpha}{2})} \leq c(1+\mu) t^{\frac{1-\alpha}{2}} \sup_\Omega \sup_{\tau<t} |\nabla \mathbf{u}(\xi,\tau)|. \tag{86}$$

The following proposition is a consequence of (81), (85), (86).

Proposition 7.2. *If* $\mathbf{u}$, $\mathbf{u}'$ *satisfy (84) and* μ *is so small that* $|\mathcal{A}\mathbf{n}_0|, |\mathcal{A}'\mathbf{n}_0| \geq c$, *then the difference* $\mathbf{n} - \mathbf{n}' = \frac{\mathcal{A}\mathbf{n}_0}{|\mathcal{A}\mathbf{n}_0|} - \frac{\mathcal{A}'\mathbf{n}_0}{|\mathcal{A}'\mathbf{n}_0|}$ *satisfies the inequalities*

$$\sup_{\tau<t} |\mathbf{n} - \mathbf{n}'|_\Gamma^{(1+\alpha)} \leq c \int_0^t |\mathbf{u} - \mathbf{u}'|_\Omega^{(2+\alpha)} \, d\tau ,$$

$$\sup_\Gamma [\mathbf{n} - \mathbf{n}']_{(0,t)}^{(\frac{1+\alpha}{2})} \leq c t^{\frac{1-\alpha}{2}} \sup_{\Omega\times(0,t)} \left| \nabla \mathbf{u}(\xi,\tau) - \nabla \mathbf{u}'(\xi,\tau) \right| . \tag{87}$$

In particular,

$$\sup_{\tau<t} |\mathbf{n} - \mathbf{n}_0|_\Gamma^{(1+\alpha)} \leq c \int_0^t |\mathbf{u}|_\Omega^{(2+\alpha)} \, d\tau ,$$

$$\sup_\Gamma [\mathbf{n} - \mathbf{n}_0]_{(0,t)}^{(\frac{1+\alpha}{2})} \leq c t^{\frac{1-\alpha}{2}} \sup_{\Omega\times(0,t)} |\nabla \mathbf{u}(\xi,\tau)| . \tag{88}$$

Similar estimates can be obtained for the coefficients of the Laplace–Beltrami operator $\Delta(t)$. We recall that in the local coordinates η_1, η_2 the Laplace–Beltrami operator is defined by

$$\Delta(t) f = \frac{1}{\sqrt{g}} \sum_{\gamma,\beta=1}^2 \frac{\partial}{\partial \eta_\beta} g^{\beta\gamma} \sqrt{g} \frac{\partial f}{\partial \eta_\gamma} = \sum_{\beta,\gamma=1}^2 g^{\beta\gamma} \frac{\partial^2 f}{\partial \eta_\beta \, \partial \eta_\gamma} + \sum_{\gamma=1}^2 h_\gamma \frac{\partial f}{\partial \eta_\gamma} ,$$

where $g = \det(g_{\beta\gamma})_{\beta,\gamma=1,2}$,

$$g_{\beta\gamma} = \frac{\partial X}{\partial \eta_\beta} \cdot \frac{\partial X}{\partial \eta_\gamma} ,$$

$g^{\beta\gamma}$ are the elements of the inverse matrix $(g_{\gamma\beta})^{-1}$, and

$$h_\gamma = \frac{1}{\sqrt{g}} \sum_{\beta=1}^2 \frac{\partial}{\partial \eta_\beta} (g^{\beta\gamma} \sqrt{g}) .$$

We cover Γ by the sets $\Gamma_k = \{\xi \in \Gamma : |\xi - \xi_k| < d\}$ where d is sufficiently small number and $\xi_k \in \Gamma$ $(k = 1, ..., M)$, take as local coordinates Cartesian coordinates $\eta_1 = \eta_1^{(k)}$, $\eta_2 = \eta_2^{(k)}$ on the tangential planes to Γ at the points ξ_k, and draw the coordinate axes $\eta_3^{(k)}$ parallel to $n_0(\xi_k)$. Let

$$\eta_3^{(k)} = \phi^{(k)}(\eta_1^{(k)}, \eta_2^{(k)}), \quad |\eta_1^{(k)}|^2 + |\eta_2^{(k)}|^2 \le d^2$$

be equation of Γ in the neighbourhood of ξ_k. In the coordinates system $\eta_1^{(k)}$, $\eta_2^{(k)}$, $\eta_3^{(k)}$

$$X(\xi, t) = \left(\eta_1 + \int_0^t \omega_1(\eta, \tau)\, d\tau, \ \eta_2 + \int_0^t \omega_2(\eta, \tau)\, d\tau, \ \phi + \int_0^t \omega_3(\eta, \tau)\, d\tau \right)$$

$$= \xi + \int_0^t \omega(\eta, \tau)\, d\tau \ ,$$

where $\xi = (\eta_1, \eta_2, \phi(\eta_1, \eta_2))$ and $\omega = \mathbf{u}(\eta_1, \eta_2, \phi, t)$ (we omit the superscript (k)). Hence,

$$g_{\gamma\beta} = g_{\gamma\beta}^{(0)} + \frac{\partial \xi}{\partial \eta_\gamma} \cdot \int_0^t \frac{\partial \omega}{\partial \eta_\beta}\, d\tau + \frac{\partial \xi}{\partial \eta_\beta} \cdot \int_0^t \frac{\partial \omega}{\partial \eta_\gamma}\, d\tau$$
$$+ \int_0^t \frac{\partial \omega}{\partial \eta_\beta}\, d\tau \cdot \int_0^t \frac{\partial \omega}{\partial \eta_\gamma}\, d\tau \ , \tag{89}$$

where

$$g_{\gamma\beta}^{(0)} = \frac{\partial \xi}{\partial \eta_\gamma} \cdot \frac{\partial \xi}{\partial \eta_\beta} \ ,$$

and

$$\dot{g}_{\gamma\beta} = \frac{\partial \xi}{\partial \eta_\gamma} \cdot \frac{\partial \omega}{\partial \eta_\beta} + \frac{\partial \xi}{\partial \eta_\beta} \cdot \frac{\partial \omega}{\partial \eta_\gamma} + \frac{\partial \omega}{\partial \eta_\gamma} \cdot \int_0^t \frac{\partial \omega}{\partial \eta_\beta}\, d\tau + \frac{\partial \omega}{\partial \eta_\beta} \cdot \int_0^t \frac{\partial \omega}{\partial \eta_\gamma}\, d\tau \ .$$

The following proposition is evident.

Proposition 7.3. *Let $\mathbf{u}$ and $\mathbf{u}'$ be two vector fields satisfying (84) with μ so small that*

$$g \ge c > 0, \quad g' \ge c > 0 \ ,$$

where $g = \det(g_{\gamma\beta})$, $g' = \det(g'_{\gamma\beta})$ and $g_{\gamma\beta}$, $g'_{\gamma\beta}$ are functions (89) corresponding to $\mathbf{u}$ and $\mathbf{u}'$, respectively. For the coefficients of the operators $\Delta(t)$ and $\Delta'(t)$ the following estimates hold

$$\sup_{\tau < t} \left| g^{\gamma\beta}(\cdot, \tau) - (g^{\gamma\beta})' \right|_{\Gamma_k}^{(1+\alpha)} + \sup_{\tau < t} \left| h_\gamma(\cdot, \tau) - h'_\gamma(\cdot, \tau) \right|_{\Gamma_k}^{(\alpha)} \le c \int_0^t |\mathbf{u} - \mathbf{u}'|_{\Gamma_k}^{(2+\alpha)}\, d\tau \ ,$$

$$\sup_{\tau < t} \left| \dot{g}^{\gamma\beta}(\cdot, \tau) - (\dot{g}^{\gamma\beta})' \right|_{\Gamma_k}^{(1+\alpha)} + \sup_{\tau < t} \left| \dot{h}_\gamma(\cdot, \tau) - \dot{h}'_\gamma(\cdot, \tau) \right|_{\Gamma_k}^{(\alpha)} \le c \sup_{\tau < t} |\mathbf{u} - \mathbf{u}'|_{\Gamma_k}^{(2+\alpha)} \ ,$$

$$\tag{90}$$

in particular,

$$\sup_{\tau<t}\left|g^{\gamma\beta}(\cdot,\tau)-g_0^{\gamma\beta}\right|_{\Gamma_k}^{(1+\alpha)} + \sup_{\tau<t}\left|h_\gamma(\cdot,\tau)-h_{0\gamma}\right|_{\Gamma_u}^{(\alpha)} \le c\int_0^t |u|_{\Gamma_k}^{(2+\alpha)}\,d\tau\ ,$$

$$\sup_{\tau<t}|\dot{g}^{\gamma\beta}(\cdot,\tau)| + \sup_{\tau<t}|\dot{h}_\gamma(\cdot,\tau)| \le c\sup_{\tau<t}|u|_{\Gamma_k}^{(2+\alpha)}\ .$$

$$(91)$$

As a consequence of inequalities (85)–(88), (90), (91), we obtain the estimates of the expressions (83). Let $(\mathbf{u}, q)$ and $(\mathbf{u}', q')$ satisfy the conditions

$$N_t(\mathbf{u}, q) = \sup_{\tau<t}|\mathbf{u}_t(\cdot,t)|_\Omega^{(\alpha)} + \sup_{\tau<t}|\mathbf{u}(\cdot,t)|_\Omega^{(2+\alpha)} + \sup_{\tau<t}|q(\cdot,t)|_\Omega^{(1+\alpha)} \le \mu\ ,$$

$$N_t(\mathbf{u}', q) \le \mu\ ,$$

$$\mathbf{u}|_{t=0} = \mathbf{u}'|_{t=0}\ .$$

Then

$$\sup_{\tau<t}\left|F_1(\mathbf{u}, q)-F_1(\mathbf{u}', q')\right|_\Omega^{(\alpha)} \le \nu\left(\sup_{\tau<t}\left|(\nabla_u^2-\nabla_{u'}^2)\,\mathbf{u}\right|_\Omega^{(\alpha)} + \sup_{\tau<t}\left|\nabla_{u'}^2(\mathbf{u}-\mathbf{u}')\right|_\Omega^{(\alpha)}\right)$$

$$+ \sup_{\tau<t}\left|(\nabla_u - \nabla_{u'})\,q\right|_\Omega^{(\alpha)} + \sup_{\tau<t}\left|\nabla_{u'}(q - q')\right|_\Omega^{(\alpha)}$$

$$\le ct\mu N_t(\mathbf{u}-\mathbf{u}', q-q')\ ,$$

$$\sup_{\tau<t}\left|F_3(\mathbf{u}) - F_2(\mathbf{u}')\right|_\Omega^{(1+\alpha)} + \sup_{\tau<t}\left|\frac{\partial}{\partial t}\left(h(\mathbf{u}) - h(\mathbf{u}')\right)\right|_\Omega^{(\alpha)}$$

$$\le \sup_{\tau<t}\left|(\nabla_u - \nabla_{u'})\cdot\mathbf{u}\right|_\Omega^{(1+\alpha)} + \sup_{\tau<t}\left|\nabla_{u'}\cdot(\mathbf{u} - \mathbf{u}')\right|_\Omega^{(1+\alpha)}$$

$$+ \sup_{\tau<t}\left|\left(\frac{\partial \mathcal{A}^T(\mathbf{u})}{\partial t} - \frac{\partial \mathcal{A}^T(\mathbf{u}')}{\partial t}\right)\mathbf{u}\right|_\Omega^{(\alpha)} + \sup_{\tau<t}\left|\frac{\partial \mathcal{A}^T(\mathbf{u}')}{\partial t}(\mathbf{u} - \mathbf{u}')\right|_\Omega^{(\alpha)}$$

$$+ \sup_{\tau<t}\left|\left(\mathcal{A}^T(\mathbf{u}) - \mathcal{A}^T(\mathbf{u}')\right)\mathbf{u}_t\right|_\Omega^{(\alpha)} + \sup_{\tau<t}\left|(I - \mathcal{A}^T(\mathbf{u}'))\,(\mathbf{u}_t - \mathbf{u}_t')\right|_\Omega^{(\alpha)}$$

$$\le c\,(t + t^{1/2})\,\mu\,N_t(\mathbf{u} - \mathbf{u}')$$

$$(92)$$

(we have used the inequality

$$
\sup_{\tau < t} \left| \frac{\partial \mathcal{A}^T(\mathbf{u})}{\partial t} - \frac{\partial \mathcal{A}^T(\mathbf{u}')}{\partial t} \right|_{\Omega}^{(\alpha)} \le c \sup_{\tau < t} \left| \nabla(\mathbf{u} - \mathbf{u}') \right|_{\Omega}^{(\alpha)} \le c t^{1/2} N_t(\mathbf{u} - \mathbf{u}') ,
$$

$$
N_t(\mathbf{u} - \mathbf{u}') \equiv N_t(\mathbf{u} - \mathbf{u}', 0) \quad) .
$$

Further we have

$$
\sup_{\tau < t} \left| F_3(\mathbf{u}) - F_3(\mathbf{u}') \right|_{\Gamma}^{(1+\alpha)} + \sup_{\tau < t} \left| F_4(\mathbf{u}, q) - F_4(\mathbf{u}', q') \right|_{\Gamma}^{(1+\alpha)}
$$

$$
\le c t \mu N_t(\mathbf{u} - \mathbf{u}', q - q') ,
$$

$$
\sup_{\Gamma} \left[F_3(\mathbf{u}) - F_3(\mathbf{u}') \right]_{(0,t)}^{(\frac{1+\alpha}{2})}
$$

$$
\le c \left(t \mu \sup_{\Gamma} [\nabla(\mathbf{u} - \mathbf{u}')]_{(0,t)}^{(\frac{1+\alpha}{2})} + t^{\frac{1-\alpha}{2}} \sup_{\Gamma \times (0,t)} |\nabla \mathbf{u}| \sup_{\Gamma \times (0,t)} |\nabla(\mathbf{u} - \mathbf{u}')| \right)
$$

$$
\le c t \mu N_t(\mathbf{u} - \mathbf{u}') ,
$$

$$
\sup_{\tau < t} \left| F_6(\mathbf{u}) - F_6(\mathbf{u}') \right|_{\Gamma}^{(\alpha)} \le c t \mu N_t(\mathbf{u} - \mathbf{u}') .
$$

(93)

Finally, since $\mathbf{n}_0 \cdot \frac{\partial \xi}{\partial n_\nu} = 0$, the lower order terms of $\dot{\Delta}(\tau) \xi$ do not give any contribution into $F_s(\mathbf{u})$, and we have

$$
\sup_{\tau < t} \left| F_5(\mathbf{u}) - F_5(\mathbf{u}') \right|_{\Gamma}^{(1+\alpha)} \le c \int_0^t |\mathbf{u} - \mathbf{u}'|_{\Omega}^{(2+\alpha)} d\tau \le c t N_t(\mathbf{u} - \mathbf{u}') .
$$

In particular, if $\mathbf{u}' = 0$, $q' = 0$, then

$$
\sup_{\tau < t} |F_1(\mathbf{u}, q)|_{\Omega}^{(\alpha)} + \sup_{\tau < t} |F_2(\mathbf{u})|_{\Omega}^{(1+\alpha)} + \sup_{\tau < t} |F_3(\mathbf{u})|_{\Gamma}^{(1+\alpha)}
$$

$$
+ \sup_{\tau < t} |F_4(\mathbf{u})|_{\Gamma}^{(1+\alpha)} + \sigma \sup_{\tau < t} |F_6(\mathbf{u})|_{\Gamma}^{(\alpha)} \le c t N_t^2(\mathbf{u}, q) ,
$$

(94)

$$
\sup_{\tau < t} |F_5(\mathbf{u})|_{\Gamma}^{(1+\alpha)} \le c \int_0^t N_t(\mathbf{u}) d\tau \le c t N_t(\mathbf{u}) ,
$$

(95)

and, if $\mathbf{u}|_{t=0} = \mathbf{v}_0(\xi)$, then

$$\sup_{\tau < t} \left| \frac{\partial}{\partial t} \mathbf{h}(\mathbf{u}) \right|_{\Omega}^{(\alpha)} \leq c \left(\sup_{\Omega \times (0,t)} |\nabla \mathbf{u}(\xi, \tau)| \, |\mathbf{u}(\xi, \tau)| + t \, N^2(\mathbf{u}) \right)$$

$$\leq c \left(\sup_{\Omega} |\nabla \mathbf{v}_0(\xi)| \cdot |\mathbf{v}_0(\xi)| + t^{\frac{1+\alpha}{2}} N(\mathbf{u}) \sup |\mathbf{v}_0(\xi)| + t \, N^2(\mathbf{u}) \right),$$

(96)

$$\sup_{\Gamma} [F_3(\mathbf{u})]_{(0,t)}^{(\frac{1+\alpha}{2})} \leq c \left(t^{\frac{1-\alpha}{2}} \sup_{\Gamma \times (0,t)} |\nabla \mathbf{u}(\xi, t)|^2 + t \, N^2(\mathbf{u}) \right)$$

$$\leq c t^{\frac{1-\alpha}{2}} \sup_{\Gamma} |\nabla \mathbf{v}_0(\xi)|^2 + c t \, N^2(\mathbf{u}) .$$

Now, we can give the proof of Theorems 5.1 and 5.2. For definiteness, we prove the second theorem; the same arguments (with $\sigma = 0$) provide the proof of Theorem 5.1.

Proof of Theorem 5.2. We write the problem (82) in an equivalent form

$$(\mathbf{u}, q) = \mathcal{F}(\mathbf{u}, q) + (\mathbf{u}_0, q_0) \equiv \mathcal{H}(\mathbf{u}, q) ,$$

(97)

where $(\mathbf{u}_0, q_0)$ is a solution of the problem (47) with $\mathbf{f} = 0$, $g = 0$, $B = 0$, $\mathbf{b} = \sigma H_0 \, \mathbf{n}_0$ and $\mathcal{F}(\mathbf{u}, q) \equiv (\mathbf{w}, r)$ is a solution of the same problem with $\mathbf{f} = F_1(\mathbf{u}, q)$, $\mathbf{g} = F_2(\mathbf{u})$, $\mathbf{v}_0 = 0$, $\mathbf{b} = F_3(\mathbf{u}) + \mathbf{n}_0 (F_4(\mathbf{u}, q) + \sigma F_5(\mathbf{u}))$, $B = \sigma F_6(\mathbf{u})$. The solution of the equation (97) is sought in the space $C_\alpha(\Omega \times (0, T))$ with the norm equal to $N_T(\mathbf{u}, q)$.

We show that $\mathcal{H}(\mathbf{u}, q)$ is a contraction operator in a subset of a certain ball

$$N_{T_0}(\mathbf{u}, q) \leq \mu, \quad T_0 \ll 1 ,$$

of this space, whose elements satisfy the initial condition $\mathbf{u}|_{t=0} = \mathbf{v}_0(\xi)$. Indeed, by virtue of (94)–(96), for arbitrary element of the above subset we have

$$\|\mathcal{H}(\mathbf{u}, q)\|_{C_\alpha} \leq N_{T_0}(\mathbf{w}, r) + N_{T_0}(\mathbf{u}_0, q_0)$$

$$\leq N_{T_0}(\mathbf{u}_0, q_0) + c_1 \left(\sup_{\Omega} |\nabla \mathbf{v}_0(\xi)| \, |\mathbf{v}_0(\xi)| + T_0^{\frac{1-\alpha}{2}} \sup |\nabla \mathbf{v}_0(\xi)|^2 \right)$$

$$+ c_2 \, N_{T_0}(\mathbf{u}, q) \left(T_0 + T_0^{\frac{1+\alpha}{2}} \sup_{\Omega} |\mathbf{v}_0(\xi)| + T_0 \, N(\mathbf{u}, q) \right)$$

and, if

$$\mu \geq 2 \left(N_{T_0}(\mathbf{u}_0, q_0) + c_1 \left(\sup_{\Omega} |\nabla \mathbf{v}_0(\xi)| \, |\mathbf{v}_0(\xi)| + T_0^{\frac{1-\alpha}{2}} \sup_{\Gamma} |\nabla \mathbf{v}_0(\xi)|^2 \right) \right)$$

and

$$c_2 \left(T_0 + T_0^{\frac{1+\alpha}{2}} \sup_{\Omega} |\mathbf{v}_0(\xi)| + T_0 \, \mu \right) \leq \frac{1}{2} ,$$

(98)

then $\mathcal{H}(\mathbf{u}, q)$ belongs to the same ball

$$\|\mathcal{H}(u,q)\|_{C_\alpha(\Omega)} \leq \mu \, .$$

Further, by virtue of (92), (93), for arbitrary $(\mathbf{u}, q)$, $(\mathbf{u}', q')$ in this ball such that $\mathbf{u}|_{t=0} = \mathbf{u}'|_{t=0} = \mathbf{v}_0$ there holds

$$\left\| \mathcal{H}(\mathbf{u}, q) - \mathcal{H}(\mathbf{u}', q') \right\|_{C_\alpha(\Omega)} \leq c_3 \left(T_0 + T_0 \, \mu \right) N_{T_0}(\mathbf{u} - \mathbf{u}', q - q') \, .$$

So, if

$$c_3 \, T_0 \left(1 + \mu \right) < 1 \, ,$$

then $\mathcal{H}(\mathbf{u}, q)$ is a contraction operator and equation (96) has a solution in the above ball. The uniqueness of the solution $(\mathbf{u}, q) \in C_\alpha(\Omega \times (0, T_0))$ of the problem (82) follows from the inequalities (51), (93)–(95). The theorem is proved. ∎

Remark 7.1. Theorem 5.2 holds for arbitrary domain $\Omega \subset \mathbb{R}^3$ with a smooth boundary. If Ω is homeomorphic to the ball, it has a sense to introduce a new pressure $p' = p - \frac{2}{R_0}$ where $R_0 = (\frac{|\Omega|}{4\pi})^{1/3}$, and then instead of (38) we shall have

$$\sup_{t<T_0} |\mathbf{v}(\cdot, t)|_{\Omega_t}^{(\alpha)} + \sup_{t<T_0} |\mathbf{v}(\cdot, t)|_{\Omega_t}^{(2+\alpha)} + \sup_{t<T_0} \left| p(\cdot, t) - \frac{2}{R_0} \right|_{\Omega_t}^{(1+\alpha)}$$

$$\leq c \left(|\mathbf{v}_0|_\Omega^{(2+\alpha)} \left(1 + |\mathbf{v}_0|_\Omega^{(2+\alpha)} \right) + \sigma \left| H_0 + \frac{2}{R_0} \right|_\Gamma^{(1+\alpha)} \right) \, .$$

The functions $(\mathbf{u}_0, q_0)$ in (96) will be a solution of (47) with $\mathbf{b} = \sigma \mathbf{n}_0 (H_0 + \frac{2}{R_0})$, and slightly more accurate estimates of $F_5(\mathbf{u})$ will lead, instead of (98), to the inequality

$$c \, e^{cT_0} \left(T_0^{\frac{1+\alpha}{2}} \sup_\Omega |\mathbf{v}_0(\xi)| + T_0 \, \mu \right) \leq \frac{1}{2} \, ,$$

which shows that $T_0 \to \infty$ as the norms of $\mathbf{v}_0$ and $H_0 + \frac{2}{R_0}$ tend to zero (i.e. as the domain Ω_0 approaches a ball).

8 Proof of Theorem 5.3

We start with the following auxiliary propositions.

Proposition 8.1. *Let G be a bounded domain in $\mathbb{R}^3$ satisfying the cone condition. For arbitrary vector field $\mathbf{u} \in W_2^1(G)$ the Korn inequality*

$$\|\mathbf{u}\|_{W_2^1(G)}^2 \leq c_1 \|S(\mathbf{u})\|_{L_2(G)}^2 + c_2 \left(\left| \int_G \mathbf{u} \, dx \right|^2 + \left| \int_G \left[\mathbf{u} \times (\mathbf{x} - \mathbf{x}_0) \right] dx \right| \right) \quad (99)$$

holds.

Korn's inequality was proved under different assumptions on the domain G and on the vector fields $u(x)$ by many authors (see, for instance, [16, 19, 20]). We use the variant of this inequality obtained in [35].

Proposition 8.2. *Assume that a classical solution of the problem (35) with $\sigma = 0$ and with $\mathbf{v}_0(x)$ satisfying the conditions (36) is defined for $t \in [0, T]$, and let the Korn inequality (99) be satisfied for $G = \Omega_t$ with the constant c_1 independent of t. Then*

$$\|\mathbf{v}(\cdot, t)\|^2_{L_2(\Omega_t)} \leq e^{-2bt} \|\mathbf{v}_0\|^2_{L_2(\Omega)} \ . \tag{100}$$

Proof. Multiplication of the equation $(35)_1$ by $\mathbf{v}(x, t)$ and integration over Ω_t gives

$$\frac{1}{2}\frac{d}{dt}\|\mathbf{v}(\cdot, t)\|^2_{L_2(\Omega)} + \frac{\nu}{2}\|S(\mathbf{v})\|^2_{L_2(\Omega)} = 0 \ .$$

The vector field $\mathbf{v}$ satisfies the condition

$$\int_{\Omega_t} \mathbf{v} \cdot \eta(x)\, dx = \int_{\Omega} \mathbf{v}_0 \cdot \eta\, dx = 0 \ ,$$

where $\eta = \mathbf{a} + \mathbf{b} \times \mathbf{x}$ is an arbitrary vector field of rigid displacements (which can be easily verified by multiplication of (5.5_1) by η and integration). Hence,

$$\int_{\Omega_t} \mathbf{v}\, dx = \int_{\Omega_t} [\mathbf{v} \times \mathbf{x}]\, dx = 0 \ ,$$

and by virtue of (99),

$$\frac{d}{dt}\|\mathbf{v}\|^2_{L_2(\Omega)} + 2b\|\mathbf{v}\|^2_{L_2(\Omega)} \leq 0$$

with $2b = \nu c^{-1}$. This inequality implies (100). ∎

Remark 8.1. It follows from the Proposition 8.2 that an arbitrary solution of (35). defined in an infinite time interval decays exponentially, if (99) holds with the constant independent of t.

To complete the proof of Theorem 5.3, we should get estimates similar to (100) for the Hölder norms of the solution of the problem (35).

Proposition 8.3. *Assume that the solution of the problem (35), with $\sigma = 0$ is defined for $t \in (0, T)$ and*

$$N'_T(\mathbf{v}, p) = \sup_{t<T} |\mathbf{v}_t(\cdot, t)|^{(\alpha)}_{\Omega_t} + \sup_{t<T} |\mathbf{v}(\cdot, t)|^{(2+\alpha)}_{\Omega_t} + \sup_{t<T} |p(\cdot, t)|^{(1+\alpha)}_{\Omega_t} \leq \mu \ .$$

Then there holds the inequality

$$\begin{aligned}
\sup_{\tau \in (t_0 - \tau_0, t_0)} |\mathbf{v}_t(\cdot, \tau)|^{(\alpha)}_{\Omega_\tau} &+ \sup_{\tau \in (t_0 - \tau_0, t_0)} |\mathbf{v}(\cdot, \tau)|_{\Omega_\tau} + \sup_{\tau \in (t_0 - \tau_0, t_0)} |p(\cdot, \tau)|^{(1+\alpha)}_{\Omega_\tau} \\
&\leq c(\tau_0) \sup_{\tau \in (t_0 - 2\tau_0, t_0)} \|\mathbf{v}(\cdot, \tau)\|_{L_2(\Omega)} \ ,
\end{aligned} \tag{101}$$

where $t_0 \in (0, T)$ and $\tau_0 \in (0, \frac{t_0}{2})$ is a number depending on μ.

Proof. Let $\zeta_\lambda(t)$ be a smooth monotone function of t equal to zero for $t < t_0 - 2\tau_0 + \frac{\lambda}{2}$, to one for $t > t_0 - 2\tau_0 + \lambda$, $\lambda \in (0, \tau_0)$, and such that

$$\left|\frac{d\zeta_\lambda(t)}{dt}\right| \le c\lambda^{-1} .$$

The functions $\mathbf{w} = \mathbf{v}\,\zeta_\lambda$, $r = p\,\zeta_\lambda$ satisfy the equation

$$\mathbf{w}_t - \nu\,\nabla^2\mathbf{w} + (\mathbf{v}\cdot\nabla)\,\mathbf{w} + \nabla r = \mathbf{v}\,\zeta_\lambda' ,$$
$$\nabla\cdot\mathbf{w} = 0, \quad x \in \Omega_t, \quad t > t_0 - 2\tau_0 ,$$
$$\mathbf{w}|_{t=t_0-2\tau_0} = 0 ,$$
$$T(\mathbf{w}, q)\,\mathbf{n}|_{\Gamma_t} = 0 .$$

(102)

We introduce the Lagrangean coordinates, according to the formula

$$\mathbf{x} = \xi + \int_{t_0-2\tau_0}^t \mathbf{u}(\xi,\tau)\,d\tau \equiv X(\xi,t), \quad t > t_0 - 2\tau_0 ,$$

where $\mathbf{u}(\xi,t) = \mathbf{v}(X(\xi,t),t)$, and write (102) in the form

$$\mathbf{w}_t - \nu\,\nabla^2\mathbf{w} + \nabla r = \mathbf{u}\,\zeta_\lambda' + \mathcal{F}_1(\mathbf{w}, q, \mathbf{u}) ,$$
$$\nabla\cdot\mathbf{w} = \mathcal{F}_2(\mathbf{w}, \mathbf{u}), \quad \xi \in \Omega', \quad t > t_0 - 2\tau_0 ,$$
$$\Pi_0 S(\mathbf{w})\,\mathbf{n}_0|_{\Gamma'} = \mathcal{F}_3(\mathbf{w}, \mathbf{u}) ,$$
$$\mathbf{n}_0 \cdot T(\mathbf{w}, r)\,\mathbf{n}_0|_{\Gamma'} = \mathcal{F}_4(\mathbf{w}, \mathbf{u}) ,$$
$$\mathbf{w}|_{t=t_0-2\tau_0} = 0 ,$$

where

$$\mathcal{F}_1(\mathbf{w}, q, \mathbf{u}) = \nu\,(\nabla_u^2 - \nabla^2)\,\mathbf{w} - (\nabla_u - \nabla)r ,$$
$$\mathcal{F}_2(\mathbf{w}, \mathbf{u}) = (\nabla - \nabla_u)\cdot\mathbf{w} = \nabla\cdot(I - A^T)\,\mathbf{w} ,$$
$$\mathcal{F}_3(\mathbf{w}, \mathbf{u}) = \Pi_0\Big(\Pi_0 S(\mathbf{w})\,\mathbf{n}_0 - \Pi S_u(\mathbf{w})\,\mathbf{n}\Big)\Big|_\Gamma ,$$
$$\mathcal{F}_4(\mathbf{w}, \mathbf{u}) = \nu\Big(\mathbf{n}_0\cdot S(\mathbf{w})\,\mathbf{n}_0 - \mathbf{n}\cdot S_u(\mathbf{w})\,\mathbf{n}\Big)\Big|_\Gamma ,$$

$\Omega' = \Omega_{t_0-2\tau_0}$, $\Gamma' = \partial\Omega'$, $\mathbf{n}_0$ is the normal to Γ, Π_0 and Π have the same meaning as in Section 7 (we have used the same notations for the functions $\mathbf{w}$ and r expressed in Eulerian and Lagrangean coordinates). By virtue of inequalities (50), (86), (87) and of the fact that $\mathbf{w}|_{t=t_0-2\tau_0} = 0$, there holds the estimate

$$N'_{t_0}(\mathbf{w}, r) \equiv \sup_{t_0 - 2\tau_0 < \tau < t_0} |\mathbf{w}_\tau(\cdot, \tau)|^{(\alpha)}_{\Omega'} + \sup_{t_0 - 2\tau_0 < \tau < t_0} |\mathbf{w}(\cdot, \tau)|^{(2+\alpha)}_{\Omega'}$$

$$+ \sup_{t_0 - 2\tau_0 < \tau < t_0} |r(\cdot, \tau)|^{(1+\alpha)}_{\Omega'} \tag{103}$$

$$\leq c\left(\tau_0 \, \mu \, N'_{t_0}(\mathbf{w}, r) + \lambda^{-1} \sup_{t_0 - 2\tau_0 < \tau < t_0} |\mathbf{v}(\cdot, \tau)|^{(\alpha)}_{\Omega}\right).$$

We subject τ_0 to the requirement

$$c\,\tau_0\,\mu \leq \frac{1}{2}$$

and use the interpolation inequality

$$|\mathbf{v}(\cdot, \tau)|^{(\alpha)}_{\Omega} \leq \theta^2 \, |\mathbf{v}(\cdot, \tau)|^{(2+\alpha)}_{\Omega'} + c\,\theta^{-\alpha - 3/2} \, \|\mathbf{v}(\cdot, \tau)\|_{L_2(\Omega)},$$

valid for arbitrarily small θ. Setting $\theta = \lambda^{1/2}\,\varepsilon^{1/2}$, we easily see that (103) implies

$$f(\lambda) \leq c_1 \, \varepsilon \, f\left(\frac{\lambda}{2}\right) + K, \tag{104}$$

where

$$f(\lambda) = \lambda^{\alpha + 5/2} \left(\sup_{\tau \in I_\lambda} |\mathbf{u}_\tau(\cdot, \tau)|^{(\alpha)}_{\Omega'} + \sup_{\tau \in I_\lambda} |\mathbf{u}(\cdot, \tau)|^{(2+\alpha)}_{\Omega'} + \sup_{\tau \in I_\lambda} |q(\cdot, \tau)|^{(1+\alpha)}_{\Omega'}\right),$$

$$K = c(\varepsilon) \sup_{t_0 - 2\tau_0 < t < t_0} \|\mathbf{u}(\cdot, \tau)\|^2_{L_2(\Omega')}, \qquad I_\lambda = (-t_0 - 2\tau_0 + \lambda,\, t_0).$$

Taking $\varepsilon = \frac{1}{2c_1}$ and making iterations we easily obtain

$$f(\lambda) \leq 2K, \tag{105}$$

which yields (101), if we fix $\lambda = \tau_0$. The proposition is proved. ∎

It follows from (101) and (100) that

$$|\mathbf{v}_t(\cdot, t_0)|^{(\alpha)}_{\Omega_{t_0}} + |\mathbf{v}(\cdot, t_0)|^{(2+\alpha)}_{\Omega_{t_0}} + |p(\cdot, t_0)|^{(2+\alpha)}_{\Omega_{t_0}} \leq c(\tau_0)\, e^{-2t_0} \, \|\mathbf{v}_0\|_{L_2(\Omega)}.$$

If $c(\tau_0) \, \|\mathbf{v}_0\|_{L_2(\Omega)} \leq \mu$ (this inequality is satisfied, if the number ε in (39) is sufficiently small), then we obtain a uniform estimate of the norm $N_t(\mathbf{v}, p)$ and by a repeated application of the local existence theorem we can extend the solution into an infinite time interval $t > 0$. The theorem is proved. ∎

Due to the exponential decay of the solution, the transformation X is well defined for all $t > 0$. The free boundary Γ_t has a limit

$$\Gamma_\infty = X(\Gamma, \infty) \in C^{2+\alpha}.$$

Examples show (see [9]) that in the case $\beta \neq 0$ Γ_t can have no definite limit as $t \to \infty$.

9 Scheme of the proof of Theorem 5.4

Theorem 5.4 is proved according to the same scheme as Theorem 5.3: at first the energy inequality is established, and then local estimate of the type (100) is obtained for the functions

$$\mathbf{w}(x,t) = \mathbf{v}(x,t) - \mathbf{v}_\infty(x) = \mathbf{v}(x,t) + \frac{\beta}{I_\infty}\,\mathbf{h}(x)\;,$$

$$s(x,t) = p(x,t) - p_\infty(x) = p(x,t) - \frac{\beta^2}{2\,I_\infty^2}\,(x_1^2 + x_2^2) - p_1\;,$$

where β and p_1 are the same constants as in Section 1,

$$\mathbf{h}(x) = (-x_2, x_1, 0)\;,$$

$$I_\infty = \int_{\Omega_\infty} (x_1^2 + x_2^2)\,dx\;,\qquad \Omega_\infty = \left\{ |x| < R_\infty\left(\frac{x}{|x|}\right) \right\}\;.$$

It is easily seen that $\mathbf{w}$, s satisfy the relations

$$\mathbf{w}_t + (\mathbf{v}\cdot\nabla)\,\mathbf{w} - \frac{\beta}{I_\infty}\,(\mathbf{w}\cdot\nabla)\,\mathbf{h} - \nu\,\nabla^2\mathbf{w} + \nabla s = 0\;,$$

$$\nabla\cdot\mathbf{w} = 0\;,\qquad x\in\Omega_t,\ t>0\;,$$

$$\mathbf{w}|_{t=0} = \mathbf{v}_0(x) - \mathbf{v}_\infty(x) \equiv \mathbf{w}_0(x)\;,\qquad x\in\Omega\;,\tag{106}$$

$$T(\mathbf{w},s)\,\mathbf{n} = \sigma\mathbf{n}\left(H + \frac{\beta^2\,|\mathbf{h}|^2}{2\,I_\infty^2} + p_1\right)\;,$$

$$V_n = \mathbf{w}\cdot\mathbf{n} - \frac{\beta}{I_\infty}\,\mathbf{h}\cdot\mathbf{n}\;,\qquad x\in\Gamma_t\;.$$

In addition, we have

$$\int_{\Omega_t} [\mathbf{w}\times x]\,dx = \frac{\beta}{I_\infty}\sum_{j=1}^{2}\mathbf{e}_j\int_{\Omega_t} x_3\,x_j\,dx + \mathbf{e}_3\,\beta\left(1 - \frac{I_t}{I_\infty}\right)$$

$$= \frac{\beta}{5\,I_\infty}\sum_{j=1}^{2}\mathbf{e}_j\int_{S_1}\omega_3\,\omega_j\left(R^5(\omega,t) - R_\infty^5(\omega)\right)\,d\omega \tag{107}$$

$$- \frac{\beta}{5\,I_\infty}\,\mathbf{e}_3\int_{S_1}\left(R^5(\omega,t) - R_\infty^5(\omega)\right)(\omega_1^2 + \omega_2^2)\,d\omega\;,$$

where

$$I_t = \int_{\Omega_t} |\mathbf{h}|^2\,dx = \frac{1}{5}\int_{S_1}(\omega_1^2 + \omega_2^2)\,R^5(\omega,t)\,d\omega\;,$$

moreover, if we assume that the barycenter of Ω coincides with the origin and $\int_\Omega \mathbf{v}_0 \, dx = 0$, then

$$\int_{\Omega_t} \mathbf{w} \, dx = \frac{\beta}{I_\infty} \int_{\Omega_t} \mathbf{h}(x) \, dx = 0 . \tag{108}$$

We present here only main ideas of the proof of Theorem 5.4 which is more technical than the proof of Theorem 5.3. The details will be given in a forthcoming paper of the author and Prof. M. Padula.

Proposition 9.1. *Assume that a classical solution of the problem (106) with $\sigma > 0$ is defined for $t \in (0, T)$, the free boundary is given by equation (44), and the Korn inequality (99) for $G = \Omega_t$ holds with the constant independent of t. Moreover, let the momentum β be small enough and let*

$$\sup_{S_1 \times (0,T)} |R(\omega, t) - R_\infty(\omega)| + \sup_{S_1 \times (0,T)} |\nabla_{S_1}(R - R_\infty)| \le \delta \ll 1 . \tag{109}$$

Then

$$\|\mathbf{w}(\cdot, t)\|^2_{L_2(\Omega_t)} + \|R(\cdot, t) - R_\infty(\cdot)\|^2_{W_2^1(S_1)}$$
$$\le c e^{-2bt} \left(\|\mathbf{w}_0\|^2_{L_2(\Omega_t)} + \|R(\cdot, 0) - R_\infty\|^2_{W_2^1(S_1)} \right) , \tag{110}$$

with the constants b and c independent of T.

Scheme of the proof. Multiplication of equation $(106)_1$ by $\mathbf{w}$ and integration over Ω_t leads to

$$\frac{1}{2} \frac{d}{dt} \|\mathbf{w}(\cdot, t)\|^2_{L_2(\Omega_t)} + \frac{\nu}{2} \|S(\mathbf{w})\|^2_{L_2(\Omega_t)}$$
$$= \sigma \int_{\Gamma_t} \left(H + \frac{\beta^2 |\mathbf{h}(x)|^2}{2 I_\infty^2} + p_1 \right) \mathbf{w} \cdot \mathbf{n} \, dS$$
$$= \sigma \int_{S_1} \left(H[R] - H[R_\infty] \right) R^2 R_t \, d\omega \tag{111}$$
$$+ \frac{\sigma \beta^2}{2 I_\infty^2} \int_{S_1} (\omega_1^2 + \omega_2^2) \left(R^2(\omega, t) - R_\infty^2(\omega) \right) R \sqrt{g} \, \mathbf{w} \cdot \mathbf{n}|_{|x|=R(\omega,t)} d\omega$$
$$- \sigma \frac{\beta}{I_\infty} \int_{S_1} \left(H[R] - H[R_\infty] \right) R^2 \sqrt{\omega_1^2 + \omega_2^2} \frac{d}{d\varphi} (R - R_\infty) d\omega \equiv I_1 + I_2 + I_3,$$

where $g = \sqrt{R^2 + |\nabla_{S_1} R|^2}$ and φ is the angle of rotation about the x_3-axis (we have used the relation $\mathbf{h} \cdot \mathbf{n}|_{\Gamma_t} = \sqrt{x_1^2 + x_2^2} \, n_\varphi$ and the fact that R_∞ is independent of φ). As shown in [42],

$$I_1 + I_2 = -\frac{d}{dt} E_1(R - R_\infty) + J_1(R - R_\infty) ,$$

where

$$E_1(R - R_\infty) = \frac{1}{2} \int_{S_1} |\nabla_{S_1}(R - R_\infty)|^2 \frac{R}{\sqrt{g}} \, d\omega - \int_{S_1} (R - R_\infty)^2 \, d\omega$$

$$+ \frac{1}{2} \int_{S_1} \left| \nabla_\omega R_\infty \cdot \nabla_\omega (R - R_\infty) \right|^2 \frac{R}{g^{3/2}} \, d\omega$$

$$+ \int_{S_1} \nabla_\omega R_\infty \cdot \nabla_\omega (R - R_\omega) \left(\frac{R}{\sqrt{g}} - \frac{R^2}{R_\infty \sqrt{g_\infty}} \right) d\omega$$

and

$$|J_1| \leq c(\delta + \beta^2) \left(\|R - R_\infty\|_{W_2^1(S_1)} \|\mathbf{w}\|_{L_2(\Gamma_t)} + \|R - R_\infty\|_{W_2^1(S_1)}^2 \right). \quad (112)$$

The integral I_3 can also be transformed by integration by parts and estimated by $c\beta\delta \|R - R_\infty\|_{W_2^1(S_1)}^2$ with the constant c depending on the second derivatives of R. Finally, the Korn inequality and equations (107), (108) give

$$\|S(\mathbf{w})\|_{L_2(\Omega_t)}^2 \geq c_1 \|\mathbf{w}\|_{W_2^1(\Omega_t)}^2 - c_2 \beta^2 \|R - R_\infty\|_{L_2(S_1)}^2.$$

Hence, relation (111) implies

$$\frac{d}{dt} \left(\frac{1}{2} \|\mathbf{w}\|_{L_2(\Omega_t)}^2 + E_1(R - R_\infty) \right) + \frac{\nu c}{2} \|\mathbf{w}\|_{W_2^1(\Omega_t)}^2 + J \leq 0, \quad (113)$$

with J satisfying (112).

Now, following the idea of M. Padula [25], we obtain an additional estimate of $\|R - R_\infty\|_{W_2^1(S_1)}^2$. We introduce an auxiliary vector field $\mathbf{V}(x, t)$ such that

$$\nabla \cdot \mathbf{V}(x, t) = 0, \quad x \in \Omega_t,$$

$$\mathbf{V} \cdot \mathbf{n}|_{\Gamma_t} = \frac{r - \bar{r}}{R\sqrt{g}} \equiv f(x, t), \quad (114)$$

where $r = R(\frac{x}{|x|}, t) - R_\infty(\frac{x}{|x|})$, $\bar{r} = \frac{1}{4\pi} \int_{S_1} r(\omega, t) \, d\omega$, and we require that $\mathbf{V}$ should satisfy the inequalities

$$\|\mathbf{V}\|_{W_2^1(\Omega_t)} \leq c \|f\|_{W_2^{1/2}(\Gamma_t)} \leq c \|R - R_\infty\|_{W_2^1(S_1)},$$

$$\|\mathbf{V}_t\|_{L_2(\Omega_t)} \leq c \left(\|f_t\|_{L_2(S_1)} + \|f\|_{L_2(\Gamma_t)} \right) \quad (115)$$

$$\leq c \left(\|\mathbf{w} \cdot \mathbf{n}\|_{L_2(\Gamma_t)} + \|R - R_\infty\|_{W_2^1(S_1)} \right)$$

(we recall that $\frac{R R_t}{\sqrt{g}} = \mathbf{w} \cdot \mathbf{n} - \frac{\beta}{I_\infty} \mathbf{h} \cdot \mathbf{n}$ and $\mathbf{h} \cdot \mathbf{n} = \sqrt{x_1^2 + x_2^2} \, n_\varphi$). Since f satisfies the necessary condition

$$\int_{\Gamma_t} f(x,t)\, ds = \int_{S_1} (r - \bar{r})\, d\omega = 0 \ ,$$

the vector field $\mathbf{V}(x,t)$ with the properties (114) exists, and inequalities (115) can be satisfied. Multiplying $(106)_1$ by $\mathbf{V}$ and integrating we obtain

$$\frac{d}{dt} \int_{\Omega_t} \mathbf{w} \cdot \mathbf{V}\, dx - \int_{\Omega_t} \mathbf{w} \cdot \left(\mathbf{V}_t + (\mathbf{v} \cdot \nabla)\mathbf{V} \right) dx$$

$$+ \frac{\nu}{2} \int_{\Omega_t} S(\mathbf{w}):S(\mathbf{V})\, dx + E_2(R - R_\infty) = 0 \ , \tag{116}$$

where

$$E_2(R - R_\infty) = -\int_{S_1} \left(H[R] - H[R_\infty] \right) (r - \bar{r})\, d\omega$$

$$- \frac{\beta^2}{2\, I_\infty^2} \int_{S_1} (R^2(\omega, t) - R_\infty^2(\omega)) \, (\omega_1^2 + \omega_2^2) \, (r - \bar{r})\, d\omega \ .$$

We add (113) and (116) multiplied by a small parameter γ, to obtain

$$\frac{d}{dt} \left(\frac{1}{2} \|\mathbf{w}\|_{L_2(\Omega_t)} + \gamma \int_{\Omega_t} \mathbf{w} \cdot \mathbf{V}\, dx + E_1(R - R_\infty) \right)$$

$$+ \frac{\nu c}{2} \|\mathbf{w}\|^2_{W_2^1(\Omega_t)} + J + \gamma E_2(R - R_\infty) + \frac{\gamma\nu}{2} \int_{\Omega_t} S(\mathbf{w}):S(\mathbf{V})\, dx \tag{117}$$

$$- \gamma \int_{\Omega} \mathbf{w} \cdot (\mathbf{V}_t + (\mathbf{v} \cdot \nabla)\mathbf{V})\, dx \le 0.$$

It may be shown that the differences

$$E_1(R - R_\infty) - \frac{1}{2} E_0(R - R_0) \quad \text{and} \quad E_2(R - R_\infty) - E_0(R - R_\infty) \ ,$$

where

$$E_0(R - R_\infty) = \int_{S_1} \left(|\nabla_{S_1}(R - R_\infty)|^2 - 2\, (R - R_\infty)^2 \right) d\omega$$

satisfy the inequality (112) and that

$$E_0(R - R_\infty) \ge 4\, \|R - R_\infty\|^2_{W_2^1(S_1)} - c_1 \left| \int_{S_1} (R - R_\infty)\, d\omega \right|^2$$

$$- c_2 \sum_{j=1}^{3} \left| \int_{S_1} (R - R_\infty)\, \omega_j\, d\omega \right|^2 \ ,$$

so that $E_0(R - R_\infty)$ is positive definite for small β and δ (see [34, 37]). Hence, E_1 and E_2 are also positive definite and there exists such constant $b > 0$ that

$$(2\,b)^{-1}\left(\frac{\nu\,c}{2}\,\|\mathbf{w}\|^2_{W^1_2(\Omega_t)} + J + \gamma\,E_2(R - R_\infty) + \right.$$

$$\left. + \frac{\gamma\,\nu}{2}\int_{\Omega_t} S(\mathbf{w}) : S(\mathbf{V})\,dx - \gamma\int_\Omega \mathbf{w}\cdot(\mathbf{V}_t + (\mathbf{v}\cdot\nabla)\mathbf{V})\,dx\right)$$

$$\geq \frac{1}{2}\,\|\mathbf{w}\|^2_{L_2(\Omega_t)} + \gamma\int_\Omega \mathbf{w}\cdot\mathbf{V}\,dx + E_1(R - R_\infty) \quad =: \ \mathcal{E}(t)$$

$$\geq c_1\,\|\mathbf{w}\|^2_{L_2(\Omega_t)} + c_2\,\|R - R_\infty\|^2_{W^1_2(S_1)}\,,$$

if small parameters γ, δ, β are chosen in an appropriate way. Then (117) implies

$$\frac{d}{dt}\mathcal{E}(t) + 2\,b\,\mathcal{E}(t) \leq 0$$

and, as a consequence, we obtain (110).

Proposition 9.2. *Let the solution of the problem (31) be defined for* $t \in (0, T)$ *and let*

$$N_T(\mathbf{w}, s) \leq \mu\,.$$

Then

$$\sup_{t_0-\tau_0<\tau<t_0}|\mathbf{w}_\tau(\cdot,\tau)|^{(\alpha)}_{\Omega_\tau} + \sup_{t_0-\tau_0<\tau<t_0}|\mathbf{w}(\cdot,\tau)|^{(2+\alpha)}_{\Omega_\tau} +$$

$$+ \sup_{t_0-\tau_0<\tau<t_0}|s(\cdot,t)|^{(1+\alpha)}_{\Omega_\tau} + \sup_{t_0-\tau_0<\tau<t_0}|R(\cdot,\tau) - R_\infty|^{(3+\alpha)}_{S_1}$$

$$\leq c(\tau_0)\left(\sup_{t_0-2\tau_0<\tau<t_0}\|\mathbf{w}(\cdot,\tau)\|_{L_2(\Omega_\tau)} + \sup_{t_0-2\tau_0<\tau<t_0}\|R(\cdot,\tau) - R_\infty\|_{W^1_2(S_1)}\right),$$

$$(118)$$

where $t_0 \in (0, T)$ *and* $\tau_0 < \frac{t_0}{2}$ *is a certain small number depending on* μ.

Scheme of the proof. We introduce the Lagrangean coordinates which are related to the Eulerian coordinates $x \in \Omega_t$ as follows

$$\frac{d\mathbf{x}(\xi, t)}{dt} = \mathbf{v}(x(\xi, t), t) = -\frac{\beta}{I_\infty}\,\mathbf{h}(x(\xi, t)) + \mathbf{w}(x(\xi, t), t)\,,$$

$$t \geq t'_0 = t_0 - 2\,\tau_0\,, \qquad \mathbf{x}(\xi, t'_0) = \xi\,.$$

Integrating this equation we arrive at

$$\mathbf{x}(\xi, t) = \mathcal{Z}(t)\,\mathbf{y}(\xi, t) \equiv X(\xi, t)\,,$$

where

$$\mathcal{Z}(t) = \begin{pmatrix} \cos\beta'(t-t_0) & -\sin\beta'(t-t_0) & 0 \\ \sin\beta'(t-t_0) & \cos\beta'(t-t_0) & 0 \\ 0 & 0 & 1 \end{pmatrix} ,$$

$$\beta' = -\frac{\beta}{I_\infty} ,$$

$$\mathbf{y}(\xi,t) = \xi + \int_{t_0'}^{t} \mathcal{Z}^{-1}(\tau)\,\mathbf{u}(\xi,\tau)\,d\tau , \qquad \mathbf{u}(\xi,t) = \mathbf{w}(X(\xi,t),t) .$$

In these coordinates the problem (106) has the form

$$\mathbf{u}_t + \beta'(\mathbf{u}\cdot\nabla_u)\mathbf{h}(X) - \nu\,\nabla_u^2\mathbf{u} + \nabla_u q = 0 ,$$

$$\nabla_u \cdot \mathbf{u} = 0 , \qquad \xi \in \Omega_{t_0}' \equiv \Omega' , \quad t > t_0 ,$$

$$\mathbf{u}|_{t=t_0'} = \mathbf{w}(\xi,t_0) ,$$

$$T_u(\mathbf{u},q)\,\mathbf{n} = \sigma\,\Delta(t)\,X(\xi,\tau) + p_\infty(X)\,\mathbf{n} , \qquad \xi \in \Gamma' \equiv \partial\Omega' ,$$

where

$$q(\xi,t) = s(X(\xi,t),t) , \qquad \nabla_u = A\nabla , \qquad A = \mathcal{Z}\,B ,$$

B is the matrix of cofactors of

$$b_{kj}(\xi,t) = \frac{\partial y_k}{\partial \xi_j} = \delta_{kj} + \sum_{m=1}^{3} \int_{t_0'}^{t} Z^{km}(\tau)\,\frac{\partial u_m(\xi,\tau)}{\partial \xi_j}\,d\tau ,$$

Z^{km} are elements of the inverse matrix $\mathcal{Z}^{-1}$.

Following the proof of Proposition 8.3, we introduce the functions $\mathbf{u}_\lambda(\xi,t) = \zeta_\lambda(t)\,\mathbf{u}(\xi,t)$, $q_\lambda(\xi,t) = \zeta_\lambda(t)\,q(\xi,t)$. They satisfy the relations

$$\mathbf{u}_{\lambda t} + \beta'(\mathbf{u}_\lambda \cdot \nabla_u)\,\mathbf{h}(X) - \nu\,\nabla_u^2\mathbf{u}_\lambda + \nabla_u q_x = \mathbf{u}\,\zeta_\lambda' ,$$

$$\nabla_u \cdot \mathbf{u}_\lambda = 0 ,$$

$$\mathbf{u}_\lambda|_{t=t_0'} = 0 ,$$

$$\sqcap_0 \sqcap S_u(\mathbf{u}_\lambda)\,\mathbf{n}|_{\Gamma'} = 0 ,$$

$$\mathbf{n}_0 \cdot T_u(\mathbf{u}_\lambda, q_\lambda)\,\mathbf{n} - \sigma\,\mathbf{n}_0 \cdot \int_0^t \Delta(\tau)\,\mathbf{u}_\lambda(\xi,\tau)\,d\tau|_{\Gamma'} = b(\xi,t) + \int_{t_0'}^{t} B(\xi,\tau)\,d\tau ,$$

$$b = (I - \mathcal{Z}(t))\,\mathbf{n}_0 \cdot T_u(\mathbf{u}_\lambda, q_x)\,\mathbf{n} ,$$

$$B(\xi,t) = \zeta_\lambda'(t)\,\mathcal{Z}\mathbf{n}_0 \cdot T_u(\mathbf{u},q)\,\mathbf{n} + \sigma\,\zeta_\lambda\,\mathbf{n}_0 \cdot \dot{\Delta}(t)\,\xi$$

$$+ \sigma\,\zeta_\lambda\mathbf{n}_0 \cdot \dot{\Delta}(t) \int_{t_0'}^{t} \mathcal{Z}^{-1}(\tau)\,\mathbf{u}(\xi,\tau)d\tau + \sigma\,\zeta_\lambda(t)\frac{d}{dt}(p_\infty(X)\,\mathbf{n}_0 \cdot \mathbf{n}) .$$

The last equation is obtained from

$$\zeta_\lambda\Big(\mathbf{n}_0\cdot T_u\mathbf{n} - \sigma\,\Delta(t)\,X\Big) - \int_{t_0'}^{t}\zeta_\lambda'(\tau)\Big(\mathbf{n}_0\cdot T_u\mathbf{n} - \sigma\,\Delta(\tau)\,X\Big)\,d\tau$$

$$= \sigma\,\zeta_\lambda\,p_\infty(X)\,\mathbf{n}_0\cdot\mathbf{n} - \sigma\int_{t_0'}^{t}\zeta_\lambda'(\tau)\,p_\infty(X)\,\mathbf{n}_0\cdot\mathbf{n}\,d\tau\,,$$

by integration by parts in some terms.

Making use of the inequality (51) and of the smallness of τ_0, we obtain the following estimate of $(\mathbf{u}_\lambda, q_\lambda)$

$$\sup_{t_0'<\tau<t}|\mathbf{u}_{\lambda\tau}(\cdot,\tau)|_{\Omega'}^{(\alpha)} + \sup_{t_0'<\tau<t}|\mathbf{u}_\lambda(\cdot,\tau)|_{\Omega}^{(2+\alpha)} + \sup_{t_0'<\tau<t}|q_\lambda(\cdot,\tau)|_{\Omega'}^{(1+\alpha)}$$

$$\le c\lambda^{-1}\Bigg(\sup_{t_0'+\frac{\lambda}{2}<\tau<t}|\mathbf{u}(\cdot,\tau)|_{\Omega'}^{(1+\alpha)} + \sup_{t_0'+\frac{\lambda}{2}<\tau<t}|\mathbf{u}(\cdot,\tau)|_{\Omega}^{(1+\alpha)} + \sup_{t_0'+\frac{\lambda}{2}<\tau<t}|q(\cdot,\tau)|_{\Gamma'}^{(\alpha)}\Bigg)$$

$$(119)$$

(cf. the proof of (61) in [42]). We estimate the norm of q using the boundary condition in (106), i.e.,

$$s = 2\nu\,\mathbf{n}\cdot\frac{\partial\mathbf{w}}{\partial n} - \sigma\Big(H - H_\infty + \frac{\beta'^2}{2}(R^2 - R_\infty^2)\frac{x_1^2 + x_2^2}{|x|^2}\Big), \qquad (120)$$

and interpolation inequalities, as follows

$$|s(\cdot,t)|_{\Gamma_t}^{(\alpha)} \le c\Big(|\mathbf{w}|_{\Gamma_t}^{(1+\alpha)} + |R - R_\infty|_{S_1}^{(2+\alpha)}\Big)$$

$$\le \theta_1\Big(|\mathbf{w}|_{\Gamma_t}^{(2+\alpha)} + |R - R_\infty|_{S_1}^{(3+\alpha)}\Big)$$

$$+ c\theta_1^{-\frac{5}{2}-\alpha}\Big(\|R - R_\infty\|_{W_2^1(S_1)} + \|\mathbf{w}\|_{L_2(\Omega_t)}\Big)\,.$$

Here $\theta_1 > 0$ is arbitrarily small. Since linearized equation (120) (with respect to $R - R_\infty$) is elliptic, there holds the estimate

$$|R - R_\infty|_{S_1}^{(3+\alpha)} \le c\Big(|s|_{\Gamma_t}^{(1+\alpha)} + |\mathbf{w}|_{\Omega_t}^{(2+\alpha)} + \|R - R_\infty\|_{L_2(S_1)}\Big)\,. \qquad (121)$$

Taking $\theta_1 = \lambda\varepsilon'$, $\varepsilon' \ll 1$, we easily obtain from (119) inequality similar to (104)

$$f(\lambda) = c\varepsilon'f\Big(\frac{\lambda}{2}\Big) + K\,,$$

where

$$f(\lambda) = \lambda^{\alpha+\frac{7}{2}}\Bigg(\sup_{\tau\in I_\lambda}|\mathbf{u}_\tau(\cdot,\tau)|_{\Omega'}^{(\alpha)} + \sup_{\tau\in I_\lambda}|\mathbf{u}(\cdot,\tau)|_{\Omega'}^{(2+\alpha)} + \sup_{\tau\in I_\lambda}|q(\cdot,\tau)|_{\Omega'}^{(1+\alpha)}\Bigg),$$

$$K = c(\varepsilon')\Bigg(\sup_{t_0-2\tau_0<\tau<t_0}\|\mathbf{u}(\cdot,t)\|_{L_2(\Omega)} + \sup_{t_0-2\tau<\tau<t_0}\|R - R_\infty\|_{W_2^1(S_1)}\Bigg)\,.$$

It implies (105) and, by virtue of (121), also the desired estimate (118). In addition, we have an estimate in a small interval $(0, 2\tau_0)$

$$N_{2\tau_0}(\mathbf{w}, s) \le c\left(|\mathbf{w}_0|_\Omega^{(2+\alpha)} + \left|R(\cdot, 0) - R_\infty(\cdot)\right|_{S_1}^{(3+\alpha)}\right).$$

This inequality, together with (110) and (118), yields a uniform estimate for $N_T(\mathbf{w}, s)$ which makes it possible to extend the solution of the problem (106) to an infinite time interval $t > 0$ and to complete the proof of Theorem 5.4.

References

[1]   G. Allain, Small time existence for the Navier–Stokes equations with a free surface, Appl. Math. Optim., 16 (1987), 37–50.

[2]   B.V. Bazalii, The Stefan problem, Dokl. Akad. Nauk Ukrain. SSR Ser. A, 11 (1986), 3–7.

[3]   B.V. Bazalii and S.P. Degtiarev, On the classical solvability of the multidimensional Stefan problem for the convective notion of viscous incompressible fluid, Mat. Sb., 132 (1987), 3–19.

[4]   J.T. Beale, Large-time regularity of viscous surface waves, Arch. Rational Mech. Anal., 84 (1984), 307–352.

[5]   J.T. Beale and T. Nishida, Large-time behaviour of viscous surface waves, Lecture Notes in Num. Appl. Anal., 8 (1985), 1–14.

[6]   G.I. Bizhanova, Solution in a weighted Hölder function space of the multidimensional two-phase Stefan and Florin problems for second order parabolic equation in a bounded domain, Algebra Anal., 7 (2) (1995), 46–76.

[7]   G.I. Bizhanova and V.A. Solonnikov, Solvability of an initial-boundary value problem for second order parubolic equation with time derivative in the boundary condition in a weighted Hölder function space, Algebra Anal., 5 (1) (1993), 109–142.

[8]   G.I. Bizhanova and V.A. Solonnikov, On free boundary problems for the second order parabolic equations, Algebra Anal., 12 (6) (2000), 98–139.

[9]   V.O. Bytev, Non-steady motion of a ring of a viscous incompressible liquid with free boundaries, J. Appl. Math. Techn. Phys., 3 (1970), 82–88.

[10]   X. Chen and F. Reitich, Local existence and uniqueness of solutions of the Stefan problem with surface tension and kinetic undercooling, J. Math. Anal. Appl., 164 (1992), 350–362.

[11]   X. Chen, J. Hong and F. Yi, Existence, uniqueness and regularity of classical solutions of the Mullin–Sekerka problem, , Comm. Part. Diff. Equat., 21 (1996), 1705–1727.

[12]   I.V. Denisova, Problem of the motion of two viscous incompressible fluids separated by a closed free interface, Acta Appl. Math., 37 (1994), 31–40.

[13]   J. Escher and G. Simonett, Classical solutions of multi-dimensional Hele–Shaw models, SIAM J. Math. Anal., 28 (1997), 1028–1047.

[14]   J. Escher and G. Simonett, Moving surfaces and abstract parubolic evolution equations, in "Topics in Nonlinear Analysis" – The Herbert Anmann anniversary volume, Progress in nonlinear differential equations and their applications, 35 (1999), 184–210.

[15] A. Friedman, Variational Principles and Free Boundary Problems, Wiley, New York, 1982.

[16] K.O. Friedrichs, On the boundary value problems of the theory of elasticity and Korn's inequality, Ann. of Math., 48 (1947), 441–471.

[17] E.I. Hanzawa, Classical solutions of the Stefan problem, Tohoku Math. J., 33 (1981), 297–335.

[18] D. Kinderlehrer and L. Nirenberg, The smoothness of the free boundary in the one phase Stefan problem, Comm. Pure Appl. Math., 31 (1978), 257–282.

[19] V.A. Kondratiev and O.A. Oleinik, Hardy's and Korn's type inequalities and their applications, Rend. Mat. Appl., 7 (1990), 641–666.

[20] A. Korn, Über einige Ungleichungen, welche in der Theorie der elastischen und elektrischen Schwingungen eine Rolle spielen, Bull. Intern. Cracovie Akad. (Cl. Sci. Math. Nat.), (1909), 705–724.

[21] S. Luckhaus, Solutions for the two-phase Stefan problem with Gibbs–Thomson law for the melting temperature, Euro. J. Appl. Math., 1 (1990), 101–111.

[22] A.M. Meirmanov, On a classical solution of multidimensional Stefan problem for quasilinear parabolic equations, Matem. Sb., 112 (1980), 170–192.

[23] A.M. Meirmanov, The Stefan Problem, De Gruyter, Berlin, 1992 (Russian edition: Novosibirsk, 1986).

[24] I.Sh. Mogilevskii and V.A. Solonnikov, On the solvability of an evolution free boundary problem for the Navier–Stokes equations in the Hölder spaces of functions, in "Math. Problems Related to Navier–Stokes Equations", Series on Advances in Math. for Appl. Sciences, 11 (1992), 105–181.

[25] M. Padula, On the exponential stability of the rest state of a viscous compressible fluid, J. Math. Fluid Mech., 1 (1999), 62–77.

[26] E.V. Radkevich, Solvability of general nonstationary problems with free boundary, in "Some Applications of Functional Analysis to the Problems of Mathematical Physics", AN SSSR, Sibirsk. Otdel., Inst. Mat., Novosibirsk (1986), 85–111.

[27] E.V. Radkevich, On conditions for the existence of a classical solution of the modified Stefan problem (the Gibbs–Thomson law), Matem. Sb., 183 (2) (1993), 77–101.

[28] J.F. Rodrigues, The variational inequality approach to the one-phase problem, Acta Appl. Math., 8 (1987), 1–35.

[29] J.F. Rodrigues, V.A. Solonnikov and F. Yi, On a parabolic system with time derivative in the boundary conditions and related free boundary problems, Math. Ann., 315 (1999), 61–95.

[30] V.A. Solonnikov, Estimates of solutions of nonstationary linearized system of the Navier–Stokes equations, Trudy Math. Inst. Steklov, 64 (1964), 213–317.

[31] V.A. Solonnikov, Estimates of solutions of initial-boundary value problem for a linear nonstationary system of the Navier–Stokes equations, Zapiski Nauchn. Semin. LOMI, 59 (1976), 178–254.

[32] V.A. Solonnikov, On the solvability of the second initial-boundary value problem for the linear time-dependent system of the Navier–Stokes equations, Zapiski Nauchn. Semin. LOMI, 69 (1977), 200–218.

[33] V.A. Solonikov, Solvability of the problem on a motion of a viscous incompressible liquid bounded by a free surface, Izvestia Acad. Nauk USSR Ser. Matem., 41 (6) (1977), 1388–1427.

[34] V.A. Solonnikov, On nonstationary motion of a finite liquid mass bounded by a free surface, Zapiski Nauchn. Semin. LOMI, 152 (1986), 137–157.

[35] V.A. Solonnikov, On the transient motion of an isolated mass of a viscous incompressible fluid, Izvestia Acad. Nauk USSR Ser. Matem., 51 (5) (1987), 1065–1087.

[36] V.A. Solonnikov, On the evolution of an isolated volume of viscous incompressible capillary fluid for large values of time, Vestnik Leningrad Univ. Ser. I, 3 (1987), 49–55.

[37] V.A. Solonnikov, On nonstationary motion of a finite isolated mass of self-gravitating fluid, Algebra Anal., 1 (1) (1989), 207–249.

[38] V.A. Solonnikov, On initial-boundary value problem for the Stokes system arising in the study of a free boundary problem, Trudy Mat. Inst. Steklov, 188 (1990), 150–188.

[39] V.A. Solonnikov, Estimates of solutions of the second initial-boundary value problem for the Stokes system in the space of functions with Hölder continuous derivatives with respect to spatial variables, Zapiski Nauchn. Semin. POMI, 259 (1999), 254–279.

[40] V.A. Solonnikov, On the justification of quasistationary approximation in the problem of motion of a viscous capillary drop, Interfaces and Free Boundaries, 1 (2) (1999), 125–174.

[41] V.A. Solonnikov, Initial-boundary value problem for generalized Stokes system in the half-space, Zap. Nauchn. Semin. POMI, 271 (2000), 224–275.

[42] V.A. Solonnikov, On the solvability of a quasistationary problem for the Navier–Stokes equations, Adv. Math. Sci. Appl., 11 (1) (2001).

[43] V.A. Solonnikov and E.V. Frolova, L_p-theory for the Stefan problem, Zap. Nauchn. Semin. POMI, 243 (1997), 299–323.

[44] V.A. Solonnikov and A. Tani, Evolution free boundary problem for equation of motion of viscous compressible barotropic liquid, Proceedings Oberwolfach, 1991, Lect. Notes in Math., 1530 (1992), 30–55.

[45] A. Visintin, Models of Phase Transitions, Birkhauser, Boston, 1996.

[46] F. Yi, A note on the classical solution of a two dimensional superconductor free boundary problem, European J. Appl. Math., 7 (1996), 113–118.

[47] F. Yi and J. Liu, Vanishing specific heat for the classical solutions of a multidimensional Stefan problem with kinetic condition, Quart. Appl. Math., 57 (1999), 661–672.

Variational and Dynamic Problems for the Ginzburg-Landau Functional

Halil Mete Soner*

Department of Mathematics
Koç University
RumeliFener Yolu Sariyer
80910 Istanbul, Turkey
msoner@ku.edu.tr

Table of Contents

1 Introduction

In these notes, I survey several results on the Ginzburg-Landau functional

$$\hat{I}^\epsilon(u) := \int_U E^\epsilon(u) \ dx, \quad E^\epsilon(u) := \frac{\epsilon}{2}|\nabla u|^2 + \frac{1}{4\epsilon}[1 - |u|^2]^2,$$

where U is an open, bounded subset of $\mathbb{R}^n$ with smooth boundary, and $\epsilon > 0$ is a small parameter.

---

* Partially supported by the National Science Foundation grant DMS-98-17525 and a NATO grant CRG 971561

This functional is a simpler version of the Ginzburg-Landau functional for superconductivity. The model for superconductivity has two fields; the complex valued order parameter u and the vector valued magnetic potential A. The functional $\hat{I}^\epsilon$ is obtained by setting A to zero and by appropriate scaling. In superconductivity, the length of the order parameter $|u|$ is proportional to the density of superconducting electrons. Hence the zeroes of u, called *vortices*, are the places where superconductivity is lost. The parameter ϵ corresponds to the inverse of the Ginzburg-Landau parameter κ and a number of type II superconducting materials have large κ justifying the asymptotic regime considered in the notes. The lecture notes of Rubinstein [23] provides a good introduction.

The starting point of these notes is the seminal work of Bethuel, Brezis and Hélein [6] which gives a detailed asymptotic description of the minimizers u^ϵ of $\hat{I}^\epsilon$ as ϵ tends to zero when $U \subset \mathbb{R}^2$. To briefly describe this result, let u^ϵ be the minimizer of $\hat{I}^\epsilon$ among all functions u satisfying a given boundary data $u = g$. Since as ϵ tends to zero the second term in E^ϵ forces the solution to have length one, it is natural to assume that g takes values in the unit circle S^1. In the complex notation, g admits the representation $g = e^{i\phi}$ for some possibly multi-valued function ϕ. Then, the boundary condition is

$$u(x) = g(x) = e^{i\phi(x)}, \quad x \in \partial U. \tag{1}$$

The degree of g around the origin (or the winding number) is an important parameter. Using the local representation $g = e^{i\phi}$ and the fact that $U \subset \mathbb{R}^2$, the degree can be computed by the following formula

$$\deg(g; \partial U) = \frac{1}{2\pi} \int_{\partial U} \nabla \phi \cdot \mathbf{t} \, d\mathcal{H}^1(x), \tag{2}$$

where $\mathbf{t}$ is the unit tangent and $\int_{\partial U} \cdots d\mathcal{H}^1$ is the line integral around ∂U.

If the boundary data g has zero degree, then ϕ is single valued. It is then relatively straightforward to show that u^ϵ converges strongly to the smooth function $u = e^{i\varphi}$ where φ is unique harmonic function in U satisfying the boundary condition $\varphi = \phi$. Hence, the interesting the case is

$$d := \deg(g; \partial U) \neq 0.$$

Then, by topological considerations, any function satisfying the boundary and in particular the minimizer u^ϵ must have d zeroes. This fact makes the problem interesting as each zero carries an energy of at least $\pi \ln(1/\epsilon)$. To see this consider the problem with $U = B_R := \{|x| < R\}$, and $g(x) = e^{iN\theta}$ where $\theta(x)$ is the angle between x and the x-axis and N is an integer. Consider a test function

$$v^\epsilon(x) = f(|x|/\epsilon) \, e^{iN\theta},$$

with some smooth, positive function f satisfying $f(0) = 0$, $f(r) = 1$ for all $r \geq 1$. By calculus,

$$\hat{I}^\epsilon(v^\epsilon) = N^2\pi \ln(1/\epsilon) + O(1).$$

The N^2 term indicates that it is better to have N distinct zeroes of degree one, instead of less zeroes with higher degree. Hence the minimizer u^ϵ is expected to have d distinct zeroes $a^\epsilon := (a_1^\epsilon, \dots, a_d^\epsilon)$, again called vortices, each having degree one. The minimum energy $\hat{I}^\epsilon(u^\epsilon)$ behaves like $d\pi \ln(1/\epsilon)$ and the asymptotic behavior of u^ϵ is determined by the location of the zeroes a^ϵ. Bethuel, Brezis and Hélein, obtained the location of the zeroes by calculating the next term in the minimum energy, which they call the *renormalized energy*. The renormalized energy $W(a^\epsilon)$ is a function of the location of the zeroes, and it has a representation in terms of the solution of the Laplace equation with point sources at a^ϵ, or equivalently the canoniacl hramonic as defined in [6]. Then, the minimum energy has the form

$$\hat{I}^\epsilon(u^\epsilon) = d\,\pi\,\ln(1/\epsilon) + W(a^\epsilon) + o(1). \tag{3}$$

In view of this expansion of the minimum energy, it is clear that a^ϵ converges to a minimum of the renormalized energy W. Once the location of the zeroes is determined it is possible to calculate the limit of u^ϵ. We refer to [6] for more information.

The chief difference between the problem considered here and the model for superconductivity is the boundary condition. In superconductivity, Neumann condition is given and the vortices are formed by an exogenous forcing term which is the applied magnetic field. While in the above problem vortices created by the Dirichlet data. For this reason, local results which do not refer to a particular boundary condition are more useful for our understanding of the model for superconductivity (we refer to the recent paper of Sandier and Serfaty [27], and the references therein for infromation on mathematical results on the model for superconductivity.) The Gamma limit is such a result. For the scalar valued functions, the Gamma limit of the Ginzburg-Landau functional, with a different rescaling, is proved by Modica and Mortola [21, 21], and by Modica [20]. A brief definition of the Gamma limit is given in §4.

In view of the above calculations, we consider the Gamma limit of the rescaled Ginzburg-Landau functional

$$I^\epsilon(u) := \frac{\hat{I}^\epsilon(u)}{\ln(1/\epsilon)} = \frac{1}{\ln(1/\epsilon)} \int_U E^\epsilon(u)\,dx.$$

In §4 below we will show that the Gamma limit of I^ϵ is equal to

$$I(u) := \begin{cases} |Ju|(U), & \text{if } u \in B2V(U; S^1), \\ +\infty, & \text{otherwise}, \end{cases}$$

where Ju is the Jacobian of u and weakly it is given by (see §2.1)

$$Ju := \frac{1}{2}\,\nabla \times j(u), \qquad j(u) := u \times Du = \det(u; Du),$$

S^1 is the unit circle, and $B2V$ is the set of all functions whose weak Jacobian is a Radon measure. The weak definition of the Jacobian in higher dimensions is discussed in §5, and BnV with a general n is defined properly in §4.1. This class of functions and its properties are studied in the two papers of the author with Jerrard [14, 15]. The set BnV, called *functions of bounded n variations* is related to the classical BV space and to the *Cartesian currents* of Giaquinta, Modica and Soucek [10, 11].

It is shown in [14] and also in §4.1 below that for $u \in B2V(U; S^1)$, the Jacobian has a special structure

$$Ju = \pi \sum_i k_i \, \delta_{a_i},$$

for some points $\{a_i\} \subset U$ and integers k_i. Here δ_{a_i} is the Dirac measure located at a_i. This is interpreted as encoding the location and the topological singularities (or zeroes) of u. Moreover, for $u \in B2V(U; S^1)$,

$$|Ju|(U) = \sum_i |k_i|.$$

Hence, the Gamma limit $I(u)$ counts the zeroes of u with multiplicity.

This Gamma limit is proved in several steps. The first step is a "local" energy lower bound of the form

$$\int_U E^\epsilon(u) \, dx \geq \pi \, \ln(1/\epsilon) \, \deg(u; \partial U) - C,$$

for some constant C. There are problems with this estimate as it is stated. The difficulty comes from the possible zeroes of u near or on ∂U. We will prove two such results, Theorem 2.1, and Theorem 2.2. They are local in nature, especially the second one. The proof technique is an elegant covering argument of Jerrard [12]. To explain this method clearly, we first prove it under slightly restrictive set of assumptions first to prove Theorem 2.1. We then modify this technique to obtain the sharper lower bound Theorem 2.2.

A corollary of this lower bound is a compactness result for the Jacobian. This estimate bounds the Jacobian by the Ginzburg-Landau energy, and yields a compactness result for the Jacobian. Indeed for any sequence u^ϵ satisfying

$$\sup_\epsilon I^\epsilon(u^\epsilon) < \infty,$$

the Jacobians Ju^ϵ are compact in dual $(C^{0,\alpha})^*$ norm for every $\alpha > 0$. Hence, on a subsequence Ju^ϵ converge to a distribution J not in a norm slightly weaker than the weak* topology of Radon measures. Although this convergence is not in the space of measures, we will show that the resulting distribution J is indeed a measure of a special form. For $B2V(U; \mathbb{R}^2)$ with $U \subset \mathbb{R}^m$ with $m = 2$, this is proved in §3.2, and for $m \geq 3$ it is stated in §5.

Then the Gamma limit is proved in §4 as a result of the lower bound and the compactness of the Jacobian.

The compactness of the Jacobian is also a useful tool in the analysis of the dynamic problems. To motivate the study of the evolution problems in this context, let us consider an experiment in superconductivity. In this experiment the vortices are formed by an external magnetic field. Then the magnetic field is turned off and the material turns back to superconducting state. To understand these transition from the vortex state to the superconducting state, both the parabolic

$$u_t - \Delta u = \frac{u}{\epsilon^2} [1 - |u|^2], \qquad t > 0, \ x \in I\!\!R^n, \tag{4}$$

and the Schrödinger

$$i \, u_t - \Delta u = \frac{u}{\epsilon^2} [1 - |u|^2], \qquad t > 0, \ x \in I\!\!R^n, \tag{5}$$

equations are proposed. The question then is to obtain evolution of the vortices that are forced into the system via the initial data. Mathematically this is achieved by studying the small ϵ asymptotics of the above equations for an initial data u_0^ϵ which contains several vortices with degree ± 1. Asymptotic expansion techniques are used by Neu, Rubinstein and E to derive these equations; see the lecture notes of Rubinstein [23]. Since (4) is the gradient flow for $\hat{I}^\epsilon$, in view of the expansion (3), as ϵ tends to zero, the vortices should satisfy the gradient flow for the potential W. Indeed, let $a(t) = (a_1(t), \dots, a_N(t))$ be the limit of vortices $a^\epsilon(t \, \ln(1/\epsilon))$. Then,

$$\frac{d}{dt} a(t) = -\nabla W(a(t)), \tag{6}$$

in the case of (4). Note that we need to speed up the dynamics by a factor of $\ln(1/\epsilon)$. For the Schrödinger equation, in the original time scale, we get the Hamiltonian dynamics. These results are rigorously proved in several papers. For the parabolic flow, in [18] Lin proved that the speed of vortices is $1/\ln(1/\epsilon)$ when the vortices all have same sign. The mixed vortex case, which is the relevant one in the experiment outline above, was proved in [16] by Jerrard and the author. First rigorous derivation of the vortex equation is also given in [16]. For $U = I\!\!R^2$, an explicit form of (6) is avaliable

$$\frac{d}{dt} a_k(t) = 2 \sum_{j=1}^{N} d_j d_k \frac{(a_k(t) - a_j(t))}{|a_k(t) - a_j(t)|^2},$$

where d_k is the degree of u^ϵ around a_k for small ϵ, and by hypothesis d_k is equal to ± 1. Note that the solutions of the above equation behaves like charged particles with a logarithmic potential; vortices with same degree expel each other while ones with opposite degree attract with a force proportional to the inverse of the distance between the vortices.

In $\mathbb{R}^n$, the set of zeroes of u^ϵ is a codimension two set, as ϵ tends to zero we obtain geometric equations for these sets; called vortex lines in $\mathbb{R}^3$. As expected from results on scalar version of (4), the limiting vortex line moves by mean curvature flow. We refer to [23, 24] and the references therein for the formal derivation of these equations. First rigorous results for the vector Ginzburg-Landau equation are [17] and [4].

In §6 we prove the convergence when there exists a smooth solution $\{\Gamma_t\}_{t\in[0,T]}$ of the codimension two mean curvature flow. The main idea set forward in [17] is to consider the limiting measures

$$\mu_t := \text{weak}^* \text{ limit of } \mu_t^\epsilon,$$

where

$$\mu_t^\epsilon(V) := \frac{1}{|\ln 1/\epsilon|} \int_V E^\epsilon(u^\epsilon(t,x))\, dx.$$

In Theorem 6.1, under appropriate assumptions on the initial data, we will show that

$$\text{support } \mu_t = \Gamma_t, \qquad \mu_t \geq \pi\, \mathcal{H}^{n-2} \lfloor \Gamma_t,$$

where $\mathcal{H}^{n-2} \lfloor \Gamma_t$ is the Hausdorff measure restricted to Γ_t, i.e., it is the surface area measure on the surface Γ_t. Moreover, the limit J_t of the Jacobians $Ju^\epsilon(t,\cdot)$ satisfies

$$|J_t| = \pi\, \mathcal{H}^1 \lfloor \Gamma_t. \tag{7}$$

To prove the inclusion support $(\mu_t) \subset \Gamma_t$, we use the energy identities and a Pohazaev type inequality as in [17]. The idea is to estimate the time derivative of

$$\alpha(t) := \int_{\mathbb{R}^n} \eta(t,x)\, \mu_t(dx),$$

when the test function η is the square distance function of Γ_t. Since $\{\Gamma_t\}_{t\in[0,T]}$ solves the mean curvature flow, the square distance function η satisfies

$$\nabla \eta_t = \nabla \Delta \eta, \qquad \text{on } \Gamma_t.$$

We use this and the other properties of the square distance function to prove that $\alpha(t) = 0$ for $t \in [0,T]$. This proves the inclusion support $(\mu_t) \subset \Gamma_t$.

The opposite inclusion is proved by studying the Jacobian. In view of the energy estimate

$$\int_{\mathbb{R}^n} E^\epsilon(u^\epsilon(T,x))\, dx + \int_0^T \int_{\mathbb{R}^n} |u_t^\epsilon(t,x)|^2\, dx\, dt = \int_{\mathbb{R}^n} E^\epsilon(u^\epsilon(0,x))\, dx,$$

our compactness result implies that $J_t^\epsilon := Ju^\epsilon(u^\epsilon(t,\cdot))$ is compact. Let J_t be a limit of J_t^ϵ. Then, by the previous inclusion we know that the support of J_t is in Γ_t. Moreover, by the weak formulation of the Jacobian (see §2.1), the Jacobian is divergence free. Since Γ_t is smooth manifold with no boundary, this implies that the density of the Jacobian on Γ_t is constant. We then show that this constant is equal to π for all $t \in [0,T]$, proving (7). Also, in view of the compactness result, Theorem 5.2, the energy measure dominates the Jacobian measure. Hence, the support of μ_t is equal to Γ_t.

Acknowledgments. I would like to thank the organizers, Professors Colli and Rodrigues, of the CIM/CIME Euro-Summer School for giving me the oppurtunity to put together several results obtained in different papers into these lecture notes, and also for a very productive Summer School.

2 Energy Lower Bounds

In this section we prove an energy lower bound in terms of the topological degree. This bound is local in nature and is a key step in the proof of the Gamma limit as well as the dynamical properties of the vortices. Local energy lower bounds were proved by covering arguments by Jerrard [12] and Sandier [25]. Here, we follow the technique developed by Jerrard to prove these estimates. In the next subsection, we give a brief and a formal discussion of this technique. Then we will give the precise statement and the proof of the lower bound.

Let U is a bounded open subset of $I\!R^2$, and $u \in H^1(U; I\!R^2)$ is a function that we have assumed (without loss of generality) to be smooth. The goal is to find an energy estimate of the form

$$\int_U E^\epsilon(u) \; dx \geq \pi \; \ln(1/\epsilon) \deg(u; \partial U) - C,$$

for some constant C, independent of ϵ and u. We want this to hold for all $u \in H^1(U; I\!R^2)$ and $\epsilon \in (0,1]$. However, there are problems with this estimate as it is stated. The function u may have a zero on the boundary of U. Then, the degree of u around ∂U is not even be defined. Also, when u has a zero very close to the boundary, most of the energy could be outside the domain U. These possibilities indicate that we have make either an assumption about the boundary behavior of u, or to modify the statement of the lower bound. The latter is better suited for the later use of these estimate as it makes the lower bound a "local" one. However the statement of this local lower bound is rather technical. To explain the main idea we first outline the proof under assumptions on the boundary behavior. Remarkably, the proof technique of this "easier" lower bound carried over the more technical one with very little change.

We start with a brief discussion of the degree.

2.1 Degree and Jacobian

In our arguments, we will use the degree and the Jacobian repeatedly. For that reason we recall their definition.

Let $V \subset U \subset \mathbb{R}^2$ and $|u| \neq 0$ on ∂V. Then, u admits a local representation $u = |u|e^{i\phi}$ on ∂V, and the degree of u around ∂V is given by

$$\deg(u; \partial V) = \frac{1}{2\pi} \int_{\partial V} \nabla\phi \cdot \mathbf{t} \, d\mathcal{H}^1(x),$$

where as in the Introduction, $\mathbf{t}$ is the unit tangent and $\int_{\partial V} \cdots d\mathcal{H}^1$ is the line integral around ∂V. For future reference, we note that

$$\deg(u; \partial V) = \deg(u/|u|; \partial V).$$

The Jacobian Ju of u satisfies

$$Ju = \frac{1}{2} \nabla \times j(u),$$

where

$$j(u) = u \times Du = \det(u; Du).$$

Hence by the Stokes' theorem

$$\int_V Ju \, dx = \int_{\partial V} j(u) \cdot \mathbf{t} \, d\mathcal{H}^1. \tag{8}$$

Generalizations to the case when $U \subset \mathbb{R}^n$ will be discussed later.

For $u = |u|e^{i\varphi}$, we directly calculate that $j(u) = |u|^2 \nabla\varphi$. Hence,

$$\nabla\varphi = j(v) = j(u)/|u|^2, \qquad v = u/|u|,$$

and by the Stokes' theorem

$$\deg(u; \partial V) = \frac{1}{2\pi} \int_{\partial V} \frac{j(u) \cdot \mathbf{t}}{|u|^2} \, d\mathcal{H}^1 = \frac{1}{2\pi} \int_{\partial V} j(v) \cdot \mathbf{t} \, d\mathcal{H}^1, \tag{9}$$

for all u which do not vanish on ∂V.

2.2 Covering argument

In this subsection we outline a covering technique developed by Jerrard to prove energy lower bounds [12]. Similar techniques were also used by Sandier [25]. To simplify the presentation, we assume that

$$|u(x)| > \frac{1}{2}, \quad \text{whenever } x \in U_{r_0}, \quad U_{r_0} := \{ x \in U \mid dist(x, \partial U) \geq r_0\}, \tag{10}$$

for some constant $r_0 > 0$. We also assume that

$$\deg(u; \partial U) \neq 0.$$

Theorem 2.1 (Jerrard [12]). *There exists a constant C, such that for all $\epsilon \in (0, 1]$, and for all $u \in H^1$ satisfying above assumptions,*

$$\int_U E^\epsilon(u) \, dx \geq \pi \, \ln(r_0/\epsilon) - C.$$

A more general result which do not assume (10) will be proved later in this section.

We introduce some notation and definitions, taken from [12]. We let S denote the set on which $|u|$ is small, that is,

$$\{x \in U \ : \ |u(x)| \leq 1/2\}. \tag{11}$$

We define the *essential* part S_E of S to be

$$S_E := \cup \{\text{components } S_i \text{ of } S \ : \ \deg(u; \partial S_i) \neq 0\}. \tag{12}$$

For any subset $V \subset U$ such that $\partial V \cap S_E \neq \emptyset$, we define the generalized degree

$$\mathrm{dg}(u; \partial V) := \sum \left\{ \deg(u; \partial S_i) \mid \text{components } S_i \text{ of } S_E \text{ such that } S_i \subset V \right\}. \tag{13}$$

Notice that if u is nonzero on the boundary of V, the generalized degree agrees with the degree of u around ∂V. Hence the generalization of the degree is only relevant when u has zeroes on ∂V. But in this case, we could remove these zeroes by slightly modifying u and with small change in the energy. Hence, in view of the Ginzburg-Landau energy these zeroes are removable and this justifies the definition of S_E and the generalized degree.

In view of (10),

$$U_{r_0} \cap S_E \neq \emptyset,$$

and by the definition of the generalized degree and the assumption that the degree of u around ∂U is nonzero,

$$\mathrm{dg}(u; \partial U_r) = \deg(u; \partial U_r) = \deg(u; \partial U) \neq 0, \quad \forall \, r \in (0, r_0].$$

Our strategy for proving Theorem 2.1 will be to find a collection of balls with a good lower bound for the Ginzburg-Landau energy on each ball. We then show that the sum of the radii of the balls is bounded below by $r_0/2$, hence obtaining a lower bound for the total Ginzburg-Landau energy in terms of this quantity. This will be done in several steps.

1. First cover of S_E.

We find the collection of balls by starting from an initial collection of small balls that cover S_E^ϵ, then letting these balls grow by expanding them and combining them. The first step is thus to establish the existence of the initial collection of small balls. This is the content of

Proposition 2.1. *There is a collection of closed, pairwise disjoint balls* $\{B_i^*\}_{i=1}^k$ *with radii* r_i^* *such that*

$$S_E \subset \cup_{i=1}^k B_i^*, \tag{14}$$

$$r_i^* \geq \epsilon \qquad \forall i, \tag{15}$$

$$\int_{B_i^* \cap U} E^\epsilon(u) dx \geq \frac{c_0}{\epsilon} r_i^*. \tag{16}$$

Proof. This is proved in [12]. Let $\{S_i\}_{\{i=1,\dots,N\}}$ be the disjoint components of S_E. Choose $x_i \in S_i$ for each i, and set

$$\rho_i := \inf\{r > 0 \ : \ \partial B_r(x_i) \cap S_i \neq \emptyset\}, \quad r_i := \max\{\epsilon \ ; \ \rho_i\}.$$

Set $B_i := B_{r_i}(x_i)$ so that in view of the definition of r_i

$$S_i \subset B_i \cap U.$$

Note that $|u| = 1/2$ on ∂S_i, and

$$E^\epsilon(u) \geq \frac{1}{2}|Du|^2 \geq |Ju|.$$

By (8) and (9),

$$\int_{B_i \cap U} E^\epsilon(u) \ dx \geq \left| \int_{S_i} Ju \ dx \right|$$

$$\geq \left| \int_{\partial S_i} j(u) \cdot t d\mathcal{H}^1 \right|$$

$$\geq \frac{1}{2\pi} |\deg(u; \partial S_i)|$$

$$\geq \frac{1}{2\pi}.$$

Moreover,

$$\partial B_r(x_i) \cap S_i \neq \emptyset, \quad \forall \, r \in (0, \rho_i].$$

This means that for every $r \in (0, \rho_i]$, there is a point $x^* \in \partial B_r(x_i)$ such that $|u(x^*)| < 1/2$. Due to the potential term $(1 - |u|^2)^2/4\epsilon^2$ term in the energy E^ϵ, near x^*, is large and in Lemma 2.3 below we will show that

$$\int_{\partial B_r(x_i) \cap U} E^\epsilon(u) \, d\mathcal{H}^1 \geq C \, \frac{1}{\epsilon}, \qquad \forall \, r \in [\epsilon, \rho_i],$$

for some constant C. Assume that $r_i \geq \epsilon$, and integrate this estimate over $[\epsilon, r_i]$. The result is

$$\int_{B_i \cap U} E^\epsilon(u) \, dx \geq C \frac{(r_i - \epsilon)^+}{\epsilon}.$$

Combining the two estimates,

$$\int_{B_i \cap U} E^\epsilon(u) \, dx \geq C \, \max\left\{ \frac{(r_i - \epsilon)^+}{\epsilon} \; ; \; 1 \right\} \geq \frac{C}{2} \frac{r_i}{\epsilon}.$$

The balls constructed above may not be disjoint. If two or more of these balls intersect, they can be combined into larger balls, relabeling as necessary. One can use the Besicovitch Covering Theorem to control the overlap and show that the larger balls still satisfy (16). The details of this argument appear in [12].   □

2. Annulus estimate.

In the previous step, we did not attempt to make the covering as large as possible. In particular, they could be off the size ϵ and when we add them we will not get the desired energy estimate.

In this step, we obtain an estimate which will be used when we extend the balls in our covering.

Suppose that $x^* \in U$, and $\epsilon \leq r_0 < r_1$ satisfy

$$[B_{r_1}(x^*) \setminus B_{r_0}(x^*)] \cap S_E = \emptyset,$$

and

$$dg(u; \partial B_{r_1}(x^*)) \neq 0.$$

Then, for all $r \in [r_0, r_1]$,

$$dg(u; \partial B_r(x^*)) = dg(u; \partial B_{r_1}(x^*)) \neq 0.$$

Set

$$\lambda^\epsilon(r) = \min_{m \in [0,1]} \left[\frac{m^2 \pi}{r} + \frac{(1-m)^2}{c_0 \epsilon} \right], \qquad \Lambda^\epsilon(r) := \int_0^r \lambda^\epsilon(s) \wedge \frac{c_1}{\epsilon} ds, \qquad (17)$$

for certain constants c_0, c_1 whose choice is discussed below.

Lemma 2.1. *There are constants c_0 and c_1 so that*

$$\int_{B_{r_1}(x_0) \setminus B_{r_0}(x_0)} E^\epsilon(u) dx \geq [\Lambda^\epsilon(r_1) - \Lambda^\epsilon(r_0)]. \qquad (18)$$

Proof. This is Proposition 3.2, [12]. The key estimate is

$$\int_{\partial B_r(x^*) \cap U} E^\epsilon(u) \, d\mathcal{H}^1 \geq \lambda^\epsilon(r), \quad \forall r \in [r_0, r_1]. \tag{19}$$

We then obtain (18) by integrating (19) over $r \in [r_0, r_1]$.
    Set

$$m := \min_{\partial B_r(x^*)} \{ |u(x)| \}.$$

Then, as in Lemma 2.3 below, we can prove that

$$\int_{\partial B_r(x^*) \cap U} E^\epsilon(u) \, d\mathcal{H}^1 \geq \frac{(1-m)^2}{c_0 \epsilon}.$$

For $u = |u| e^{i\varphi}$,

$$E^\epsilon(u) = \frac{1}{2} |u|^2 |\nabla \varphi|^2 + \frac{1}{2} |\nabla |u||^2.$$

Since

$$|j(u)| = |u|^2 |\nabla \varphi|,$$

$$E^\epsilon(u) = \frac{1}{2} \frac{|j(u)|^2}{|u|^2} + \frac{1}{2} |\nabla |u||^2 \geq \frac{1}{2} |u|^2 \, |j(u/|u|)|^2 \geq \frac{m^2}{2} \, |j(u/|u|)|^2.$$

Suppose that $m > 0$. Then, for $r \in [r_0, r_1]$, $\deg(u; \partial[B_r^* \cap U]) = \mathrm{dg}(u; \partial[B_r^* \cap U]) \neq 0$, and

$$\int_{\partial[B_r(x^*) \cap U]} E^\epsilon(u) \, d\mathcal{H}^1 \geq \frac{m^2}{2} \int_{\partial[B_r(x^*) \cap U]} |j(u/|u|)|^2 \, d\mathcal{H}^1$$

$$\geq \frac{m^2}{4\pi r} \left| \int_{\partial[B_r(x^*) \cap U]} j(u/|u|) \, d\mathcal{H}^1 \right|^2$$

$$= \frac{\pi \, m^2}{r} \, |\deg(u; \partial[B_r(x^*) \cap U]|^2$$

$$\geq \frac{\pi \, m^2}{r}.$$

Combining the two preceeding estimates we obtain (19).   □

3. Properties of λ^ϵ.

The following elementary estimates are proved in Propositions 3.1 and 3.2 in [12]

$$\Lambda^\epsilon(r_1 + r_2) \leq \Lambda^\epsilon(r_1) + \Lambda^\epsilon(r_2), \tag{20}$$

$$s \mapsto \frac{1}{s} \Lambda^\epsilon(s) \quad \text{is non increasing}, \quad \frac{1}{s} \Lambda^\epsilon(s) \leq \frac{c_1}{\epsilon} \quad \forall s \tag{21}$$

and $\Lambda^\epsilon(r) \geq \pi \ln(r/\epsilon) - c_2$ for some constant c_2. Also, clearly, $\lambda^\epsilon(r) \leq \pi/r$, and therefore, by redefining c_2 if necessary,

$$|\Lambda^\epsilon(r) - \pi \ln(r/\epsilon)| \leq c_2 \qquad \forall\, r \geq \epsilon. \tag{22}$$

4. Amalgamation.

Our next result is Lemma 3.1 in [12]. It is used below when we allow the small balls to grow and merge, to form large balls. For the sake of completeness, here we state it and give its short proof.

Lemma 2.2. *Given any finite collection of closed balls in $\mathbb{R}^k$, say $\{C_i\}_{i=1}^N$, we can find a collection $\{\tilde{C}_i\}_{i=1}^{\tilde{N}}$ of pairwise disjoint balls such that*

$$\bigcup_{i=1}^N C_i \subset \bigcup_{i=1}^{\tilde{N}} \tilde{C}_i,$$

$$\sum_{C_j \subset \tilde{C}_i} diam\, C_j = diam\, \tilde{C}_i,$$

$\tilde{N} \leq N$, *with strict inequality unless* $\{C_i\}_{i=1}^N$ *is pairwise disjoint.*

Proof. Replace pairs of intersecting balls C_i, C_j by larger single balls $\tilde{C}$ such that $C_i \cup C_j \subset \tilde{C}$ and diam $\tilde{C} =$ diam $C_i +$ diam C_j, continuing until a pairwise disjoint collection is reached. This collection has the stated properties. □

5. Cover of S_E.

We next show that, starting from the initial collection of balls, we can let them grow in such a way that each ball continues to satisfy a good lower bound.

As above let $\{B_i^*\}$ denote the balls found in Proposition 2.1, with radii r_i^* and generalized degree $d_i^* := \mathrm{dg}(u; \partial[B_i^* \cap U])$. Define

$$\sigma^* := \min\{\, r_i^* \mid d_i^* \neq 0 \,\}.$$

The idea is to extend the balls with the smallest radius with nonzero degree until they hit each other or the boundary of U. When they hit each other we use the amalgamation lemma and continue the process until one of the balls with nonzero degree hits the boundary of U. However, here we follow the presentation of Sandier and Serfaty [27]. Although this is slightly more technical than the outline procedure, it extends very easily to more general situations.

Proposition 2.2. *For every* $\sigma \geq \sigma^*$, *there exists a collection of disjoint, closed balls* $\mathcal{B}(\sigma) = \{B_k^\sigma\}_{k=1}^{k(\sigma)}$ *satisfying* $r_k^\sigma \geq \epsilon$,

$$S_E \subset \cup_k B_k^\sigma , \tag{23}$$

$$\int_{U \cap B_k^\sigma} E^\epsilon(u) \, dx \geq \frac{r_k^\sigma}{\sigma} \Lambda^\epsilon(\sigma) , \tag{24}$$

$$r_k^\sigma \geq \sigma \qquad whenever \ B_k^\sigma \cap \partial U = \emptyset \ , and \ d_k^\sigma \neq 0, \tag{25}$$

where r_k^σ *is the radius and* $d_k^\sigma = dg(u; \partial[B_k^\sigma \cap U])$ *is the generalized degree.*

Proof. Let C be the set of all $\sigma \geq \sigma^*$ for which such a collection exists.

1. We first claim that $\sigma^* \in C$. Indeed $\{B_k^*\}$ be the collection of balls constructed in Proposition 2.1. Set $\mathcal{B}(\sigma^*) := \{B_k^*\}$. The definition 17 of Λ^ϵ easily implies that $\Lambda^\epsilon(\sigma)/\sigma \leq c_1/\epsilon$ forall σ, so Proposition 2.1 implies that this collection satisfies (23) and (24). Also, (25) is satisfied due to the definition of σ^*.

2. In step we will show that C is closed. Let $\{\sigma^n\}_n$ be a sequence in C and suppose that σ^n converges to σ_0 as n tends to infinity. Since the balls are disjoint, and their radii are at least ϵ, the total number of balls $k(\sigma_n)$ is uniformly bounded in n. Therefore by passing to a subsequence we may assume that $k(\sigma_n)$ is equal to a constant k_0 independent of n. By passing to a further subsequence, we may assume that the radii $r_k^{\sigma^n}$ and the centers $a_k^{\sigma^n}$ converge to r_k^0 and a_k^0, respectively, for each $k \leq k_0$. Let $B_{k,0}$ be the closed ball centered at a_k^0 with radius r_k^0. It is clear that this collection of balls satisfies (23), (24), and (25). If the balls are disjoint , we set $\mathcal{B}(\sigma_0) := \{B_{k,0}\}_{k=1}^{k_0}$. If they are not disjoint, we apply the amalgamation process outlined in Lemma 2.2. Let $\{B_j^{\sigma_0}\}$ be the resulting balls and $r_j^{\sigma_0}$ be their radius. Then, by Lemma 2.2,

$$r_j^{\sigma_0} = \sum_{B_{k,0} \subset B_j^{\sigma_0}} r_{k,0}. \tag{26}$$

Since $\{B_{k,0}\}$ satisfies (24), this implies that

$$\int_{U \cap B_j^{\sigma_0}} E^\epsilon(u) \, dx \geq \sum_{B_{k,0} \subset B_j^{\sigma_0}} \int_{U \cap B_{k,0}} E^\epsilon(u) \, dx$$

$$\geq \sum_{B_{k,0} \subset B_j^{\sigma_0}} \frac{r_{k,0}}{\sigma_0} \Lambda^\epsilon(\sigma_0)$$

$$= \frac{r_j^{\sigma_0}}{\sigma_0} \Lambda^\epsilon(\sigma_0) .$$

Hence, $\mathcal{B}(\sigma_0) := \{B_j^{\sigma_0}\}$ satisfies (24). Moreover,

$$|d_j^{\sigma_0}| = \left| \sum_{B_{k,0} \subset B_j^{\sigma_0}} d_{k,0} \right| \leq \sum_{B_{k,0} \subset B_j^{\sigma_0}} |d_{k,0}|.$$

Hence, if $B_j^{\sigma_0} \cap \partial U = \emptyset$ and $d_j^{\sigma_0} \neq 0$, then $B_{k,0} \cap \partial U = \emptyset$ for all $B_{k,0} \subset B_j^{\sigma_0}$ and at least one $d_{k^*,0} \neq 0$. Since $B_{k^*,0}$ satisfies (25),

$$r_j^{\sigma_0} \geq r_{k^*,0} \geq \sigma_0.$$

This implies that the balls in the collection $\mathcal{B}(\sigma_0)$ satisfy (25).

3. Suppose that $\sigma_1 \in C$. We will show that there is $\delta > 0$ such that $[\sigma_1, \sigma_1 + \delta] \subset C$. Indeed, let

$$K_1 := \{ k \mid B_k^{\sigma_1} \cap \partial U = \emptyset \text{ and } d_k^{\sigma_1} \neq 0 \},$$

and set

$$s_1 := \min_{k \in K_1} \{ r_k^{\sigma_1} \}.$$

By (25), $\sigma_1 \leq s_1$. If this inequality is strict, we set $\mathcal{B}(\sigma) = \mathcal{B}(\sigma_1)$ for all $\sigma \in [\sigma_1, s_1]$. It is clear that this collection of balls satisfies (23), and (25). Also (24) follows from (21). So let us assume that $s_1 = \sigma_1$, and let

$$K_2 := \{ k \in K_1 \mid s_1 = r_k^{\sigma_1} \}.$$

For $\sigma \geq \sigma_1$, set

$$r_k^\sigma := \begin{cases} r_k^{\sigma_1}, & \text{if } k \notin K_2, \\ \dfrac{\sigma}{\sigma_1} r_k^{\sigma_1}, & \text{if } k \in K_2. \end{cases}$$

Let B_k^σ be the closed ball with radius r_k^σ with the same center as $B_k^{\sigma_1}$ and let $\mathcal{B}(\sigma)$ be the collection of these balls. Since $\{B_k^{\sigma_1}\}_k$ are disjoint closed sets, there is $\delta_1 > 0$ such that for all $\sigma \in [\sigma_1, \sigma_1 + \delta_1]$ B_k^σ's are disjoint and

$$K_1(\sigma) := \{ k \mid B_k^\sigma \cap \partial U = \emptyset \text{ and } d_k^\sigma \neq 0 \} = K_1.$$

Then, for $k \in K_2$,

$$\frac{r_k^\sigma}{\sigma} = \frac{r_k^{\sigma_1}}{\sigma_1} = 1,$$

and for $k \notin K_2$,

$$\frac{r_k^\sigma}{\sigma} = \frac{\sigma_1}{\sigma} \frac{r_k^{\sigma_1}}{\sigma_1}.$$

Since for $k \notin K_2$, $r_k^{\sigma_1} > \sigma_1$, there is $0 < \delta \leq \delta_1$ such that (25) is satisfied by the collection $\mathcal{B}(\sigma)$. Since $r_k^\sigma \geq r_k^{\sigma_1}$, (23) is also satisfied.

To verify (24), we observe that for $k \notin K_2$, $B_k^\sigma = B_k^{\sigma_1}$ and (24) is satisfied in light of (21). If, however, $k \in K_2$, then

$$d_k^\sigma = d_k^{\sigma_1}, \qquad r_k^\sigma = \sigma, \tag{27}$$

and

$$[B_k^\sigma \setminus B_k^{\sigma_1}] \cap S^E = \emptyset .$$

Then by (18),

$$\int_{B_k^\sigma} E^\epsilon(u) \, dx = \int_{B_k^{\sigma_1}} E^\epsilon(u) \, dx + \int_{B_k^\sigma \setminus B_k^{\sigma_1}} E^\epsilon(u) \, dx$$

$$\geq \frac{r_k^{\sigma_1}}{\sigma_1} \Lambda^\epsilon(\sigma_1) + [\Lambda^\epsilon(r_k^\sigma) - \Lambda^\epsilon(r_k^{\sigma_1})]$$

$$= \Lambda^\epsilon(r_k^{\sigma_1}) + [\Lambda^\epsilon(r_k^\sigma) - \Lambda^\epsilon(r_k^{\sigma_1})]$$

$$= \Lambda^\epsilon(r_k^\sigma)$$

$$= \frac{r_k^\sigma}{\sigma} \Lambda^\epsilon(\sigma) .$$

Here we repeatedly used the identities (27) and the fact that $B_k^{\sigma_1}$ satisfies (24). Hence $\mathcal{B}(\sigma)$ also satisfies (24) for all $\sigma \in [\sigma_1, \sigma_1 + \delta]$.

4. We have shown that C is closed set including σ^* and for every $\sigma \in C$, there exists $\delta > 0$ such that $[\sigma, \sigma + \delta] \subset C$. Hence, $C = [\sigma^*, \infty)$. $\square$

6. Proof of Theorem 2.1

Let σ^* be as in the previous Lemma and let r_0 be as in (10).

1. First suppose that $r_0 < 2\sigma^*$. The opposite inequality will be treated later in the proof.

Consider the balls $\{B_k^*\}$ constructed in Proposition 2.1. Set

$$K^* := \{ k \mid \text{and } d_k^* \neq 0 \}.$$

Recall that $d_k^* = \mathrm{dg}(u; \partial[B_k^* \cap U])$. Since

$$0 \neq \deg(u; \partial U) = \sum_k \mathrm{dg}(u; B_k^*),$$

K^* is nonempty. Then, by (16) and the definition of σ^*,

$$\int_U E^\epsilon(u) \, dx \geq \sum_k \int_{U \cap B_k^*} E^\epsilon(u) \, dx \geq \sum_k \frac{c_1}{\epsilon} r_k^*$$

$$\geq \frac{c_1 \sigma^*}{\epsilon} \sum_{k \in K^*} 1 \geq c_1 \frac{r_0}{2\epsilon} |K^*| \geq c_1 \frac{r_0}{2\epsilon}$$

$$\geq \Lambda^\epsilon\left(\frac{r_0}{2}\right).$$

In view of (22), this gives the desired lower bound.

2. We now assume that $r_0 \geq 2\sigma^*$. Set $\bar{\sigma} := r_0/2$ and consider the collection of balls $\mathcal{B}(\bar{\sigma})$ provided by Proposition 2.2. Let

$$\bar{K} := \{\, k \mid d_k^{\bar{\sigma}} \neq 0 \,\}.$$

Since $\deg(u; \partial U) \neq 0$, $\bar{K}$ is nonempty. We claim that

$$r_k^{\bar{\sigma}} \geq \bar{\sigma} = \frac{r_0}{2}, \qquad \forall\, k \in \bar{K}.$$

Indeed if $B_k^{\bar{\sigma}} \cap \partial U = \emptyset$ for some $k \in \bar{K}$, then this follows from (25). Suppose that $B_k^{\bar{\sigma}} \cap \partial U \neq \emptyset$ for some $k \in \bar{K}$. Since $d_k^{\bar{\sigma}} \neq 0$, $B_k^{\bar{\sigma}}$ contains a zero of u, and by (10),

$$B_k^{\bar{\sigma}} \cap [U \setminus U_{r_0}] \neq \emptyset.$$

Since, by assumption, $B_k^{\bar{\sigma}} \cap \partial U \neq \emptyset$, this implies that the diameter of $B_k^{\bar{\sigma}}$ is at least r_0, proving the claim.

3. By the previous step

$$\sum_{k \in \bar{K}} r_k^{\bar{\sigma}} \geq \bar{\sigma} = \frac{r_0}{2}.$$

Hence by (24),

$$\begin{aligned}
\int_U E^\epsilon(u)\, dx &\geq \sum_{k \in \bar{K}} \int_{U \cap B_k^{\bar{\sigma}}} E^\epsilon(u)\, dx \\
&\geq \sum_{k \in \bar{K}} r_k^{\bar{\sigma}} \frac{\Lambda^\epsilon(\bar{\sigma})}{\bar{\sigma}} \\
&\geq \Lambda^\epsilon(\bar{\sigma}) \,.
\end{aligned}$$

$$\square$$

2.3 Main lower bound

In this section, we prove a generalization of the lower bound. This sharper lower bound is taken from [13] and its proof is very similar to that of Theorem 2.1. Here we repeat the arguments of the previous section with minor modifications for the sake of completeness.

Let U be a bounded open subset of $\mathbb{R}^2$, and $u \in H^1(U; \mathbb{R}^2)$ be a function that we have assumed (without loss of generality) to be smooth. In addition, ϕ is a nonnegative Lipschitz test function that vanishes on ∂U.

Throughout this section we will use the notation

$$T = \|\phi\|_\infty = \max_U \phi(x). \tag{28}$$

Given $\phi \in C_c^{0,1}(U)$ we use the notation

Reg(ϕ) for the set

$$\{ t \in [0, T] : \ \partial\Omega(t) = \phi^{-1}(t), \ \partial\Omega(t) \text{ is rectifiable}, \ \mathcal{H}^1(\partial\Omega(t)) < \infty \}. \tag{29}$$

The coarea formula implies that $\text{Reg}(\phi)$ is a set of full measure. For every $t \in \text{Reg}(\phi)$, $\partial\Omega(t)$ is a union of finite Jordan curves $\Gamma_i(t)$, i.e.,

$$\partial\Omega(t) = \cup_i \Gamma_i(t) , \qquad \forall\, t \in \text{Reg}(\phi).$$

In particular, this holds for almost every t. For $t \in \text{Reg}(\phi)$ we define

$$\Gamma(t) = \cup \left\{ \text{ components } \Gamma_i(t) \text{ of } \partial\Omega(t) \mid \min_{x \in \Gamma_i(t)} |u(x)| > 1/2 \right\}. \qquad (30)$$

We also define $\gamma(t) = \partial\Omega(t) \setminus \Gamma(t)$,

$$\gamma(t) = \cup \left\{ \text{ components } \Gamma_i(t) \text{ of } \partial\Omega(t) \mid \min_{x \in \Gamma_i(t)} |u(x)| \le 1/2 \right\}. \qquad (31)$$

When we want to indicate explicitly the dependence of $\Gamma(t)$ on ϕ and u, we will write $\Gamma_{\phi,u}(t)$.

We will also use the notation

$$t_\epsilon := \epsilon \|\nabla\phi\|_\infty. \qquad (32)$$

For any positive integer d, let

$$D_d^\epsilon := \{ t \in \text{Reg}(\phi) : t \ge t_\epsilon, \Gamma(t) \text{ is nonempty, and } |\deg(u; \Gamma(t))| \ge d \}. \qquad (33)$$

The main result of this section is the following theorem. We follow very closely arguments introduced in [12].

Theorem 2.2 (Jerrard & Soner [13]). *If $u : U \to \mathbb{R}^2$ is a smooth function and ϕ is a nonnegative Lipschitz function such that $\phi = 0$ on ∂U, then for any positive integer d,*

$$\int_{spt(\phi)} E^\epsilon(u) \ge d\Lambda^\epsilon \left(\frac{|D_d^\epsilon|}{2d\|\nabla\phi\|_\infty} \right).$$

For any $t_2 > t_1$ the ratio $(t_2 - t_1)/\|\nabla\phi\|_\infty$ is a lower bound for the distance between $\partial\Omega(t_2)$ and $\partial\Omega(t_1)$. This explains the role of $\|\nabla\phi\|_\infty$ in the estimate.

Similar results were proven in [12] under more or less the assumption that D_d^ϵ is an interval; and in [7] in the case $d = 1$. Related results have also appeared in Sandier [26].

Note that the case covered in the statement of the theorem, $\{x : \phi(x) > 0\} \subset U$, can be reduced to the case $\{x : \phi(x) > 0\} = U$, if we replace U by $\tilde{U} := \{x \in U : \phi(x) > 0\}$. So we will henceforth assume for notational simplicity that this holds, so that $spt(\phi) = \tilde{U}$.

For the proof of Theorem 2.2, we define

$$S_E^\epsilon := \cup\{\text{components } S_i \text{ of } S_E \ : \ S_i \subset \Omega(t_\epsilon)\}. \tag{34}$$

If $x \in \Omega(t_\epsilon)$ and $y \in \partial U$, then

$$|x - y| \, \|\nabla\phi\|_\infty \ge |\phi(x) - \phi(y)| = |\phi(x)| \ge t_\epsilon = \epsilon\|\nabla\phi\|_\infty.$$

In particular,

$$\text{dist}(x, \partial U) \ge \epsilon \qquad \text{for all } x \in S_E^\epsilon. \tag{35}$$

Note also that if $V \subset \Omega(t_\epsilon)$ and $\partial V \cap S_E = \emptyset$, then

$$\text{dg}(u; \partial V) := \sum \{\deg(u; \partial S_i) \mid \text{components } S_i \text{ of } S_E^\epsilon \text{ such that } S_i \subset\subset V\}.$$

In other words, for such sets V we can ignore $S_E \setminus S_E^\epsilon$ when computing $\text{dg}(u; \partial V)$. In the proof of Theorem 2.2 below we will always be concerned with subsets $V \subset \Omega(t_\epsilon)$, so this will always be the case.

Our strategy for proving Theorem 2.2 is very similar to the method we used to prove Theorem 2.1. We will first find a collection of balls such we have a good lower bound for the Ginzburg-Landau energy on each ball. We then show that the sum of the radii of the balls is bounded below by $|D_d^\epsilon|/(2\|\nabla\phi\|_\infty)$, hence obtaining a lower bound for the total Ginzburg-Landau energy in terms of this quantity. We start with a technical step.

1. If $\gamma(t) \ne \emptyset$.

In this step, we prove an estimate in the case when one of the level sets intersects with the zero set. In this case, $|u|$ falls below $1/2$ on $\gamma(t)$ and we expect the Ginzburg-Landau energy to be large on $\gamma(t)$. The following technical lemma proves this under the assumption that $\gamma(t)$ is not too small.

Lemma 2.3. *Suppose that*

$$\mathcal{H}^1(\gamma(t)) \ge \epsilon \ .$$

Then

$$\int_{\partial\Omega(t)} E^\epsilon(u) \, d\mathcal{H}^1 \ge \frac{1}{25\epsilon} \ . \tag{36}$$

Proof. This is very similar to Lemma 2.3 in [12]. Fix a connected component $\Gamma_i(t)$ of $\gamma(t)$ and set $\rho := |u|$ and

$$\gamma_i := \int_{\Gamma_i(t)} \frac{1}{2}|\nabla\rho|^2 \, d\mathcal{H}^1 \ .$$

By the definition (31) of $\gamma(t)$ there is a point $x_{min} \in \Gamma_i(t)$ such that $\rho(x_{min}) \le 1/2$. Parametrize $\Gamma_i(t)$ by arclength so that

$$\Gamma_i(t) = \{\, x(s) \mid s \in [0, G_i] \}\,, \qquad G_i := \mathcal{H}^1(\Gamma_i(t))\,,$$

with $x_{\min} = x(0) = x(G_i)$. Then since $|\dot{x}(s)| = 1$,

$$\rho(x(s)) = \rho(x(0)) + \int_0^s \nabla\rho(x(r)) \cdot \dot{x}(r)\ dr$$

$$\leq \frac{1}{2} + s^{1/2} \left(\int_0^s |\nabla\rho(x(r))|^2 dr \right)^{1/2}$$

$$\leq \frac{1}{2} + \sqrt{\gamma_i\, s} \leq \frac{3}{4}\,,$$

provided that $s \leq \sigma_i := [G_i \wedge 1/(16\gamma_i)]$. Then, for $s \in [0, \sigma_i]$, $(1 - \rho^2(x(s)))^2/4 \geq 1/25$. Therefore,

$$\int_{\Gamma_i(t)} E^\epsilon(u)\ d\mathcal{H}^1 \geq \gamma_i + \int_{\Gamma_i(t)} \frac{1}{4\epsilon^2}(1 - \rho^2)^2\ d\mathcal{H}^1$$

$$\geq \gamma_i + \frac{\sigma_i}{25\,\epsilon^2}\,.$$

By calculus,

$$\gamma_i + \frac{\sigma_i}{25\,\epsilon^2} = \gamma_i + \frac{G_i \wedge (1/4\gamma_i)}{25\,\epsilon^2} \geq \frac{1}{25\epsilon} \left[\frac{G_i}{\epsilon} \wedge 5 \right]\,.$$

Thus

$$\int_{\Gamma_i(t)} E^\epsilon(u)\ d\mathcal{H}^1 \geq \frac{1}{25\epsilon} \left[\frac{G_i}{\epsilon} \wedge 5 \right]\,.$$

Since

$$\mathcal{H}^1(\gamma(t)) = \sum_{\{i \mid \Gamma_i(t) \text{ is a component of } \gamma(t)\}} \mathcal{H}^1(\Gamma_i(t)) = \sum_i G_i \geq \epsilon\,,$$

we can sum over components $\Gamma_i(t)$ of $\gamma(t)$ to conclude that

$$\int_{\partial\Omega(t)} E^\epsilon(u)\ d\mathcal{H}^1 \geq \frac{1}{25\epsilon}\,.$$

$\square$

2. First Cover.

This is very similar to Step 1 of the previous subsection.

We find the collection of balls by starting from an initial collection of small balls that cover S_E^ϵ, then letting these balls grow by expanding them and combining them. The first step is thus to establish the existence of the initial collection of small balls. This is the content of

Proposition 2.3. *There is a collection of closed, pairwise disjoint balls* $\{B_i^*\}_{i=1}^k$ *with radii* r_i^* *such that*

$$S_E^\epsilon \subset \cup_{i=1}^k B_i^*, \tag{37}$$

$$r_i^* \geq \epsilon \qquad \forall i, \tag{38}$$

$$\int_{B_i^* \cap U} E^\epsilon(u)dx \geq \frac{c_0}{\epsilon}r_i^* \geq \Lambda^\epsilon(r_i^*). \tag{39}$$

This is essentially proved in the previous subsection; Proposition 2.1 and in Proposition 3.3 in [12].

Proposition 2.3 differs from Proposition 2.1 in that in the latter, S_E appears in place of S_E^ϵ in the counterpart of (37).

3. Cover of S_E^ϵ.

This step is similar to Step 5 of the previous section.

Starting from the initial collection of balls, we can let them grow in such a way that each ball continues to satisfy a good lower bound.

As above let $\{B_i^*\}$ denote the balls found in Proposition 2.3, with radii r_i^* and generalized degree $d_i^* := dg(u; \partial[B_i^* \cap U])$. Define

$$\sigma^* := \min\{ \frac{r_i^*}{|d_i^*|} \mid d_i^* \neq 0 \} .$$

Proposition 2.4. *For every* $\sigma \geq \sigma^*$, *there exists a collection of disjoint, closed balls* $\mathcal{B}(\sigma) = \{B_k^\sigma\}_{k=1}^{k(\sigma)}$ *satisfying* $r_k^\sigma \geq \epsilon$,

$$S_E^\epsilon \subset \cup_k B_k^\sigma , \tag{40}$$

$$\int_{U \cap B_k^\sigma} E^\epsilon(u) \, dx \geq \frac{r_k^\sigma}{\sigma} \Lambda^\epsilon(\sigma) , \tag{41}$$

$$r_k^\sigma \geq \sigma|d_k^\sigma| \qquad \text{whenever } B_k^\sigma \cap \partial U = \emptyset , \tag{42}$$

where r_k^σ *is the radius and* d_k^σ *is the generalized degree.*

The proof of this proposition is very similar to that of Proposition 2.2. The only difference is we consider the ratio $r_k^\sigma/|d_k^\sigma|$ to decide which balls to expand.

Proof. Let C be the set of all $\sigma \geq \sigma^*$ for which such a collection exists.

1. We first claim that $\sigma^* \in C$. Indeed $\{B_k^*\}$ be the collection of balls constructed in Proposition 2.3. Set $\mathcal{B}(\sigma^*) := \{B_k^*\}$. The definition (17) of Λ^ϵ easily implies that $\Lambda^\epsilon(\sigma)/\sigma \leq c_1/\epsilon$ forall σ, so Proposition 2.3 implies

that this collection satisfies (40) and (41). Also, (42) is satisfied due to the definition of σ^*.

2. In step we will show that C is closed. Let $\{\sigma^n\}_n$ be a sequence in C and suppose that σ^n converges to σ_0 as n tends to infinity. Since the balls are disjoint, and their radii are at least ϵ, the total number of balls $k(\sigma_n)$ is uniformly bounded in n. Therefore by passing to a subsequence we may assume that $k(\sigma_n)$ is equal to a constant k_0 independent of n. By passing to a further subsequence, we may assume that the radii $r_k^{\sigma^n}$ and the centers $a_k^{\sigma^n}$ converge to r_k^0 and a_k^0, respectively, for each $k \leq k_0$. Let $B_{k,0}$ be the closed ball centered at a_k^0 with radius r_k^0. It is clear that this collection of balls satisfies (40), (41), and (42). If the balls are disjoint , we set $\mathcal{B}(\sigma_0) := \{B_{k,0}\}_{k=1}^{k_0}$. If they are not disjoint, we apply the amalgamation process outlined in Lemma 2.2. Let $\{B_j^{\sigma_0}\}$ be the resulting balls and $r_j^{\sigma_0}$ be their radius. Then, by Lemma 2.2,

$$r_j^{\sigma_0} = \sum_{B_{k,0} \subset B_j} r_{k,0} \geq \sum_{B_{k,0} \subset B_j} \sigma_0 |d_{k,0}|. \tag{43}$$

Since $\{B_{k,0}\}$ satisfies (41), this implies that

$$\int_{U \cap B_j} E^\epsilon(u) \, dx \geq \sum_{B_{k,0} \subset B_j} \int_{U \cap B_{k,0}} E^\epsilon(u) \, dx$$

$$\geq \sum_{B_{k,0} \subset B_j} \frac{r_{k,0}}{\sigma_0} \Lambda^\epsilon(\sigma_0)$$

$$\geq \frac{r_j^{\sigma_0}}{\sigma_0} \Lambda^\epsilon(\sigma_0) \ .$$

Hence, $\mathcal{B}(\sigma_0) := \{B_j^{\sigma_0}\}$ satisfies (41). Moreover,

$$|d_j^{\sigma_0}| = \left| \sum_{B_{k,0} \subset B_j} d_{k,0} \right| \leq \sum_{B_{k,0} \subset B_j} |d_{k,0}|,$$

and this together with (43) implies that the balls in the collection $\mathcal{B}(\sigma_0)$ satisfy (42).

3. Suppose that $\sigma_1 \in C$. We will show that there is $\delta > 0$ such that $[\sigma_1, \sigma_1 + \delta] \subset C$. Indeed, let K_1 be the set of indices k such that $B_k^{\sigma_1} \cap \partial U = \emptyset$ and set

$$s_1 := \min_{k \in K_1} \frac{r_k^{\sigma_1}}{|d_k^{\sigma_1}|} \ .$$

By (42), $\sigma_1 \leq s_1$. If this inequality is strict, we set $\mathcal{B}(\sigma) = \mathcal{B}(\sigma_1)$ for all $\sigma \in [\sigma_1, s_1]$. It is clear that this collection of balls satisfies (40), and (42). Also (41) follows from (21). So let us assume that $s_1 = \sigma_1$, and let $K_2 \subset K_1$ be the indices k which minimize the ratio $r_k^{\sigma_1}/d_k^{\sigma_1}$. For $\sigma \geq \sigma_1$, set

$$r_k^\sigma := \begin{cases} r_k^{\sigma_1} \,, & \text{if } k \notin K_2 \,, \\[2mm] \dfrac{\sigma}{\sigma_1} \, r_k^{\sigma_1} \,, & \text{if } k \in K_2 \,. \end{cases}$$

Let B_k^σ be the closed ball with radius r_k^σ with the same center as $B_k^{\sigma_1}$ and let $\mathcal{B}(\sigma)$ be the collection of these balls. Since $\{B_k^{\sigma_1}\}_k$ are disjoint closed sets, there is $\delta_1 > 0$ such that for all $\sigma \in [\sigma_1, \sigma_1 + \delta_1]$ B_k^σ's are disjoint and

$$K_1(\sigma) := \{\, k \mid B_k^\sigma \cap \partial U = \emptyset \,\} = K_1 \,.$$

Then, for $k \in K_2$,

$$\frac{r_k^\sigma}{\sigma} = \frac{r_k^{\sigma_1}}{\sigma_1} = |d_k^{\sigma_1}| = |d_k^\sigma| \,,$$

and for $k \notin K_2$,

$$\frac{r_k^\sigma}{\sigma} = \frac{\sigma_1}{\sigma} \frac{r_k^{\sigma_1}}{\sigma_1} \,.$$

Since for $k \notin K_2$, $r_k^{\sigma_1}/\sigma_1 > |d_k^{\sigma_1}|$, there is $0 < \delta \le \delta_1$ such that (42) is satisfied by the collection $\mathcal{B}(\sigma)$. Since $r_k^\sigma \ge r_k^{\sigma_1}$, (40) is also satisfied.

To verify (41), we observe that for $k \notin K_2$, $B_k^\sigma = B_k^{\sigma_1}$ and (41) is satisfied in light of (21). If, however, $k \in K_2$, then

$$d_k^\sigma = d_k^{\sigma_1} \,, \qquad\qquad r_k^\sigma = \sigma|d_k^\sigma| \,, \qquad\qquad (44)$$

and

$$[B_k^\sigma \setminus B_k^{\sigma_1}] \cap S^E = \emptyset \,.$$

Then by Lemma 2.1

$$\begin{aligned} \int_{B_k^\sigma} E^\epsilon(u) \, dx &= \int_{B_k^{\sigma_1}} E^\epsilon(u) \, dx + \int_{B_k^\sigma \setminus B_k^{\sigma_1}} E^\epsilon(u) \, dx \\ &\ge \frac{r_k^{\sigma_1}}{\sigma_1} \Lambda^\epsilon(\sigma_1) + |d_k^{\sigma_1}| \left[\Lambda^\epsilon\left(\frac{r_k^\sigma}{|d_k^\sigma|}\right) - \Lambda^\epsilon\left(\frac{r_k^{\sigma_1}}{|d_k^{\sigma_1}|}\right) \right] \\ &= |d_k^{\sigma_1}| \, \Lambda^\epsilon\left(\frac{r_k^{\sigma_1}}{|d_k^{\sigma_1}|}\right) + |d_k^{\sigma_1}| \left[\Lambda^\epsilon\left(\frac{r_k^\sigma}{|d_k^\sigma|}\right) - \Lambda^\epsilon\left(\frac{r_k^{\sigma_1}}{|d_k^{\sigma_1}|}\right) \right] \\ &= |d_k^{\sigma_1}| \, \Lambda^\epsilon\left(\frac{r_k^\sigma}{|d_k^\sigma|}\right) \\ &= \frac{r_k^\sigma}{\sigma} \Lambda^\epsilon(\sigma) \,. \end{aligned}$$

Here we repeatedly used the identities (44) and the fact that $B_k^{\sigma_1}$ satisfies (41). Hence $\mathcal{B}(\sigma)$ also satisfies (41) for all $\sigma \in [\sigma_1, \sigma_1 + \delta]$.

4. We have shown that C is closed set including σ^* and for every $\sigma \in C$, there exists $\delta > 0$ such that $[\sigma, \sigma + \delta] \subset C$. Hence, $C = [\sigma^*, \infty)$. $\square$

We are now ready for the

Proof of Theorem 2.2. Set $R := |D_d^\epsilon|/(2\|\nabla\phi\|_\infty)$ and $\bar\sigma := R/d$. Let σ^* be as in the previous Lemma. We suppose that D_d^ϵ is nonempty as there is nothing to prove otherwise .

  1. First suppose that $\bar\sigma < \sigma^*$. The opposite inequality will be treated later in the proof.

Consider the balls $\{B_k^*\}$ constructed in Proposition 2.3. By (39) and the definition of σ^*,

$$\int_U E^\epsilon(u)\,dx \geq \sum_k \int_{U\cap B_k^*} E^\epsilon(u)\,dx$$

$$\geq \sum_k \frac{c_1}{\epsilon}\, r_k^* \geq \frac{c_1\sigma^*}{\epsilon} \sum_k |d_k^*| \geq c_1 \frac{R}{d\epsilon} \sum_k |d_k^*|\ .$$

Let $t_0 \in D_d^\epsilon$. Then the definition (33) of D_d^ϵ implies that $d \leq |\deg(u;\Gamma(t_0))|$ and by definition, (30), $|u| > 1/2$ on $\Gamma(t_0)$. Hence $d \leq |\mathrm{dg}(u;\Gamma(t_0))|$. Moreover, by (13) and (37),

$$d \leq |\mathrm{dg}(u;\Gamma(t_0))| \leq \sum_{\{k\,:\,B_k^*\cap\Omega(t_0)\neq\emptyset\}} |d_k^*| \leq \sum_k |d_k^*|\ .$$

Hence, by (21),

$$\int_U E^\epsilon(u)\,dx \geq c_1 \frac{R}{d\epsilon}\, d \geq d\Lambda^\epsilon\left(\frac{R}{d}\right)\ ,$$

which is what we needed to prove.

  2. We now assume that $\bar\sigma \geq \sigma^*$. Consider the collection of balls $\mathcal{B}(\bar\sigma)$ provided by Proposition 2.4. Assume towards a contradiction that

$$\sum_k r_k^{\bar\sigma} < R\ . \tag{45}$$

Set

$$C := \{\, t \in (0,\|\phi\|_\infty) \mid \Gamma(t) \cap [\cup_k B_k^{\bar\sigma}] \neq \emptyset \,\}\ .$$

The definitions imply that $C \subset \cup_k \phi(B_k^{\bar\sigma})$, and, as a consequence,

$$|C| \leq 2\|\nabla\phi\|_\infty \sum_k r_k^{\bar\sigma} < 2\|\nabla\phi\|_\infty R = |D_d^\epsilon|\ .$$

Hence $D_d^\epsilon \setminus C \neq \emptyset$.

  3. Let $t_0 \in D_d^\epsilon \setminus C$. The definition of D_d^ϵ implies that $|\mathrm{dg}(u;\Gamma(t_0))| = |\deg(u;\Gamma(t_0)| \geq d$. On the other hand, the definition of C implies that $\Gamma(t_0)\cap(\cup_k B_k^{\bar\sigma}) = \emptyset$, so (40) and the additivity of the degree yield

$$d \leq |dg(u; \Gamma(t_0))| \leq \sum_{\{k \,:\, B_k^{\bar{\sigma}} \subset \Omega(t_0)\}} |d_k^{\bar{\sigma}}|$$

$$\leq \sum_{\{k \,:\, B_k^{\bar{\sigma}} \cap \partial U = \emptyset\}} |d_k^{\bar{\sigma}}|$$

$$\leq \sum_{\{k \,:\, B_k^{\bar{\sigma}} \cap \partial U = \emptyset\}} \frac{r_k^{\bar{\sigma}}}{\bar{\sigma}} \,,$$

by (42). On the other hand, by (45),

$$d = \frac{R}{\bar{\sigma}} > \sum_k \frac{r_k^{\bar{\sigma}}}{\bar{\sigma}} \,.$$

Therefore we conclude that (45) is false.

4. By the previous step $\sum_k r_k^{\bar{\sigma}} \geq R = d\bar{\sigma}$. Hence, by (41),

$$\int_U E^\epsilon(u) \, dx \geq \sum_k \int_{U \cap B_k^{\bar{\sigma}}} E^\epsilon(u) \, dx$$

$$\geq \sum_k r_k^{\bar{\sigma}} \frac{\Lambda^\epsilon(\bar{\sigma})}{\bar{\sigma}}$$

$$\geq d \, \Lambda^\epsilon(\bar{\sigma}) \,.$$

$\square$

3 Jacobian and the GL Energy

In this section, we show that the Jacobian is bounded by the GL energy. This estimate is the crucial step in a Gamma convergence result.

Results of this section are taken from [13].

3.1 Jacobian estimate

The chief result of this section is the following estimate of the Jacobian in terms of the Ginzburg Landau energy. This estimate will be the main ingredient in the compactness result. We give a more precise version of the estimate at the end of the section.

We use the notation introduced in the previous section and set

$$\mu^\epsilon(V) := \mu_u^\epsilon(V) = \frac{1}{\ln(1/\epsilon)} \int_V E^\epsilon(u) \, dx.$$

Theorem 3.1 (Jerrard & Soner [13]). *Suppose $\phi \in C_c^{0,1}(U)$ and $u \in H^1(U; \mathbb{R}^2)$. For any $\lambda \in (1, 2]$, and $\epsilon \in (0, 1]$,*

$$\left| \int_U \phi\, Ju\, dx \right| \le \pi d_\lambda \|\phi\|_\infty + \|\phi\|_{C^{0,1}} h^\epsilon(\phi, u, \lambda), \qquad (46)$$

where

$$d_\lambda = \left\lfloor \frac{\lambda}{\pi} \mu_u^\epsilon(spt(\phi)) \right\rfloor, \qquad (47)$$

$\lfloor x \rfloor$ *denotes the greatest integer less than or equal to x, and*

$$h^\epsilon(\phi, u, \lambda) \le C\epsilon^{\alpha(\lambda)}(1 + \mu_u^\epsilon(spt(\phi)))(1 + Leb^2(spt(\phi))), \qquad (48)$$

where $\alpha(\lambda) = \frac{\lambda - 1}{12\lambda}$ and C is a constant independent of u, ϕ, ϵ, λ and U.

Note that h^ϵ depends on ϕ only through the support of ϕ, and on u only through its (linear) dependence on $\mu_u^\epsilon(spt(\phi))$.

It suffices to consider nonnegative test functions, since we can decompose an arbitrary function ϕ into its positive and negative parts. So we will assume that $\phi \ge 0$.

By an approximation argument, we may also assume that u is smooth.

The main idea behind the above estimate is the following identity, which relies on the co-area formula, integration by parts, and the identity $Ju = \nabla \times j(u)/2$

$$\int_U \phi\, Ju\, dx = \frac{1}{2} \int_0^T \int_{\partial\Omega(t)} j(u) \cdot \mathbf{t}\, d\mathcal{H}^1\, dt, \qquad (49)$$

where, as before,

$$\Omega(t) = \{\, x \in U \mid \phi(x) > t\, \},$$

$$\mathbf{t} = \text{unit tangent to } \partial\Omega(t) = \frac{\nabla \times \phi}{|\nabla \times \phi|}.$$

The proof shows that

$$\int_{\partial\Omega(t)} j(u) \cdot \mathbf{t}\, d\mathcal{H}^1 \approx 2\pi \deg(u; \partial\Omega(t)),$$

for most values of t. The other main point is then to prove that the set of t such that $\deg(u; \partial\Omega(t)) > d_\lambda$ has Leb^1 measure that can be controlled by $\mu^\epsilon(spt(\phi))$. This last point is similar in spirit to results established in [7, 12, 25] for example. We use the lower bounds obtained in the previous section to achieve this.

We start the proof of Theorem 3.1 with two simple estimates.

Lemma 3.1. *For any set A,*

$$\int_A \left| \int_{\partial\Omega(t)} j(u) \cdot \mathbf{t}\, d\mathcal{H}^1 \right| dt \le \frac{|A|}{2} \int_{spt(\phi)} E^\epsilon(u)\, dx. \qquad (50)$$

For any nonnegative function f,

$$\int_0^T \int_{\partial\Omega(t)} f(x) \, d\mathcal{H}^1 \, dt \leq \|\nabla\phi\|_\infty \int_{spt(\phi)} f(x) \, dx \; . \tag{51}$$

Proof. For any $t \in \text{Reg}(\phi)$, Stokes' Theorem yields

$$\int_{\partial\Omega(t)} j(u) \cdot \mathbf{t} \, d\mathcal{H}^1 = \frac{1}{2} \int_{\Omega(t)} Ju \, dx \; .$$

Since $|Ju| \leq \frac{1}{2}|\nabla u|^2 \leq E^\epsilon(u)$, (50) follows from the above identity.
For (51), we calculate, by using the coarea formula,

$$\int_0^T \int_{\partial\Omega(t)} f \, d\mathcal{H}^1 \, dt = \int_{spt(\phi)} f \, |\nabla\phi| \, dx$$

$$\leq \|\nabla\phi\|_\infty \int_{spt(\phi)} f \, dx.$$

$\square$

We are now in a position to prove Theorem 3.1. In the proof we repeatedly absorb logarithmic factors by using the fact that, if $\beta < \alpha$, then

$$\epsilon^\alpha \ln(1/\epsilon) \leq C\epsilon^\beta,$$

for some $C = C(\alpha, \beta)$ independent of $\epsilon \in (0, 1]$.

Proof of Theorem 3.1.
1. Recall that we are writing $T = \|\phi\|_\infty$. Fix $\lambda \in (1, 2]$ and define $d_\lambda := \lfloor \frac{\lambda}{\pi}\mu^\epsilon(\text{spt}(\phi)) \rfloor$. We define sets $A, B \subset [0, T]$ by

$$B := \{ t \in \text{Reg}(\phi) \; : \; |\deg(u; \Gamma(t))| \geq d_\lambda + 1 \text{ or } \mathcal{H}^1(\gamma(t)) \geq \epsilon \}, \tag{52}$$

$$A = \text{Reg}(\phi) \setminus B. \tag{53}$$

Because almost every t belongs to $A \cup B = \text{Reg}(\phi)$, (49) implies that

$$\int_U \phi \, Ju \, dx = \frac{1}{2} \int_A \int_{\Gamma(t)} j(u) \cdot \mathbf{t} \, d\mathcal{H}^1 \, dt$$

$$+ \frac{1}{2} \int_A \int_{\gamma(t)} j(u) \cdot \mathbf{t} \, d\mathcal{H}^1 \, dt + \frac{1}{2} \int_B \int_{\partial\Omega(t)} j(u) \cdot \mathbf{t} \, d\mathcal{H}^1 \, dt$$

$$= I_{A,\Gamma} + I_{A,\gamma} + I_B. \tag{54}$$

2. **Estimate of $I_{A,\Gamma}$**
Suppose $t \in A$. On $\Gamma(t)$, $|u| \geq 1/2$ by the definition (30), and we set $v := u/|u|$, so that $j(v) = j(u)/|u|^2$, and

$$\int_{\Gamma(t)} j(v) \cdot \mathbf{t} \, d\mathcal{H}^1 = 2\pi \, \deg(u; \Gamma(t)).$$

Then

$$\int_{\Gamma(t)} j(u) \cdot \mathbf{t} \, d\mathcal{H}^1 = 2\pi \, \deg(u; \Gamma(t)) + \int_{\Gamma(t)} j(u) \, \frac{|u|^2 - 1}{|u|^2} \cdot \mathbf{t} \, d\mathcal{H}^1 \ .$$

Since $|j(u) \le |u| \, |\nabla u|$, and since $|u| \ge 1/2$ on $\Gamma(t)$, Cauchy's inequality and (51) imply that

$$\int_A \left| \int_{\Gamma(t)} j(u) \cdot \mathbf{t} \, d\mathcal{H}^1 - 2\pi \deg(u; \Gamma(t)) \right| \, dt \le \int_A \int_{\Gamma(t)} |\nabla u| \, \left| \frac{|u|^2 - 1}{|u|} \right| \, d\mathcal{H}^1$$

$$\le 4\epsilon \int_A \int_{\Gamma(t)} E^\epsilon(u) \, d\mathcal{H}^1$$

$$\le 4\epsilon \ln(1/\epsilon) \, \|\nabla \phi\|_\infty \, \mu^\epsilon(spt(\phi)) \ .$$

$$(55)$$

Clearly $A \subset [0, T]$ has measure less than $T = \|\phi\|_\infty$. Also, by definition of A, if $t \in A$ and $\Gamma(t)$ is nonempty, then $|\deg(u; \Gamma(t))| \le d_\lambda$. It follows that

$$|I_{A,\Gamma}| \le \pi \|\phi\|_\infty d_\lambda + C\epsilon^{1/2} \|\nabla \phi\|_\infty \, \mu^\epsilon(spt(\phi)). \tag{56}$$

3. Estimate of $I_{A,\gamma}$

Using Cauchy's inequality and the elementary fact that $x \le \frac{1}{b}(1 - x)^2 + (1 + \frac{b}{4})$ for all $x \in \mathbb{R}$ and $b > 0$, we have

$$|j(u)| \le |u||\nabla u| \le \frac{a}{2} \left(|\nabla u|^2 + \frac{1}{a^2}|u|^2 \right) \le \frac{a}{2} \left(|\nabla u|^2 + \frac{(1 - |u|^2)^2}{a^2 b} \right) + \frac{1}{2a}(1 + \frac{b}{4}) \ ,$$

for every $a, b > 0$. We select $a = \epsilon^\alpha$ for $\alpha \in (0, 1)$ and $b = \epsilon^{2 - 2\alpha}$ to find

$$|j(u)| \le C\epsilon^\alpha E^\epsilon(u) + C\epsilon^{-\alpha} \ . \tag{57}$$

The definition (53) of A implies that $|A| \le T = \|\phi\|_\infty$ and that $\mathcal{H}^1(\gamma(t)) < \epsilon$ for every $t \in A$, so we can take $\alpha = 1/2$ and use (51) to find

$$|I_{A,\gamma}| \le C \int_A \int_{\gamma(t)} \sqrt{\epsilon} E^\epsilon(u) d\mathcal{H}^1(x) dt \ + C \int_A \int_{\gamma(t)} \frac{C}{\sqrt{\epsilon}} d\mathcal{H}^1(dx) dt$$

$$\le C \ln(1/\epsilon) \sqrt{\epsilon} \, \mu^\epsilon(spt(\phi)) \|\nabla \phi\|_\infty \ + C\sqrt{\epsilon} \|\phi\|_\infty \ .$$

4. Estimate of I_B

To estimate I_B we prove that B has small measure. Toward this end we define

$$B_1 := \{t \in \mathrm{Reg}(\phi) \ : \ \mathcal{H}^1(\gamma(t)) \ge \epsilon\},$$

$$B_2 := \{t \in \mathrm{Reg}(\phi) \ : \ \Gamma(t) \text{ is nonempty, and } |\deg(u; \Gamma(t))| \ge d_\lambda + 1\}. \tag{58}$$

The estimate of B_2 is deferred to the end of this subsection, where we prove

Proposition 3.1. *For every* $\lambda \in (1,2]$, $\epsilon \in (0,1]$, *smooth* $u : U \to I\!\!R^2$, *and nonnegative test function* $\phi \in C_c^{0,1}(U)$,

$$|B_2| \leq C\epsilon^{1-\frac{1}{\lambda}}\|\nabla\phi\|_\infty(d_\lambda + 1) \leq C\epsilon^{1-\frac{1}{\lambda}}\|\nabla\phi\|_\infty(1 + \mu^\epsilon(spt(\phi)). \quad (59)$$

For the time being we assume this fact and use it to complete the proof of the theorem.

The measure of B_1 is easily estimated: using (51) and Lemma 2.3,

$$\frac{1}{25\epsilon}|B_1| \leq \int_{t\in B_1} \int_{\partial\Omega(t)} E^\epsilon(u)d\mathcal{H}^1 dt$$

$$\leq \|\nabla\phi\|_\infty \ln(\frac{1}{\epsilon})\mu^\epsilon(spt(\phi)). \quad (60)$$

Clearly $|B| \leq |B_1| + |B_2|$, so, by combining (60) and (59), we obtain

$$|B| \leq C\epsilon^{\frac{\lambda-1}{2\lambda}}\|\nabla\phi\|_\infty(1 + \mu^\epsilon(spt(\phi))). \quad (61)$$

Finally, we use (50) to estimate

$$|I_B| \leq C\epsilon^{\frac{\lambda-1}{3\lambda}}\|\nabla\phi\|_\infty(1 + \mu^\epsilon(spt(\phi)))\mu^\epsilon(spt(\phi)). \quad (62)$$

5. The previous three steps imply that

$$\left|\int \phi Ju \, dx\right| \leq d_\lambda\|\phi\|_\infty + \|\phi\|_{C^1}h_0^\epsilon(\phi, u, \lambda),$$

for

$$h_0^\epsilon(\phi, u, \lambda) \leq C\epsilon^{4\alpha(\lambda)}\left(1 + \mu^\epsilon(spt(\phi)) + (\mu^\epsilon(spt(\phi)))^2\right), \qquad \alpha(\lambda) = \frac{\lambda-1}{12\lambda}.$$

To complete the proof of the Theorem, note that by (51) and (57) (with $\alpha = 2\alpha(\lambda)$)

$$\left|\int \phi Ju \, dx\right| \leq \int_0^T \int_{\partial\Omega(t)} |j(u)|d\mathcal{H}^1 dt$$

$$\leq C\|\nabla\phi\|_\infty \int_{spt(\phi)} \epsilon^{2\alpha(\lambda)}E^\epsilon(u) + \epsilon^{-2\alpha(\lambda)} \, dx$$

$$\leq C\|\phi\|_{C^1}h_1^\epsilon(\phi, u, \lambda),$$

for $h_1^\epsilon = \epsilon^{\alpha(\lambda)}\mu^\epsilon(spt(\phi)) + \epsilon^{-2\alpha(\lambda)}\text{Leb}^2(spt(\phi))$. We define $h^\epsilon(\phi, u, \lambda) := \min\{h_0^\epsilon, h_1^\epsilon\}$, so that (46) clearly holds. It thus suffices to verify that (48) holds, that is,

$$h^\epsilon(\phi, u, \lambda) = \min\{h_0^\epsilon, h_1^\epsilon\} \leq C\epsilon^{\alpha(\lambda)}(1 + \mu^\epsilon(spt(\phi))(1 + \text{Leb}^2(spt(\phi))),$$

for some appropriately large constant C. This follows immediately from the definition of h_0^ϵ if $\mu^\epsilon(\text{spt}(\phi)) \leq \epsilon^{-3\alpha(\lambda)}$, and if not, it follows directly from the definition of h_1^ϵ. $\qquad\square$

Note that the result we have proved is in fact somewhat sharper than Theorem 3.1 as stated, in that it not only provides an upper bound for $\int \phi Ju$, but in fact gives an approximate value for the integral. The following corollary states a small technical modification of this sharper estimate.

Corollary 3.1. *Let U be a bounded, open subset of $\mathbb{R}^2$, and suppose that $\phi \in C_c^{0,1}(U)$ and $u \in H^1(U; \mathbb{R}^2)$. Define $Reg(\phi)$, $\Gamma(t)$ and $\gamma(t)$ as in (29), (30) and (31) respectively.*

Then for any $\lambda \in (1, 2]$ and $\epsilon \in (0, 1]$, there exists a set $A = A(\phi, u, \lambda, \epsilon) \subset (0, \|\phi\|_\infty)$ such that

$$|A| \geq \|\phi\|_\infty - C\epsilon^{\alpha(\lambda)}\|\nabla\phi\|_\infty(1 + \mu^\epsilon(spt(\phi))); \tag{63}$$

$$\Gamma(t) \text{ is nonempty, and } |deg(u; \Gamma(t))| \leq d_\lambda \qquad \forall t \in A; \text{ and} \tag{64}$$

$$\left|\int \phi Ju - \pi \int_{t \in A} deg(u; \Gamma(t))\, dt\right| \leq \|\phi\|_{C^1} h^\epsilon(\phi, u, \lambda), \tag{65}$$

where h^ϵ is defined in (48) and d_λ is defined in (47).

Proof. We cannot take A to be the set defined in (53), as we have now imposed the additional condition that $\Gamma(t) \neq \emptyset$ for $t \in A$. So we let $\tilde{A}$ be the set formerly known as A, defined in (53), and we define

$$A = \{t \in \tilde{A} \ : \ \Gamma(t) \text{ is nonempty}\}.$$

Then (64) follows from the definition of $\tilde{A}$, and (65) follows from (55). We claim moreover that $\tilde{A} \setminus A$ has measure at most $\epsilon\|\nabla\phi\|_\infty$. In view of (61) and (53), this will suffice to establish (63), and thus to complete the proof of the Corollary.

To prove our claim, note first that for every $t \in \tilde{A}$, $\mathcal{H}^1(\gamma(t)) < \epsilon$. If $t \in \tilde{A} \setminus A$, then $\Gamma(t)$ is empty, and so $\mathcal{H}^1(\phi^{-1}(t)) = \mathcal{H}^1(\gamma(t)) < \epsilon$ for all $t \in \tilde{A} \setminus A$. On the other hand, let $x_0 \in U$ be a point such that $\phi(x_0) = \|\phi\|_\infty$. If $|y - x_0| \leq \epsilon$ then $\phi(y) \geq \|\phi\|_\infty - \epsilon \|\nabla\phi\|_\infty$. It follows that $B_\epsilon(x_0) \subset \Omega(t)$ for all $t < \|\phi\|_\infty - \epsilon \|\nabla\phi\|_\infty$. Thus the isoperimetric inequality implies that $\mathcal{H}^1(\phi^{-1})(t) \geq 2\pi\epsilon$.

We conclude that if $t \in \tilde{A} \setminus A$, then $t \geq \|\phi\|_\infty - \epsilon\|\nabla\phi\|_\infty$, which proves the claim. $\qquad\square$

We now use Theorem 2.2 and the facts about Λ^ϵ to give the proof of Proposition 3.1.

Recall that for Proposition 3.1 we want to estimate the measure of a set $B_2 \subset Reg(\phi)$, and from the definition (58) of B_2 we see that

$$D^\epsilon_{d^*_\lambda} = B_2 \cap \{t \; : \; t \geq t_\epsilon\}, \quad \text{for } d^*_\lambda := \lfloor \tfrac{\lambda}{\pi} \mu^\epsilon(\mathrm{spt}(\phi)) \rfloor + 1 \geq \tfrac{\lambda}{\pi} \mu^\epsilon(\mathrm{spt}(\phi)).$$

$$(66)$$

Proof of Proposition 3.1. We need to show that

$$|B_2| \leq C\epsilon^{1-\frac{1}{\lambda}} \|\nabla\phi\|_\infty d^*_\lambda, \qquad d^*_\lambda := d_\lambda + 1.$$

Let $R := \frac{|D^\epsilon_{d^*_\lambda}|}{2\|\nabla\phi\|_\infty}$. From (66) and the definition (32) of t_ϵ it suffices to show that

$$\frac{R}{d^*_\lambda} \leq C\epsilon^{1-\frac{1}{\lambda}}.$$

We may assume that $\frac{R}{d^*_\lambda} \geq \epsilon$, as otherwise the conclusion is obvious. Then (22), Theorem 2.2, and the choice (66) of d^*_λ imply that

$$
\begin{aligned}
\ln\left(\frac{R}{d^*_\lambda}\right) &= \frac{1}{\pi}\left[\pi\ln\left(\frac{R}{\epsilon d^*_\lambda}\right) - \pi\ln\left(\frac{1}{\epsilon}\right)\right] \\
&\leq \frac{1}{\pi}\Lambda^\epsilon\left(\frac{R}{d^*_\lambda}\right) + C - \ln\left(\frac{1}{\epsilon}\right) \\
&\leq \frac{1}{\pi d^*_\lambda}\ln\left(\frac{1}{\epsilon}\right)\mu^\epsilon(\mathrm{spt}(\phi)) + C - \ln\left(\frac{1}{\epsilon}\right) \\
&\leq \left(\frac{1}{\lambda} - 1\right)\ln\left(\frac{1}{\epsilon}\right) + C.
\end{aligned}
$$

$\square$

3.2 Compactness in two dimensions

In this section we consider a sequence of functions $u^\epsilon \in H^1(U; \mathbb{R}^2)$, where U is a bounded open subset of $\mathbb{R}^2$ and the renormalized Ginzburg-Landau energy is uniformly bounded:

$$K_U := \sup_{\epsilon\in(0,1]} \mu^\epsilon(U) < \infty, \qquad \mu^\epsilon := \mu^\epsilon_{u^\epsilon}. \tag{67}$$

We will show that under this assumption, the Jacobian is compact in the dual norm $(C^{0,\beta})^*$ for every $\beta \in (0,1]$. We refer to §5 and to [13] for a compactness in higher dimensions.

We introduce the Jacobian (signed) measure

$$Ju^\epsilon(E) := \int_E \det(\nabla u^\epsilon) \; dx, \qquad E \subset U.$$

Since $\det(\nabla u^\epsilon) = \frac{1}{2}\nabla \times j(u^\epsilon)$ for $j(u^\epsilon) := u^\epsilon \times \nabla u^\epsilon$,

$$\int_{\mathbb{R}^2} \phi \, dJu^\epsilon = \frac{1}{2} \int_{\mathbb{R}^2} \nabla \times \phi(x) \cdot j(u^\epsilon)(x) \, dx \, , \qquad \forall \phi \in C^1_c(U),$$

where for a scalar function ϕ, we write $\nabla \times \phi := (\phi_{x_2}, -\phi_{x_1})$.

Theorem 3.2. *Let $\{u^\epsilon\} \subset H^1(U; \mathbb{R}^2)$ satisfy (67). Then there exists a subsequence ϵ_n converging to zero and a signed Radon measure J such that Ju^{ϵ_n} converges to J in the dual norm $\left(C_c^{0,\beta}\right)^*$ for every $\beta \in (0, 1]$. Moreover, there are $\{a_i\}_{i=1}^N \subset U$ and integers k_i such that*

$$J = \pi \sum_{i=1}^N k_i \, \delta_{a_i} \, , \qquad and \qquad |J|(U) = \pi \sum_i |k_i| \leq K_U \, .$$

Finally, if μ^ϵ converges weakly to a limit μ, then $J \ll \mu$, and $\frac{dJ}{d\mu}(x) \leq 1$ for μ almost every x.

We will first prove

Proposition 3.2. *Assume (67). Then, Ju^ϵ can be written in the form*

$$Ju^\epsilon = J_0^\epsilon + J_1^\epsilon$$

where J_0^ϵ and J_1^ϵ are signed measures such that

$$\|J_0^\epsilon\|_{(C^0)^*} \leq C, \quad and \qquad \|J_1^\epsilon\|_{(C_c^{0,1})^*} \leq C\epsilon^\alpha \qquad (68)$$

for some $\alpha > 0$ and a constant C depending only on the constant K_U in (67).

Proof.

1. In light of the assumption $\mu^\epsilon(U) \leq K$, Theorem 3.1 (with $\lambda = 2$ and $\alpha = 1/24$, for example) implies that

$$\int \phi J u^\epsilon \leq C\|\phi\|_\infty + C\epsilon^\alpha \|\nabla \phi\|_\infty \qquad \text{for all } \phi \in C_c^{0,1}(U). \qquad (69)$$

We write $\delta = \epsilon^\alpha$, and we define $U_\delta = \{x \in U : \text{dist}(x, \partial U) > \delta\}$. Let

$$\chi_\delta = \begin{cases} 1 & \text{if } x \in U_{2\delta} \\ 0 & \text{if not.} \end{cases}$$

We define $J_0^\epsilon := \chi_\delta(\eta^\delta * Ju^\epsilon)$, where η^δ is a standard mollifier with support in $B_\delta(0)$. We then define $J_1^\epsilon := Ju^\epsilon - J_0^\epsilon$.

Suppose that ϕ is a C^1 test function vanishing on ∂U, and note that

$$\int \phi \, J_0^\epsilon dx = \int \eta^\delta * (\chi_\delta \phi) Ju^\epsilon dx.$$

We write $\phi^\delta := \eta^\delta * (\chi_\delta \phi)$. It is clear that ϕ^δ is compactly supported in U, and one easily checks that

$$\|\phi^\delta\|_\infty \leq \|\chi_\delta \phi\|_\infty \leq \|\phi\|_\infty, \qquad \|\nabla\phi^\delta\|_\infty \leq \frac{C}{\delta}\|\chi_\delta\phi\|_\infty \leq \frac{C}{\delta}\|\phi\|_\infty.$$

Then (69) implies that

$$\int \phi\, J_0^\epsilon dx \leq C\|\phi\|_\infty.$$

2. We now estimate J_1^ϵ. Given $\phi \in C_0^1(U)$, write

$$\phi_1 := \min\{\phi, 2\delta\|\nabla\phi\|_\infty\}, \qquad \phi_2 := \phi - \phi_1.$$

It is clear that $\phi \leq 2\delta\|\nabla\phi\|_\infty$ in $U \setminus U_{2\delta}$, so ϕ_2 is supported in $U_{2\delta}$.
  From the definitions,

$$\int \phi_1 J_1^\epsilon\, dx = \int (\phi_1 - \eta^\delta * (\chi_\delta\phi_1)) Ju^\epsilon\, dx.$$

It is clear that

$$\|\phi_1\|_\infty \leq 2\delta\|\nabla\phi\|_\infty, \qquad \|\nabla\phi_1\|_\infty \leq \|\nabla\phi\|_\infty.$$

Similarly, $\eta^\delta * (\chi_\delta\phi_1)$ satisfies

$$\|\eta^\delta * (\chi_\delta\phi_1)\|_\infty \leq 2\delta\|\nabla\phi\|_\infty, \qquad \|\nabla\eta^\delta * (\chi_\delta\phi_1)\|_\infty \leq \frac{C}{\delta}\|\phi_1\|_\infty \leq C\|\nabla\phi\|_\infty.$$

So (69) implies that

$$\int \phi_1 J_1^\epsilon\, dx \leq C\delta\|\nabla\phi\|_\infty = C\epsilon^\alpha\|\nabla\phi\|_\infty.$$

Finally, since ϕ_2 is supported in $U_{2\delta}$,

$$\int \phi_2 J_1^\epsilon\, dx = \int (\phi_2 - \eta^\delta * (\chi_\delta\phi_2)) Ju^\epsilon\, dx = \int (\phi_2 - \eta^\delta * \phi_2) Ju^\epsilon\, dx.$$

It is easy to check that

$$\|\phi_2 - \eta^\delta * \phi_2\|_\infty \leq C\delta\|\nabla\phi\|_\infty, \qquad \|\nabla(\phi_2 - \eta^\delta * \phi_2)\|_\infty \leq C\|\nabla\phi\|_\infty.$$

So we again use (69) to conclude

$$\int \phi_2 J_1^\epsilon\, dx \leq C\delta\|\nabla\phi\|_\infty = C\epsilon^\alpha\|\nabla\phi\|_\infty.$$

$\square$

Once we have the above decomposition, the compactness of the sequence Ju^ϵ follows from soft arguments.

Lemma 3.2. *If ν is a Radon measure on U, then*

$$\|\nu\|_{(C_c^{0,\alpha})^*} \leq C\|\nu\|_{(C_c^{0,1})^*}^{\alpha}\|\nu\|_{(C_c^0)^*}^{1-\alpha}. \tag{70}$$

Proof. Since U is bounded and we are considering compactly supported functions, the Hölder seminorm is in fact a norm and is topologically equivalent to the usual $C^{0,\alpha}$ norm. So for this lemma we set

$$\|\phi\|_{C_c^{0,\alpha}(U)} := [u]_{C^{0,\alpha}} = \sup_{x \neq y} \frac{|\phi(x) - \phi(y)|}{|x - y|^\alpha}, \qquad \alpha \in (0, 1].$$

Fix $\phi \in C_c^{0,\alpha}$, and let $\tilde\phi^\epsilon = \eta^\epsilon * \phi$, where η^ϵ is a smoothing kernel and ϵ will be chosen later. Then one easily checks that

$$\|\tilde\phi^\epsilon\|_{C^{0,1}} \leq C\epsilon^{\alpha-1}\|\phi\|_{C^{0,\alpha}} := M_\epsilon, \qquad \|\phi - \tilde\phi^\epsilon\|_{C^0} \leq C\epsilon^\alpha\|\phi\|_{C^{0,\alpha}}. \tag{71}$$

In particular, $|\tilde\phi^\epsilon| \leq C\epsilon^\alpha\|\phi\|_{C^{0,\alpha}}$ on ∂U.

We next modify $\tilde\phi^\epsilon$ so that it vanishes on ∂U while continuing to satisfy the above estimates. Let

$$u(x) = \sup_{y \in \partial U} \left(\tilde\phi^\epsilon(y) - M_\epsilon|x - y|\right)^+, \qquad v(x) = \sup_{y \in \partial U} \left(\tilde\phi^\epsilon(y) + M_\epsilon|x - y|\right)^-.$$

Then one easily checks that $\tilde\phi^\epsilon = u - v$ on ∂U. Moreover, if we define $\phi^\epsilon := \tilde\phi^\epsilon - u + v$, then ϕ^ϵ satisfies the estimates in (71) and also vanishes on ∂U.

So we have

$$\begin{aligned}
\int \phi d\nu &= \int \phi^\epsilon \, d\mu + \int (\phi - \phi^\epsilon) \, d\nu \\
&\leq \|\phi^\epsilon\|_{C^{0,1}}\|\nu\|_{(C_c^{0,1})^*} + \|\phi - \phi^\epsilon\|_{C^0}\|\nu\|_{(C^0)^*} \\
&\leq C\|\phi\|_{C^{0,\alpha}} \left(\epsilon^{\alpha-1}\|\nu\|_{(C_c^{0,1})^*} + \epsilon^\alpha\|\nu\|_{(C^0)^*}\right).
\end{aligned}$$

Taking $\epsilon = \|\nu\|_{(C_c^{0,1})^*}/\|\nu\|_{(C^0)^*}$ gives the conclusion of the lemma. $\qquad\square$

Lemma 3.3. *If $\alpha > 0$, then $(C^0)^* \subset\subset (C^{0,\alpha})^*$.*

Proof. The Arzela-Ascoli Theorem implies that any sequence that is bounded on $C^{0,\alpha}$ is precompact in C^0. The lemma follows by duality.

More concretely: given a sequence of measures bounded in $(C^0)^*$, we can extract a subsequence, say μ_n that converges to a limit μ in the weak-* topology. We must show that this sequence converges in norm in $(C^{0,\alpha})^*$. If not, then we can find a sequence of functions ψ_n with $\|\psi_n\|_{C^{0,\alpha}} \leq 1$ such that

$$\int \psi_n d(\mu_n - \mu) \geq c_0 > 0, \tag{72}$$

for all n. However, the Arzela-Ascoli theorem implies that, upon passing to a subsequence, ψ_n converges to some limit ψ uniformly, whence (72) is impossible. $\qquad\square$

We now prove

Theorem 3.3. *Assume (67). Then Ju^ϵ is strongly precompact in $(C^{0,\beta})^*$ for all $\beta > 0$.*

Proof. By Proposition 3.2 we can write $Ju^\epsilon = J_0^\epsilon + J_1^\epsilon$, where the two measures on the right-hand side satisfy (68).

Fix any $\beta \in (0, 1]$. Lemma 3.3 implies that $\{J_0^\epsilon\}$ is precompact in $(C^{0,\beta})^* \subset (C_c^{0,\beta})^*$.

Also, it is clear from the definitions that

$$\|J_1^\epsilon\|_{(C^0)^*} \leq \|Ju^\epsilon\|_{L^1} + \|J_0^\epsilon\|_{(C^0)^*} \leq C\|\nabla u^\epsilon\|_{L^2}^2 + C \leq K \ln(\frac{1}{\epsilon}).$$

So together with (68) and the interpolation inequality (70) this implies that $\|J_1^\epsilon\|_{(C_c^{0,\beta})^*} \to 0$ as $\epsilon \to 0$. $\qquad\qquad\square$

Remark 3.1. The above result is sharp in the sense that Ju^ϵ need not be precompact, or even weakly precompact, in $(C^0)^*$. To see this, consider the sequence of functions

$$u^\epsilon(x, y) = (1, 0) + \epsilon^2 (\ln(\frac{1}{\epsilon}))^{1/2} (\cos(\frac{x}{\epsilon^2}), \sin(\frac{y}{\epsilon^2}))$$

on the open unit disk D in the plane. One easily verifies that $\mu^\epsilon(D) \leq C$, and that $\|Ju^\epsilon\|_{(C^0)^*} = \|Ju^\epsilon\|_{L^1} \geq c^{-1} \ln(\frac{1}{\epsilon})$. In particular, since $\|Ju^\epsilon\|_{(C^0)^*}$ is unbounded, the Uniform Boundedness Principle implies that the sequence cannot converge weakly in $(C^0)^*$.

Remark 3.2. Suppose ν^ϵ is any sequence of measures on a bounded open set $U \subset \mathbb{R}^m$, and that

$$|\nu^\epsilon|(U) \leq K \ln(\frac{1}{\epsilon}), \qquad \int \phi d\nu^\epsilon \leq C\|\phi\|_\infty + C\epsilon^\alpha \|\nabla \phi\|_\infty,$$

for some $\alpha > 0$. The arguments given above then show, with essentially no change, that $\{\nu^\epsilon\}$ is precompact in $(C^{0,\beta})^*$ for all $\beta \in (0, 1]$.

We are now in a position to give the

Proof of Theorem 3.2. Suppose $\{u^\epsilon\}_{\epsilon \in (0,1]} \subset H^1(U; \mathbb{R}^2)$ is a sequence satisfying (67). By an approximation argument, we may assume that in fact each u^ϵ is smooth. In view of Theorem 3.3, we can find a measure J and a subsequence ϵ_n such that $Ju^{\epsilon_n} \to J$ in $(C_c^{0,\beta})^*$ for every $\beta \in (0, 1]$.

1. Since μ^{ϵ_n} is a sequence of uniformly bounded, nonnegative Radon measures, we may assume upon passing to a further subsequence (still labeled ϵ_n) that there is a Radon measure μ such that

$$\mu_n := \mu^{\epsilon_n} \overset{*}{\to} \mu,$$

in the weak* topology of Radon measures in U. For $x \in U$, set

$$\Theta(x) := \lim_{r \downarrow 0} \mu\left(B_r(x) \cap U\right) .$$

We first claim that J is supported only on the points with $\Theta(x) \geq \pi$.

Indeed, suppose that $\Theta(x_0) < \pi$ at some $x_0 \in U$. Then there exists some $r_0 > 0$ and a number $\alpha < \pi$ such that

$$\mu_n(B_{r_0}(x_0)) \leq \alpha < \pi$$

for all sufficiently large n. Then Theorem 3.1 with $\lambda = \frac{a+\pi}{2a} > 1$ immediately implies that

$$\int \phi \, dJ(x) = \lim_{n \to \infty} \int \phi \, Ju^{\epsilon_n} \, dx = 0$$

for all smooth ϕ with support in $B_{r_0}(x_0)$, since $d_\lambda = 0$ for such ϕ. Thus $x_0 \notin \mathrm{spt}(J)$.

Since μ is bounded on U, there are finitely many points $\{a_i\}_i \subset U$ such that

$$\Theta(a_i) \geq \pi .$$

Therefore there are constants c_i such that the limit measure J satisfies

$$J = \pi \sum_i c_i \, \delta_{a_i} .$$

We need to prove that c_i's are integers and that $\pi|c_i| \leq \Theta(a_i)$ for all i; this will immediately imply all the remaining conclusions of Theorem 3.2.

2. Choose $r_1 \leq 1$ so that $B_{r_1}(a_1)$ does not intersect $\{a_i\}_{i>1} \cup \partial U$. We may also assume, taking r_1 smaller if necessary, that there exists some $\lambda > 1$ and an integer N_0 such that

$$d_\lambda := \lfloor \frac{\lambda}{\pi} \mu_n(B_{r_1}(a_1)) \rfloor \leq \frac{1}{\pi} \Theta(a_1) \qquad \forall n \geq N_0. \tag{73}$$

We first apply Theorem 3.1 to the function $\phi(x) := (r_1 - |x - a_1|)^+$, which is supported in $B_{r_1}(a_1)$. Let $A^n = A(\phi, u^{\epsilon_n}, \lambda, \epsilon_n)$ be the set whose existence is asserted in Theorem 3.1. Note that if $t \in A^n$, then $\Gamma_{\phi,u^{\epsilon_n}}(t)$ is nonempty, which is to say that there is a component of $\phi^{-1}(t)$ on which $\min |u| \geq 1/2$. However, $\phi^{-1}(t) = \partial B_{r_1 - t}(a_1)$ is connected, so in fact $\Gamma_{\phi,u^{\epsilon_n}}(t) = \partial B_{r_1 - t}(a_1)$ for all $t \in A^n$. So for every $t \in A^n$ and $n \geq N_0$, Theorem 3.1 and the choice of λ imply that

$$\min_{x \in \partial B_{r_1 - t}(a_1)} |u^{\epsilon_n}| \geq \frac{1}{2}, \qquad |\deg(u^{\epsilon_n}; \partial B_{r_1 - t}(a_1))| \leq d_\lambda \leq \frac{1}{\pi} \Theta(a_1).$$

It follows that for all such n there is an integer $d(n)$ such that the set

$$S_n^{d(n)} := \{r \in [0, r_1] : \min_{\partial B_r(a_i)} |u^{\epsilon_n}| > \frac{1}{2}, \deg(u^{\epsilon_n}; \partial B_r) = d(n)\}$$

has measure at least $k_0 := r_1/(3d_\lambda)$. Note also that $S_n^{d(n)}$ is open, since u^{ϵ_n} is by assumption continuous (indeed, smooth). We can therefore find an open set $\Sigma_n \subset S_n^{d(n)}$ such that $|\Sigma_n| = k_0$.

3. We now define new test functions ψ^n as follows. First let

$$f^n(r) = |[r, r_1] \cap \Sigma_n|.$$

We then define $\psi^n(x) = f^n(|x - a_1|)$. One can then check that t is a regular value of ψ^n if and only if

$$(\psi^n)^{-1}(t) = \partial B_r(a_1) \quad \text{for some } r \in \Sigma_n.$$

In particular, $\deg(u; (\psi^n)^{-1}(t)) = d(n)$ for a.e. $0 < t < \|\psi^n\|_\infty = k_0$.

One can then easily check, using Theorem 3.1, that

$$\int \psi^n J u^{\epsilon_n} \, dx = \pi d(n) k_0 + O(\epsilon^\alpha).$$

On the other hand, since the functions ψ^n are uniformly bounded in $C_c^{0,1}$ and since $J u^{\epsilon_n} \to J = \pi \sum c_i \delta_{a_i}$ in $C_c^{0,1}(U)^*$,

$$0 = \lim_n \left| \int \psi^n J u^{\epsilon_n} \, dx - \pi c_1 \psi^n(a_1) \right| = \lim_n \left| \int \psi^n J u^{\epsilon_n} \, dx - \pi c_1 k_0 \right|.$$

Comparing the last two equations, we find that $d(n) = c_1$ for all sufficiently large n. In particular, c_1 is an integer and $|c_1| \leq d_\lambda \leq \frac{1}{\pi} \Theta(a_1)$, which is what we needed to show. $\square$

4 Gamma Limit

Suppose that a sequence of functionals J_n on a Banach space X is given. We assume that they are bounded from below and by adding a constant, if necessary, we may assume that these functionals are nonnegative, i.e.,

$$J_n : X \mapsto [0, \infty].$$

The Gamma limit, J, of these functional in the topology of X is roughly given by

$$J(x) := \liminf\{ J_n(x_n) \,|\, (n, x_n) \to (\infty, x) \}.$$

More precisely, the Gamma limit is a nonnegative functional

$$J : X \mapsto [0, \infty],$$

satisfying the following two conditions

– Let $\{x_n\} \subset X$ be a convergent sequence with limit $x \in X$, i.e.,

$$\lim_{n \to \infty} \|x_n - x\|_X = 0.$$

Then,

$$\liminf_{n \to \infty} J_n(x_n) \geq J(x).$$

– For any $x \in X$, there exists a sequence $\{x_n^*\}$ satisfying

$$\lim_{n \to \infty} \|x_n^* - x\|_X = 0,$$

and

$$\lim_{n \to \infty} J_n(x_n^*) = J(x).$$

Generally an accompanying compactness result is useful. Such a compactness result states that, if for a sequence $\{x_n\} \subset X$

$$\sup_n J_n(x_n) < \infty, \tag{74}$$

then the set $\{x_n\}$ is compact in X.

An immediate corollary to a Gamma limit and to the compactness result is this: Let y_n^* be a minimizer of J_n, and they satisfy (74). Then, by compactness on a subsequence, denoted by n again, y_n converges to a point $y^* \in X$. By the first condition on the Gamma limit we know that

$$J(y^*) \leq \liminf_n J_n(y_n).$$

Let x be any point in X and let $\{x_n^*\}$ be the sequence in the second condition on the Gamma limit. Since y_n's minimize J_n,

$$\begin{aligned} J(x) &= \lim_n J_n(x_n) \\ &\geq \limsup_n J_n(y_n) \\ &\geq J(y^*). \end{aligned}$$

Hence, any limit y^* of the minimizing sequence y_n is a minimizer of the Gamma limit J. In practice, this method is used to a construct minimizer of a given functional J. Given J, we construct a "regular" functionals J_n such that the Gamma limit of this sequence is J. Then, we obtain minimizers of J_n by standard methods and then apply the above method to construct a minimizer of J. The sequence J_n is often called a *relaxation* of J.

We refer to the book of Dal Maso [8] for more information.

In this section we calculate the Gamma limit of

$$I^\epsilon(u) := \frac{1}{\ln(1/\epsilon)} \int E^\epsilon(u) \, dx.$$

First we introduce a function space which is needed in order to state the Gamma limit.

4.1 Functions of BnV

Motivated by the analysis of I_ϵ and the central role of BV in the scalar case, Jerrard and the author introduced and studied a class of functions called BnV in [14]; a short summary is provided in [15]. It turns out that the Gamma limit of the functional I^ϵ is finite only for functions $u \in B2V$. For that reason, we give a brief discussion of BnV. We refer to [14, 15] for more information.

Briefly, function $u \in W^{1,n-1}(U; \mathbb{R}^n)$, for $U \subset \mathbb{R}^m, m \geq n$, is said to belong to BnV if the weak determinants of all n by n submatrices of the gradient matrix ∇u are signed Radon measures. Here we only we give the precise definition of $B2V$ and refer to [14] for higher dimensional case.

For $U \subset \mathbb{R}^m \to \mathbb{R}^2$ with $m \geq 2$ we view the Jacobian as a measure taking values in the exterior algebra $\Lambda^2 \mathbb{R}^m$. For every n (and in particular for $n = 2$) we endow $\Lambda^n \mathbb{R}^m$ with the natural inner product structure, which we denote $(\cdot, \cdot)$, and for a multivector $v \in \Lambda^n \mathbb{R}^m$ we write $|v| = (v, v)^{1/2}$. If $u \in W^{1,1}(U; \mathbb{R}^2)$ we define

$$j(u) = \sum_{i=1}^{m} u \times u_{x_i} \, dx^i , \qquad (75)$$

and if $j(u) \in L^1_{\text{loc}}$, we define

$$Ju = \frac{1}{2} \, d \, j(u) \qquad \text{in the sense of distributions}, \qquad (76)$$

where d is the exterior derivative. Thus if $u \in H^1_{\text{loc}}$, then

$$Ju = \sum_{i<j} J^{ij}u \, dx^i \wedge dx^j = \frac{1}{2} \sum_{i,j} J^{ij}u \, dx^i \wedge dx^j,$$

where

$$J^{ij}u = -J^{ji}u = u_{x_i} \times u_{x_j} = \det(u_{x_i}, u_{x_j}).$$

For sufficiently differentiable $u : \mathbb{R}^m \to \mathbb{R}^n$ one can define in a similar way Ju as a measure taking values in $\Lambda^n \mathbb{R}^m$. We omit the most general definition as we will not need it here.

Here we give the precise definition of $B2V$. Let $U \subset \mathbb{R}^m$ with $m \geq 2$.
Definition. *We say that u belongs to $B2V(U; \mathbb{R}^2)$ if both of the conditions are satisfied*

– $j(u) \in L^1_{loc}(U; \mathbb{R}^2)$,
– Ju is a Radon measure with values in $\Lambda^2 \mathbb{R}^m$.

A priori Ju is only a distribution; we say that $u \in B2V$ if it happens to be a measure. Also there are several conditions that ensure that $j(u) \in L^1$. For instance if $u \in W^{1,1} \cap L^\infty$ $j(u) \in L^1$.

The class BnV is very closely related to the *Cartesian Currents* of Giaquinta, Modica and Soucek [10, 11]. This connection is discussed in detail in [14].

In [14] it is shown that if $u \in B2V(\mathbb{R}^m; S^1)$, then the Jacobian measures Ju is supported on an $m - 2$ dimensional rectifiable set. In particular, if $u \in B2V(U; S^1)$ and $U \subset \mathbb{R}^2$, then there are $\{a_i\} \subset U$ and integers k_i such that

$$Ju = \pi \sum_i k_i \, \delta_{a_i} .$$

This is interpreted as encoding the location and degree of the topological singularities of u.

Here we outline only the proof in the case of $u \in B2V(U; S^1)$ with $U \subset \mathbb{R}^2$, and refer to [14] for the higher dimensional result.

Lemma 4.1. *Let $U \subset \mathbb{R}^2$ and $u \in B2V(U; \mathbb{R}^2)$. Then*

$$Ju = \pi \sum_j k_j \, \delta_{a_j} , \qquad (77)$$

for finite collections of points $\{a_j\} \subset U$ and integers k_j.

Outline of Proof. In this case, Ju is a scalar valued, signed Radon measure. Hence for every $x \in U$, the following limit exists

$$d(x) := \lim_{r \downarrow 0} d_r(x), \qquad d_r(x) := Ju(B_r(x)).$$

We claim that for every $x \in U$, and for almost every $r > 0$, $d_r(x)/\pi$ is an integer. Indeed, since $Ju = \nabla \times j(u)$, for any smooth function $\phi \in C_c^\infty(U; \mathbb{R}^1)$, by integration by parts

$$\int_U \phi Ju(dx) = \int_U \nabla \phi \cdot j(u) \, dx.$$

Formally if we take ϕ to be the characteristic function of the set $B_r(x)$, we obtain

$$Ju(B_r(x)) = \int_{B_r(x))} Ju(dx)$$
$$= \frac{1}{2} \int_{\partial B_r(x)} j(u) \cdot \mathbf{t} \, d\mathcal{H}^1.$$

By approximation, we may show that the above identity holds if u is sufficiently smooth. For $u \in B2V$, in [14], the above identity is proved for almost every $r > 0$.

Moreover, since $u \in W^{1,1}(U; S^1)$, for almost every $r > 0$, $u \in W^{1,1}(B_r(x); S^1)$. Hence, for these values of r, u is absolutely continuous with values in S^1. Then, by the degree formulae discussed earlier,

$$\frac{1}{2} \int_{\partial B_r(x)} j(u) \cdot \mathbf{t} \, d\mathcal{H}^1 = \pi \, \deg(u; \partial B_r(x)).$$

Therefore $d_r(x)/\pi$ is an integer for almost every r. Since the limit of $d_r(x)$ exists, we conclude that for every $x \in$ there exists $r(x) > 0$ such that

$$d_r(x) = d(x), \qquad \forall \, r \in (0, r(x)],$$

and

$$d(x)/\pi \in \mathbb{Z}, \qquad \forall \, x \in U.$$

Since $|Ju|(U)$ is finite, by a simple covering argument these yield the desired result. □

4.2 Gamma limit of I^ϵ

Let U be an open bounded subset of $\mathbb{R}^2$ with a smooth boundary.
   In this section we study the Γ limit of the functionals

$$I_\epsilon(u) := \frac{1}{\ln(1/\epsilon)} \int_U \frac{1}{2} |\nabla u|^2 + \frac{1}{4\epsilon^2}(1 - |u|^2)^2 \, dx \, ,$$

as ϵ tends to zero and show that the limiting functional is

$$I(u) := \begin{cases} |Ju|(U) = \pi \sum_i |k_i| \, , & \text{if } u \in B2V(U; S^1) \, , \\ +\infty \, , & \text{if } u \notin B2V(U; S^1) \, . \end{cases}$$

Theorem 4.1. *The Γ limit of I_ϵ in the topology of $W^{1,1}(U; \mathbb{R}^2)$ is equal to I, i.e., for every sequence u^ϵ converging to u in $W^{1,1}(U; \mathbb{R}^2)$,*

$$\liminf_{\epsilon \to 0} I_\epsilon(u^\epsilon) \geq I(u) \, , \tag{78}$$

and for every $u \in B2V(U; S^1)$, there exist functions u^ϵ converging to u in $W^{1,1}(U; \mathbb{R}^2)$ satisfying

$$\liminf_{\epsilon \to 0} I_{\epsilon_n}(u^\epsilon) = I(u) \, . \tag{79}$$

A similar result also holds in higher dimensions [13].
   The above result is proved independently in [1] and in [13].

Remark 4.1. In view of our compactness result and our introductory discussion of Gamma Limit, the Banach space $W^{1,1}$ is not appropriate. Since we do not have a compactness result in $W^{1,1}$, and the only compactness result is for the Jacobian, it is more natural to consider the Gamma Limit in the space of equivalent classes of functions with a topology equivalent to the convergence of the Jacobian. However, here and in [13], we chose to work with $W^{1,1}$ as it is a standard Banach space.

Proof. We start with the proof of (78). Suppose that u^ϵ converges to u in $W^{1,1}(U; \mathbb{R}^2)$. We assume that

$$\liminf_\epsilon I_\epsilon(u^\epsilon) < \infty ,$$

as there would be nothing to prove otherwise.

1. By the Compactness Theorem 3.2, there exists a subsequence ϵ_n converging to zero such that the Jacobian measure Ju^{ϵ_n} converges to a Radon measure J in $(C_c^{0,\beta})^*$ for all $\beta > 0$. We claim that $J = Ju$. In particular this will show that $u \in B2V(U; S^1)$.

To simplify the notation, set $u_n := u^{\epsilon_n}$.

2. We directly estimate that

$$\left| j(u_n) - \frac{j(u_n)}{|u_n|^2 \wedge 1} \right| \le |u_n||\nabla u_n| \left| \frac{|u_n|^2 \wedge 1 - 1}{|u_n|^2 \wedge 1} \right|$$

$$= |\nabla u_n| \frac{|1 - |u_n|^2|}{|u_n|} \chi_{|u_n| \ge 1}$$

$$\le \epsilon_n \left[\frac{1}{2} |\nabla u_n|^2 + \frac{1}{2\epsilon_n^2}(1 - |u_n|^2)^2 \right] .$$

Hence,

$$\lim_{n \to \infty} \int_U \left| j(u_n) - \frac{j(u_n)}{|u_n|^2 \wedge 1} \right| dx = 0 .$$

3. Set $v_n := u_n/(|u_n|^2 \wedge 1)$ so that

$$\frac{1}{|u_n|^2 \wedge 1} j(u_n) - j(u) = v_n \times \nabla u_n - u \times \nabla u$$

$$= v_n \times (\nabla u_n - \nabla u) + (v_n - u) \times \nabla u .$$

Hence,

$$\left| \frac{1}{|u_n|^2 \wedge 1} j(u_n) - j(u) \right| \le |v_n||\nabla u_n - \nabla u| + |v_n - u||\nabla u|$$

$$\le |\nabla u_n - \nabla u| + |v_n - u||\nabla u| .$$

Since u_n converges to u in $W^{1,1}(U; \mathbb{R}^2)$, there exists a subsequence, denoted by n again, so that u_n converges to u almost everywhere. Hence $|v_n - u||\nabla u|$ converges to zero almost everywhere and also it is less than $2|\nabla u|$. So we may use the dominated convergence theorem to conclude that

$$\lim_{n \to \infty} \int_U \left| \frac{1}{|u_n|^2 \wedge 1} j(u_n) - j(u) \right| dx = 0 .$$

4. Steps 2 and 3 imply that on a subsequence $j(u_n)$ converges to $j(u)$ in L^1. Hence, Ju^{ϵ_n} converges to Ju in the sense of distributions. This implies

that $J = Ju$. Since by Theorem 3.2, J is a Radon measure, so is Ju and therefore $u \in B2V(U; \mathbb{R}^2)$. It is also clear that $|u| = 1$ almost everywhere. Hence, $u \in B2V(U; S^1)$.

5. The Jacobian estimate (46) implies that

$$\left| \int_U \phi \, Ju(dx) \right| = \lim_{n \to \infty} \left| \int_U \phi \, Ju_n(dx) \right|$$
$$\leq \lambda \|\phi\|_\infty \liminf_{n \to \infty} I^{\epsilon_n}(u_n),$$

for every $\lambda > 1$. Hence,

$$\liminf_{n \to \infty} I^{\epsilon_n}(u_n) \geq \sup \left\{ \left| \int_U \phi \, Ju(dx) \right| : \|\phi\|_\infty \leq 1 \right\}$$
$$= |Ju|(U)$$
$$= I(u).$$

This proves (78).

6. We continue by proving the Γ-limit upper bound (79). Fix $u \in B2V(U; S^1)$. As remarked above, it is shown in [14] that Ju must have the form

$$Ju = \pi \sum_j k_j \, \delta_{a_j}.$$

It suffices to show that, given any sufficiently small $\delta > 0$, there exists a sequence of functions $\{v^\epsilon\} \subset H^1(U; \mathbb{R}^2)$ such that

$$I_\epsilon(v^\epsilon) \to \pi \sum |k_j|, \qquad \limsup_\epsilon \|v^\epsilon - u\|_{W^{1,1}(U)} \leq C\delta.$$

To do this, fix some small $\delta > 0$. Let $r_0 > 0$ be a number such that the balls $\{B_{2r}(a_j)\}$ are pairwise disjoint and do not intersect ∂U, whenever $r \leq r_0$, and select some $r > 0$ such that

$$\sum_j \int_{B_{2r}(a_j)} |\nabla u| \, dx \leq \delta, \qquad r \leq \min\{r_0, \delta\}. \tag{80}$$

For any $s > 0$, let U_s denote $U \setminus \cup_j B_s(a_j)$. Demengel [9] proves that if V is an open subset of $\mathbb{R}^2$, then smooth functions taking values in S^1 are dense in the subspace $\{w \in W^{1,1}(V; S^1) : Jw = 0\}$. Since $Ju = 0$ on U_r, this implies that there exists a function $v \in C^\infty(U_r, S^1)$ such that

$$\|u - v\|_{W^{1,1}(U_r)} \leq \delta. \tag{81}$$

Demengel's proof in fact shows that we may also assume that

$$\|j(u) - j(v)\|_{L^1(U_r)} \leq \delta. \tag{82}$$

7. Clearly (80) and (81) imply that

$$\sum_j \int_r^{2r} \int_{\partial B_s(a_j)} |\nabla v(x)| \, d\mathcal{H}^1(x) ds \;=\; \sum_j \int_{B_{2r} \setminus B_r(a_j)} |\nabla v| \, dx \;\leq\; 2\delta.$$

So, for each j, we can find some number $r_j \in [r, 2r]$ such that

$$\int_{\partial B_{r_j}(a_j)} |\nabla v(x)| \, d\mathcal{H}^1(x) \leq \frac{2\delta}{r}. \tag{83}$$

We also claim that

$$\deg(v; \partial B_{r_j}(a_j)) = k_j, \tag{84}$$

if δ is sufficiently small. Indeed, since v is smooth and S^1-valued it is clear that $s \mapsto \deg(v; \partial B_s(a_j))$ is constant for $s \in [r, 2r_0]$, so we only need to verify that this constant must equal k_j. To do this, note that if ϕ is any function of the form $\phi(x) = \bar{\phi}(|x - a_j|)$ that is constant on $B_r(a_j)$ and has its support in $B_{2r_0}(a_j)$, then

$$\frac{1}{2} \int \nabla \times \phi \cdot j(v) \, dx \;=\; \frac{1}{2} \int_0^\infty \deg(v; \phi^{-1}(s)) \, ds \;=\; \pi\phi(a_j)\deg(v; \partial B_{r_j}(a_j))$$

and

$$\frac{1}{2} \int \nabla \times \phi \cdot j(u) \, dx \;=\; \int \phi \, dJu \;=\; \pi\phi(a_j)k_j.$$

If δ is small enough, (84) follows from these two identities and (82), since $\nabla \times \phi$ is supported in U_r.

8. We claim that for each j there exists smooth functions v_j^ϵ, defined in $B_{r_j}(a_j)$ such that $v_j^\epsilon(x) = v(x)$ for $x \in \partial B_{r_j}(a_j)$,

$$\int_{B_{r_j}(a_j)} |\nabla v_j^\epsilon| dx \leq C\delta, \quad \text{and} \quad \lim_{\epsilon \to 0} \frac{1}{|\ln \epsilon|} \int_{B_{r_j}(a_j)} E^\epsilon(v_j^\epsilon) dx = \pi|k_j|. \tag{85}$$

To see this, fix some j. We may assume without loss of generality that $a_j = 0$, and due to (84) we can write

$$v(x) = \exp[i(k_j\theta + \alpha_j + \psi(x))] \qquad \text{for } x \in \partial B_{r_j},$$

where α_j is a constant, ψ is a smooth, single-valued function on ∂B_{r_j}, and θ as usual satisfies $\frac{x}{|x|} = (\cos\theta, \sin\theta)$. We are identifying $\mathbb{R}^2 \cong \mathbb{C}$ in the usual way. We extend ψ to be homogeneous of degree zero on $\mathbb{R}^2 \setminus \{0\}$, and we define

$$v_j^\epsilon(x) \;=\; \exp[i(k_j\theta + \alpha_j + \frac{2|x| - r_j}{r_j}\psi(x))] \quad \text{if} \quad \frac{1}{2}r_j \le |x| \le r_j.$$

For $|x| \le \frac{1}{2}r_j$ we define $v_j^\epsilon(x)$ to be a minimizer of

$$\int_{B_{r_j/2}} E^\epsilon(w)\,dx,$$

subject to the boundary conditions $w = \exp[i(k_j\theta + \alpha_j)]$ on $\partial B_{r_j/2}$.

Since v_j^ϵ restricted to the annulus $B_{r_j}\backslash B_{r_j/2}$ is just a fixed smooth function of unit modulus, independent of ϵ, it is clear that

$$\lim_\epsilon \frac{1}{|\ln \epsilon|}\int_{B_{r_j}\backslash B_{r_j/2}} E^\epsilon(v_j^\epsilon)\,dx \;=\; \lim_\epsilon \frac{1}{|\ln \epsilon|}\int_{B_{r_j}\backslash B_{r_j/2}} \frac{1}{2}|\nabla v_j^\epsilon|^2\,dx \;=\; 0.$$

Also, using (83) one can check that

$$\int_{B_{r_j}\backslash B_{r_j/2}(a_j)} |\nabla v_j^\epsilon|dx \;\le\; C\delta.$$

Finally, the book of Bethuel, Brezis, and Hélein gives a detailed description of the asymptotics of Ginzburg-Landau energy-minimizers, and their results imply that

$$\lim_\epsilon \frac{1}{|\ln \epsilon|}\int_{B_{r_j/2}} E^\epsilon(v_j^\epsilon)\,dx = \pi|k_j|, \qquad \limsup_\epsilon \int_{B_{r_j/2}} |\nabla v_j^\epsilon|\,dx \le Cr_j \le C\delta.$$

Putting these facts together we find that the sequence $\{v_j^\epsilon\}$ has the properties specified in (85).

9. Finally we define

$$v^\epsilon(x) = \begin{cases} v(x) & \text{if } x \in U \backslash (\cup_j B_{r_j}(a_j)) \\ v_j^\epsilon(x) & \text{if } x \in B_{r_j}(a_j) \end{cases}.$$

Since v is a fixed smooth function and $|v| \equiv 1$, $\frac{1}{|\ln \epsilon|}E^\epsilon(v) = \frac{1}{|\ln \epsilon|}|\nabla v|^2$ tends to zero uniformly as $\epsilon \to 0$. Thus it is clear from (85) that

$$\lim_{\epsilon \to 0}\frac{1}{|\ln \epsilon|}\int_U E^\epsilon(v^\epsilon)dx \;=\; \sum_j \lim_{\epsilon \to 0}\frac{1}{|\ln \epsilon|}\int_{B_{r_j}(a_j)} E^\epsilon(v_j^\epsilon)dx = \pi\sum_j|k_j|.$$

Also,

$$\|u - v^\epsilon\|_{W^{1,1}(U)} \le \|u - v\|_{W^{1,1}(U_r)}$$
$$+ \sum_j(\|u\|_{W^{1,1}(B_{r_j}(a_j))} + \|v_j^\epsilon\|_{W^{1,1}(B_{r_j}(a_j))})$$
$$\le C\delta,$$

by (80), (81), and (85). So the sequence $\{v^\epsilon\}$ has all the required properties. $\qquad\square$

5 Compactness in Higher Dimensions

Now suppose U is a bounded, open subset of $\mathbb{R}^m$ with $m \geq 3$.

In this section we will show that if $\{u^\epsilon\}_{\epsilon \in (0,1]} \subset H^1(U; \mathbb{R}^2)$ is a sequence of functions such that the normalized Ginzburg-Landau energy measure $\mu^\epsilon(U)$ is uniformly bounded, then the Jacobians Ju^ϵ are precompact in $(C^{0,\beta})^*$ for all $\beta > 0$, and any limit is rectifiable. In addition, we prove that

$$|\bar{J}|(U) \leq \liminf \mu^\epsilon(U).$$

This is not a full Γ-convergence result, but it shows that the mass of the Jacobian is a reasonable candidate for the Γ-limit. We also believe that the compactness result and the upper bound for the Jacobian (ie, lower bound for the energy) are interesting and will be useful in other contexts.

We start by defining some of the terms used above. We remark that good general references for this material include Giaquinta *et. al* [10] and Simon [28].

Let $\Lambda^2 \mathbb{R}^m$, $j(u)$ and Ju be as in subsection 4.1.

A set $M \subset \mathbb{R}^m$ is said to be a k-dimensional rectifiable set if there are Lipschitz functions $f_j : \mathbb{R}^k \to \mathbb{R}^m$ and measurable subsets A_j of $\mathbb{R}^k$ such that

$$M = M_0 \cup \left(\cup_{j=1}^{\infty} f_j(A_j) \right), \qquad \mathcal{H}^k(M_0) = 0.$$

Thus, in a precise measure theoretic sense, a k-dimensional rectifiable set is not much worse than a k-dimensional Lipschitz submanifold. Rectifiable sets can also be characterized by the fact that they have k-dimensional approximate tangent spaces $\mathcal{H}^k$ almost everywhere; see [28] or [10].

Suppose that M is an oriented, rectifiable $(m-n)$-dimensional subset of $\mathbb{R}^m$, and for $\mathcal{H}^{m-n}$ almost every $x \in M$, let $\nu(x) \in \Lambda^n \mathbb{R}^m$ be the unit n-vector representing the appropriately oriented normal space to M. (It is more convenient for our purposes to work with normal spaces rather than tangent spaces.) Suppose also that $\theta : M \to \mathbb{N}$ is a $\mathcal{H}^{m-n}$-integrable function. One can define a measure J taking values in $\Lambda^n \mathbb{R}^m$ by

$$\int \phi(x) J(dx) = \int_M \phi(x) \cdot \nu(x) \theta(x) \mathcal{H}^{m-n}(dx) \qquad \forall \phi \in C^0(\mathbb{R}^m; \Lambda^n \mathbb{R}^m).$$

(86)

We say that a measure J taking values in $\Lambda^n \mathbb{R}^m$ is $(m-n)$-dimensional integer multiplicity rectifiable (or more briefly, integer multiplicity rectifiable) if it has the form (86) for some rectifiable set M and an integer-valued function θ as above.

The class of functions for which Ju is a measure is denoted $BnV(U, \mathbb{R}^n)$ and was defined and studied in [14]. In particular we prove there that if $u \in BnV(\mathbb{R}^m, S^{n-1})$ then Ju/ω_n is integer multiplicity rectifiable, where ω_n

is the volume of the unit ball in $I\!R^n$. This is deduced as a consequence of a more general rectifiability criterion which we recall here.

Let J be a measure on a subset $U \subset I\!R^m$ taking values in $\Lambda^n I\!R^m$, where $n \leq m$. We can write J in the form $J = \nu|J|$, where $|J|$ is a nonnegative Radon measure, and ν is a $|J|$-measurable function taking values in $\Lambda^n I\!R^m$ such that $|\nu(x)| = 1$ at $|J|$-a.e. $x \in U$.

Suppose that $e_1, ..., e_m$ is an orthonormal basis for $I\!R^m$. Given any point $x \in I\!R^m$, we write $y_i = x \cdot e_i$ if $i = 1, ..., m - n$; and $z_i = x \cdot e_{m-n+i}$ if $i \in 1, ..., n$. We write $I\!R_y^{m-n}$ to denote the span of $\{e_i\}_{i=1}^{m-n}$. Similarly, $I\!R_z^n = \mathrm{span}\{e_i\}_{i=m-n+1}^m$. Thus we identify points $x \in I\!R^m$ with corresponding $(y, z) \in I\!R_y^{m-n} \times I\!R_z^n$. Let $dz := dz^1 \wedge \ldots \wedge dz^n$, and let J^z denote the scalar signed measure defined by $J^z := (dz, \nu)|J|$.

We say that J^z is *locally represented by slices* $J_y(dz)$ if, given any open set $O \subset U$ of the form $O = O_y \times O_z$, with $O_y \subset I\!R_y^{m-n}$ and $O_z \subset I\!R_z^n$, there exist signed Radon measures $J_y(dz)$ on O_z for a.e. $y \in O_y$, such that

$$\int \phi J^z = \int_{O_y} \int_{O_z} \phi(y, z) J_y(dz) \, dy \tag{87}$$

for all continuous ϕ with compact support in O.

We say that a statement holds for a.e. $J_y(dz)$ if, for every open set O as above, it is valid for a.e. $y \in O_y$.

In [14] we prove the following

Theorem 5.1. *Suppose that J is a Radon measure on $U \subset I\!R^m$ taking values in $\Lambda^n I\!R^m$, and also that $dJ = 0$ in the sense of distributions. Suppose also that for every choice on an orthonormal basis $\{e_i\}_{i=1}^m$ (determining a decomposition of $I\!R^m$ into $I\!R_y^{m-n} \times I\!R_z^n$) J^z is represented locally by slices, and that for a.e. $y \in O_y$ these slices have the form*

$$J_y(dz) = \sum_{i=1}^K d_i \delta_{a_i}(dz),$$

for an integers K and d_i, and points $a_i \in O_z$.
Then J is rectifiable.

A much more general version of this result was later established by Ambrosio and Kirchheim [2]. A similar theorem in somewhat different and very general setting was proved independently (and slightly earlier) by White [33].

We will need Theorem 5.1 to prove

Theorem 5.2 (Jerrard & Soner [13]). *Let $U \subset I\!R^m$, and suppose that $\{u^\epsilon\}_{\epsilon \in (0,1]}$ is a collection of functions in $W^{1,2}(U; I\!R^2)$ such that $\mu^\epsilon(U) \leq K_U < \infty$ for all ϵ. Then there exists a subsequence $\epsilon_n \to 0$ such that*

(i) Ju^{ϵ_n} converges to a limit $\bar{J}$ in the $(C^{0,\alpha})^*$ norm for every $\alpha > 0$;

(ii) *For any choice of basis $\{e_i\}$ for $\mathbb{R}^m$ (determining a decomposition of $\mathbb{R}^m$ into $\mathbb{R}_y^{m-2} \times \mathbb{R}_z^2$), $\bar{J}^z$ is represented locally by slices $\bar{J}_y(dz)$, and for a.e. y these slices have the form $\bar{J}_y(dz) = \pi \sum_{i=1}^{K} d_i \delta_{a_i}$, with $d_i \in \mathbb{Z}$ for all i.*

(iii) *$d\bar{J} = 0$ in the sense of distributions, and $\bar{J}/\pi$ is integer multiplicity rectifiable;*

(iv) *Finally, if $\bar{\mu}$ is any weak limit of a subsequence of μ^{ϵ_n}, then $|\bar{J}| \ll \bar{\mu}$, and $\frac{d|\bar{J}|}{d\bar{\mu}} \le 1$. In particular, $|\bar{J}|(U) \le K_U$.*

Remark 5.1. For any $\bar{J}$ as above, (*iii*) and the definition of rectifiability imply that a lower density bound

$$\liminf_{r \to 0} \frac{|\bar{J}|(B_r(x))}{\mathcal{H}^{m-2}(B_r(x))} \ge \pi,$$

for $|\bar{J}|$ almost every x. Also, if $\bar{\mu}$ is as in (*iv*), then clearly the $m-2$-dimensional density of μ is greater than $m-2$-dimensional density of $\bar{J}$. In particular,

$$\liminf_{r \to 0} \frac{\bar{\mu}(B_r(x))}{\mathcal{H}^{m-2}(B_r(x))} \ge \pi,$$

for $|\bar{J}|$ almost every x.

The basic idea of the proof is to decompose a component of Ju^ϵ, for example $J^{m-1,m}u^\epsilon$, into two-dimensional slices, say $J_y^\epsilon(dz)$, and to use the two-dimensional estimates on each slice. Arguing in this fashion, it is quite easy to obtain uniform estimates for $J^{m,m-1}u^\epsilon$ in certain weak spaces, and these imply (*i*) by results of the previous section.

To prove (*ii*), it is convenient to view the sliced measures $J_y^\epsilon(dz)$ as constituting a function mapping $\mathbb{R}_y^{m-2}$ into $C_c^1(\mathbb{R}_z^2)^*$; the latter is a space that contains measures on $\mathbb{R}_z^2$ and is endowed with a rather weak topology. Claim (*i*) can be seen as assertion that the function $y \mapsto J_y^\epsilon(dz)$ is precompact in some weak sense. What one would like to do is to show that in fact $y \mapsto J_y^\epsilon(dz)$ is precompact in some stronger sense, for example in $L^1(\mathbb{R}_y^{m-2}; (C_c^1(\mathbb{R}_z^2)^*)$, so that one can extract a subsequence that converges to some limiting function $y \mapsto \bar{J}_y(dz)$ in L^1. In particular, after passing to a further subsequence we could then assume that $J_y^{\epsilon_n}(dz) \to \bar{J}_y(dz)$ for almost every y. In addition, by our two-dimensional results, for almost every y, one can find a subsequence $\epsilon_{n_m} \to 0$ (in general depending on y) such that $J_y^{\epsilon_{n_m}}(dz)$ converges to some limit that has the form sought in (*ii*). By combining these results one can hope to show that in fact $\frac{1}{\pi}\bar{J}_y(dz)$ is a sum of point masses with integer multiplicities.

The key point is then to establish some sort of strong compactness of the sequence of functions $y \mapsto J_y^\epsilon(dz)$ as $\epsilon \to 0$. We do this using the observation from [14, 15] that the total variation of $y \mapsto J_y^\epsilon(dz)$ in the $(C_c^{0,1})^*$ norm can be estimated by controlling "orthogonal" components of Ju^ϵ, which is already done in the proof of (i). Using this one can argue that the functions $y \mapsto J_y^\epsilon(dz)$ have uniformly bounded variation in $(C_c^{0,1})^*$, modulo terms that vanish in still weaker norms, and this gives the necessary strong convergence. (The terms involving weaker norms force us to work with test functions that are C^2 instead of C^1 in much of the proof.)

The remaining points follow quite directly from (ii) and the rectifiability criterion of Theorem 5.1, and from the two-dimensional results.

6 Dynamic Problems: Evolution of Vortex Filaments

The stationary results, in particular the energy lower bounds and the compactness of the Jacobian are very useful tools in the asymptotic study of the evolution problems related to I^ϵ. In these note we only consider the parabolic equation, which is the gradient flow of I^ϵ

$$u_t^\epsilon - \Delta u^\epsilon = \frac{u^\epsilon}{\epsilon^2}\,(1 - |u^\epsilon|^2), \quad t > 0, \ x \in \mathbb{R}^n, \tag{88}$$

where the unknown function

$$u^\epsilon : \mathbb{R}^n \mapsto \mathbb{R}^2$$

satisfies an initial condition

$$u^\epsilon(0, x) = u_0^\epsilon(x), \qquad x \in \mathbb{R}^n, \tag{89}$$

where u_0^ϵ is a given function.

Here we do not consider the corresponding nonlinear Schrödinger and the nonlinear heat equations. Instead we refer to the notes of Rubinstein [23] for an introduction and the description of formal results and to the paper of Colliander and Jerrard [7] for a rigorous study of the nonlinear, planar Schrödinger equation.

Also here we only consider the case $n \geq 3$. Results for $n = 2$ are outlined in the Introduction.

The asymptotic analysis of these equations initiated by the seminal paper of Rubinstein, Sternberg and Keller [24]. Much has been done for the scalar equation since then. We refer to the survey article of the author [30] and the references therein for the scalar equation, which is also called as the Cahn-Allen equation in the literature.

The vector valued Ginzburg-Landau equation with $u^\epsilon : \mathbb{R}^n \mapsto \mathbb{R}^m$ is studied first by Struwe [32]. He considered this equation as a relaxation of the heat flow for harmonic maps. Under the assumption that original (not rescaled) energy is uniformly bounded in ϵ, he obtained deep partial regularity results. A monotonicity result is one of the interesting results of [32]. Our results differ from [32] in that we only consider the case $u^\epsilon \in \mathbb{R}^2$ but we allow for singularities to form and study the time evolution of the singular structures.

Here we outline the results of [17] and [4]. This technique is also outlined in the lecture notes of the author [30] for the scalar equations. As discussed in the Introduction, in the following proof, we will use the compactness of the Jacobian to provide a shorter proof.

6.1 Energy identities

Set
$$e_\epsilon := e_\epsilon(t, x) = E^\epsilon(u^\epsilon(t, x)).$$

Recall that the energy density E^ϵ is given as
$$E^\epsilon(u) := \frac{1}{2}|\nabla u|^2 + \frac{(1 - |u|^2)^2}{4\epsilon^2}.$$

By calculus and (88),
$$(e_\epsilon)_t = -|u_t^\epsilon|^2 + \mathrm{div}\, p^\epsilon, \tag{90}$$

$$\nabla e_\epsilon = -p^\epsilon + \mathrm{div}(\nabla u^\epsilon \otimes \nabla u^\epsilon), \tag{91}$$

$$p^\epsilon := \nabla u^\epsilon\, u_t^\epsilon.$$

To localize the energy estimates, let $\eta \geq 0$ be a smooth compactly supported test function. Multiply (90) by η and (91) by $\nabla\eta$ and subtract the two identities. Then use the resulting identity to compute the time derivative of the integral of ηe_ϵ. The result is

$$\frac{d}{dt}\int \eta\, e_\epsilon = \int (\eta_t - \Delta\eta)\, e_\epsilon + D^2\eta\nabla u^\epsilon \cdot \nabla u^\epsilon - \int \eta\, |u_t^\epsilon|^2. \tag{92}$$

Although we will not use it in these notes, let us mention that if we add the two identities instead of subtracting them, we obtain the following identity

$$\frac{d}{dt}\int \eta\, e_\epsilon = \int (\eta_t + \Delta\eta)\, e_\epsilon - D^2\eta\nabla u^\epsilon \cdot \nabla u^\epsilon \tag{93}$$

$$+ \frac{|\nabla\eta \cdot \nabla u^\epsilon|^2}{\eta} - \int \eta \left| u_t^\epsilon - \frac{\nabla\eta \cdot \nabla u^\epsilon}{\eta} \right|^2.$$

In [32], Struwe used the above identity with the special choice for η

$$\eta(t, x) = \frac{1}{(4(t_0 - t))^{(n-m)/2}} \exp\left(-\frac{|x - x_0|^2}{4(t_0 - t)}\right), \qquad t < t_0,\ x \in \mathbb{R}^n,$$

where $t_0 > 0$ and $x_0 \in \mathbb{R}^n$ are arbitrary, and m is the dimension of the range of u.

The special case of (92) with $\eta \equiv 1$ yields the classical energy estimate,

$$\int e^\epsilon(t, x)\, dx + \int_0^t |u_t^\epsilon|^2\, dx dt = \int e^\epsilon(0, x)\, dx. \tag{94}$$

6.2 Mean curvature flow and the distance function

The distance and the square distance functions to a smooth manifold can be used to describe all the relevant geometric quantities. These functions were first used in [17] in the convergence proofs.

Let $\{\Gamma_t\}_{t \in [0,T]}$ be a smooth solution of co-dimension two mean curvature flow. Then the square distance function $\eta(t, x)$ satisfies

$$\eta_t - \Delta\eta = -2, \quad \text{on } \Gamma_t, \tag{95}$$

$$\nabla[\eta_t - \Delta\eta] = 0, \quad \text{on } \Gamma_t, \tag{96}$$

$$D^2\eta \leq I_n, \quad \text{on whenever it is smooth.} \tag{97}$$

The equation (96) is an observation of De Giorgi. It can be viewed as the definition of the codimension two mean curvature flow. We refer to [3] for information on the mean curvature flow in any codimension. However, (95), and (97) are the properties of any square distance function to a smooth codimension two manifold.

Since Γ_t is smooth, η is smooth in a tubular neighborhood of Γ_t. (95) and (96) imply that in this neighborhood,

$$|\eta_t - \Delta\eta + 2| \leq C \, \eta, \tag{98}$$

for some constant C. We extend η smoothly to all of $[0, T] \times \mathbb{R}^n$ so that the extension satisfies (97), (98), and

$$\eta \geq 0, \quad \text{and } \eta(t, x) = 0, \quad \text{if and only if } \ x \in \Gamma_t. \tag{99}$$

In (92) we choose η to be the square distance function, modified as above. Set

$$\alpha^\epsilon(t) := \frac{1}{\ln(1/\epsilon)} \int \eta \, e_\epsilon(t, x) \, dx,$$

so that by (92), (97) and (98),

$$\frac{d}{dt}\alpha^\epsilon(t) \leq \frac{1}{\ln(1/\epsilon)} \int [C \, \eta_t \, e_\epsilon - 2e_\epsilon + |\nabla u^\epsilon|^2] - \frac{1}{\ln(1/\epsilon)} \int |u_t^\epsilon|^2 \, dx$$

$$\leq C \, \alpha^\epsilon(t) - \frac{1}{\ln(1/\epsilon)} \int |u_t^\epsilon|^2 \, dx. \tag{100}$$

6.3 Convergence

In this subsection, we prove the convergence of the solutions of 88 to smooth solutions of the mean curvature flow. This is first proved in [17], and here we follow a new approach using the compactness result on the Jacobians. The convergence to weak solutions of the mean curvature flow is still not known. A first step in this direction is proved in [4].

Here we assume that the initial data concentrates on a smooth co-dimension two manifold Γ_0 and that there exists a smooth solution $\{\Gamma_t\}_{t\in[0,T]}$ of the codimension two mean curvature flow. then, we will prove that the energy measure concentrates on the smooth solution.

Let

$$\mu_t^\epsilon(V) := \frac{1}{\ln(1/\epsilon)} \int_V E^\epsilon(u^\epsilon(t,x))\, dx$$

be the rescaled Ginzburg-Landau energy. Precisely, we assume that u_0^ϵ is such that

$$\mu_0^\epsilon \stackrel{*}{\rightharpoonup} \pi\, \mathcal{H}^{n-2}\lfloor\Gamma_0,$$

in the weak* topology of Radon measure, where $\mathcal{H}^{n-2}\lfloor\Gamma_0$ is the $n-2$ dimensional Hausdorff measure restricted to the codimension two manifold Γ_0, i.e, it is the surface area measure of Γ_0.

A simple, consequence of this assumption is that

$$C^* := \sup_\epsilon\ \mu_0^\epsilon(\mathbb{R}^n) < \infty.$$

This together with the energy estimate (94) imply

$$\sup\{\ \mu_t^\epsilon(\mathbb{R}^n)\ |\ t\in[0,T],\ \epsilon > 0\ \} \leq\ C^*.$$

Using an argument due to Brakke (see [17] for details), this implies that there exists a subsequence, denoted by ϵ again, such that

$$\mu_t^\epsilon \stackrel{*}{\rightharpoonup} \mu_t, \qquad t\in[0,T],$$

where $\{\mu_t\}_{t\in[0,T]}$ stands for a family of nonnegative Radon measures.

Theorem 6.1. *Suppose that $\{\Gamma(t)\}_{t\in[t_0,t_1]}$ is a collection of compact sets which is a classical solution of the codimension two mean curvature flow. Let μ_t be a weak* limit of the rescaled energy measure μ_t^ϵ. Then,*

$$\mu_t \geq \pi\mathcal{H}^1\lfloor\Gamma_t, \qquad t\in[0,T],$$

and

$$\text{support } \mu_t = \Gamma_t, \qquad t\in[0,T].$$

Proof. Since each connected component of the solution can be studied separately, without loss of generality, we may assume that Γ_t is conneceted with no boundary.

1. Let $\alpha^\epsilon(t)$ be as in the previous subsection. Then by the convergence of μ_t^ϵ,

$$\lim_{\epsilon\to 0} \alpha^\epsilon(t) = \alpha(t) := \int \eta\, \mu_t(dx).$$

In view of (99) and our assumption on μ_0,

$$\alpha(0) = 0.$$

Then by (100) and the Gronwall's inequality

$$\alpha(t) \le \alpha(0) \, e^{Ct} = 0.$$

Since η is nonnegative, this implies that the support of μ_t is included in the zero set of η. Hence by (99),

$$\text{support } \mu_t \subset \Gamma_t, \qquad t \in [0, T].$$

2. Fix $t \in [0, T]$ and let $J_t^\epsilon := Ju^\epsilon(t, \cdot)$. Since the rescaled Ginzburg-Landau energy $\mu_t^\epsilon(\mathbb{R}^n)$ is uniformly bounded in ϵ, we may apply the compactness result proved earlier. Then, on a subsequence denoted by ϵ again, J_t^ϵ converges to a signed Radon measure J_t in the topology of $(C^{0,\alpha})^*$ for every $\alpha > 0$. Moreover,

$$\frac{d|J_t|}{d\mu_t} \le 1, \qquad \mu_t \text{ a.e.},$$

where $|J_t|$ is the total variation of the vector valued measure J_t and μ_t is the weak* limit of μ_t^ϵ. Hence, by step 1,

$$\text{support } J_t \subset \text{support } \mu_t \subset \Gamma_t.$$

3. Since

$$J_t^\epsilon = dj(u^\epsilon(t, \cdot)), \qquad \text{and} \qquad dJ_t^\epsilon = 0 \;\; \Rightarrow \;\; dJ_t = 0.$$

If we see J_t as a current, the above states that it has no boundary. By the previous step J_t is included in a smooth manifold Γ_t which has no boundary. In Lemma 6.1 below, we will show that these fact imply that the density of $|J_t|$ on Γ_t

$$\Theta_t := \frac{d|J_t|}{d\mathcal{H}^{n-2}\lfloor \Gamma_t}$$

is constant. We postpone the proof of this result to Lemma 6.1 and complete the proof of this theorem.

In view of our compactness result, this constant has to be an integer multiple of π, i.e.,

$$\Theta_t \equiv n_t \, \pi, \quad \mathcal{H}^{n-2}\lfloor \Gamma_t \text{ a.e.},$$

for some integer n_t.
4. It suffices to show that $n_t \equiv 1$ for $t \in [0, T]$. Indeed this implies that

$$|J_t| = \pi \, \mathcal{H}^{n-2}\lfloor \Gamma_t.$$

Since $|J_t| \le \mu_t$,

$$\mu_t \geq \pi \, \mathcal{H}^{n-2} \lfloor \Gamma_t.$$

and

$$\Gamma_t \subset \text{support } \mu_t.$$

The opposite inclusion is proved in Step 1.

5. To prove that $n_t \equiv 1$, consider the space-time Jacobian J^ϵ of u^ϵ on $[0, T] \times \mathbb{R}^n$. The space-time Ginzburg-Landau energy is

$$\mathcal{E}^\epsilon := \frac{1}{2} \left[|\nabla u^\epsilon|^2 + |u_t^\epsilon|^2 \right] + \frac{(1 - |u^\epsilon|^2)^2}{4\epsilon^2},$$

and in view of (94),

$$\sup_\epsilon \left\{ \frac{1}{\ln(1/\epsilon)} \int_0^T \int_{\mathbb{R}^n} \mathcal{E}^\epsilon \, dx \, dt \right\} < \infty.$$

Hence we may apply our compactness result to J^ϵ, concluding that on a subsequence, denoted by ϵ again, it converges to a Radon measure J^*. Moreover

$$|J^*|([0, T] \times \mathbb{R}^n) \leq \mu([0, T] \times \mathbb{R}^n),$$

where μ is the weak* limit of the space-time energy. Then,

$$\mu([0, T] \times \mathbb{R}^n) = \int_0^T \mu_t(\mathbb{R}^n) \, dt + \nu([0, T] \times \mathbb{R}^n),$$

where ν is the weak* limit of

$$\nu^\epsilon(dt \times dx) := \frac{1}{\ln(1/\epsilon)} |u_t^\epsilon|^2 \, dt \times dx.$$

Let η and α be as in Step 1. By (92), and the properties of η,

$$\frac{d}{dt} \alpha^\epsilon(t) = C \, \alpha^\epsilon(t) - \frac{1}{2 \ln(1/\epsilon)} \int_{\mathbb{R}^n} \eta(t, x) \, |u_t^\epsilon|^2 \, dx.$$

By the Gronwall's inequality

$$\alpha^\epsilon(T) \leq \alpha^\epsilon(0) \, e^{CT} - \frac{1}{2 \ln(1/\epsilon)} \int_0^T \int_{\mathbb{R}^n} e^{C(T-t)} \eta(t, x) \, |u_t^\epsilon|^2 \, dx \, dt$$

$$\leq \alpha^\epsilon(0) \, e^{CT} - \frac{1}{2 \ln(1/\epsilon)} \int_0^T \int_{\mathbb{R}^n} \eta(t, x) \, \nu^\epsilon(dt \times dx).$$

Since α^ϵ tends to zero as ϵ approaches to zero. In the limit we obtain

$$\int_0^T \int_{\mathbb{R}^n} \eta \, \nu = 0.$$

Hence, the support of ν is included in the graph Γ of $\{\Gamma_t\}_{t \in [0, T]}$

$$\Gamma := \{\ (t,x) \in [0,T] \times I\!\!R^n \mid x \in \Gamma_t\ \}.$$

Therefore,

$$\text{support } J^* \subset \Gamma.$$

Using Lemma 6.1, we conclude that

$$|J^*| = \pi\ \mathcal{H}^{n-1}\lfloor\Gamma,$$

and consequently

$$|J_t| = \pi\ \mathcal{H}^{n-2}\lfloor\Gamma_t, \qquad t \in [0,T].$$

$\square$

In the following lemma we assume J is a Radon measure of the form in Theorem 5.2. In particular it satisfies (86)

$$\int \phi(x)J(dx)\ =\ \int_M \phi(x) \cdot \nu(x)\Theta(x)\mathcal{H}^{m-2}(dx) \qquad \forall \phi \in C^0(I\!\!R^m; \Lambda^2 I\!\!R^m).$$

Lemma 6.1. *Let J be as above. Further assume that $M \subset I\!\!R^n$ be a codimension two, smooth manifold with no boundary. Then, Θ is constant on M.*

Proof. Let $\tilde{J}$ be the same as J but with $\Theta \equiv 1$, i.e.,

$$\int \phi(x)\tilde{J}(dx)\ =\ \int_M \phi(x) \cdot \nu(x)\mathcal{H}^{m-2}(dx) \qquad \forall \phi \in C^0(I\!\!R^m; \Lambda^2 I\!\!R^m).$$

Since M has no boundary and since M is smooth, we directly calculate that $d\tilde{J} = 0$. We also know that $dJ = 0$. Using these two facts we calculate that

$$0 = d\,J = d\,[\Theta\,\tilde{J}] = \Theta\,d\,\tilde{J} + d\,\Theta \wedge \tilde{J}$$
$$= d\,\Theta \wedge \tilde{J}.$$

Hence

$$\nabla_{tan}\Theta = 0, \quad \text{on } M,$$

where ∇_{tan} is the tangential derivative on M. Since M is connected, this implies that Θ is constant on M. $\square$

References

[1]   G. Alberti, S. Baldo and G. Orlandi, in preparation.
[2]   L. Ambrosio and B. Kirchheim, Currents in metric spaces, *Math. Ann.*, 318, 527-555 (2000).

[3]   L. Ambrosio and H.M. Soner, Level set approach to mean curvature flow in arbitrary codimension, *J. Diff. Geometry*, 43, 693-737 (1996).

[4]   L. Ambrosio and H.M. Soner, A measure theoretic approach to higer codimension mean curvature flow, *Ann. Scuola Norm. Sup. Pisa Cl. Sci. (4)*, 25, 27-48 (1997).

[5]   F. Bethuel, A characterization of maps in $H^1(B^3; S^2)$ which can be approximated by smooth maps, *Ann. Inst. H. Poincaré Anal. Non Linéaire*, 7, 269-286 (1990).

[6]   F. Bethuel, H. Brezis and F. Hélein, *Ginzburg Landau Vortices*, Birkhauser, New-York, 1994.

[7]   J.E. Colliander and R.L. Jerrard, Ginzburg-Landau vortices; weak stability and Schrödinger equation dynamics, *Journal d'Analyse Mathematique*, 77, 129-205 (1999).

[8]   G. Dal Maso, *An Introduction to Γ-Convergence*, Birkhauser, Boston, 1993.

[9]   F. Demengel, Une caractérisation des applications de $W^{1,p}(B^N, S^1)$ qui peuvent être approchées par des fonctions régulieres. *C. R. Acad. Sci. Paris Sér. I Math.*, 310, 553-557 (1990).

[10] M. Giaquinta, G. Modica and J. Soucek, *Cartesian Currents in the Calculus of Variations I*, Springer-Verlag, New York, 1998.

[11] M. Giaquinta, G. Modica and J. Soucek, *Cartesian Currents in the Calculus of Variations II*, Springer-Verlag, New York, 1998.

[12] R.L. Jerrard, Lower bounds for generalized Ginzburg-Landau functionals, *SIAM Math. Anal.*, 30, 721-746 (1999).

[13] R.L. Jerrard and H.M. Soner, The Jacobian and the Ginzburg-Landau energy, *Calc. Var. Partial Differential Equations*, 14, 151-191 (2002).

[14] R.L. Jerrard and H.M. Soner, Functions of bounded higher variation, *Indiana Univ. Math. J.*, 51, 645-677 (2002).

[15] R.L. Jerrard and H.M. Soner, Rectifiability of the distributional Jacobian for a class of functions, *C. R. Acad. Sci. Paris Sér. I Math.*, 329, 683-688 (1999).

[16] R.L. Jerrard and H.M. Soner, Dynamics of Ginzburg-Landau vortices, *Arch. Rational Mech. Anal.*, 142, 185-206 (1998).

[17] R.L. Jerrard and H.M. Soner, Scaling limits and regularity for a class of Ginzburg-Landau systems, *Ann. Inst. H. Poincaré Anal. Non Linéairee*, 16, 423-466 (1999).

[18] F.H. Lin, Some dynamical properties of Ginzburg-Landau vortices, *Comm. Pure Appl. Math.*, 49, 323-359 (1996).

[19] F.-H. Lin, Solutions of Ginzburg-Landau equations and critical points of the renormalized energy, *C.R. Acad. Sci. Paris, Série I*, 12, 599-622 (1995).

[20] L. Modica, The gradient theory of phase transitions and the minimal interface criterion, *Arch. Rational Mech. Anal.*, 98, 123-142 (1987).

[21] L. Modica and S. Mortola, Il limite nella Γ-convergenza di una famiglia di funzionali elliptici, *Boll. Un. Mat. Ital. A*, 14, 526-529 (1977).

[22] L. Modica and S. Mortola, Un esempio di Γ convergenza, *Boll. Un. Mat. Ital. B*, 14, 285-299 (1977).

[23] J. Rubinstein, Six lectures on superconductivity. *Boundaries, interfaces, and transitions (Banff, AB, 1995)*, 163-184, CRM Proc. Lecture Notes, 13, Amer. Math. Soc., Providence, RI, 1998.

[24] J. Rubinstein, P. Sternberg and J. Keller, Fast reaction, slow diffusion and curve shortening, *SIAM J. Appl. Math.*, 49, 116-133 (1989).

[25] E. Sandier, Lower bounds for the energy of unit vector fields and applications, *J. Funct. Anal.*, 152, 379-403 (1998).

[26] E. Sandier, Ginzburg-Landau minimizers from $I\!R^{n+1}$ into $I\!R^n$ and minimal connections, *Indiana Univ. Math. J.*, 50, 1807-1844 (2001).

[27] E. Sandier and S. Serfaty, Global minimizers for the Ginzburg-Landau functional below the first critical magnetic field, *Ann. Inst. H. Poincaré Anal. Non Linéaire*, 17, 119-145 (2000).

[28] L. Simon, *Lectures on Geometric Measure Theory*, Australian National University, 1984.

[29] H.M. Soner, Ginzburg-Landau equation and motion by mean curvature. I. Convergence; II. Development of the initial interface, *J. Geometric Analysis*, 7, 437-475; 476-491 (1995).

[30] H. M. Soner, Front propagation. *Boundaries, interfaces, and transitions (Banff, AB, 1995)*, 185-206, *CRM Proc. Lecture Notes*, 13, Amer. Math. Soc., Providence, RI, 1998.

[31] M. Struwe, On the asymptotic behavior of minimizers of the Ginzburg-Landau model in two dimensions, *Differential Integral Equations*, 7, 1613-1624 (1994).

[32] M. Struwe, On the evolution of harmonic maps in higher dimensions, *J. Diff. Geometry*, 28, 485-502 (1988).

[33] B. White, Rectifiability of flat chains, *Annals of Mathematics*, 150, 165-184 (1999).

List of Participants

1. Carlos Albuquerque, Univ. de Lisboa, Portugal
   albuquer@lmc.fc.ul.pt
2. Luigi Ambrosio, Scuola Normale Superiore, Pisa, Italy
   luigi@ambrosio.sns.it
3. Rebekka Axthelm, Univ. Freiburg, Germany
   rebekka@mathematik.uni-freigurg.de
4. Michal Benes, Czech Tech. Univ. of Prague, Czech Rep.
   benes@km1.fjfi.cvut.cz
5. Zsolt Biró, CAR Inst. Hungarian Acad. Sci., Budapest, Hungary
   zbiro@sztaki.hu
6. Elena Bonetti, Univ. Pavia, Italy
   bonetti@dimat.unipv.it
7. Antonin Chambolle, Univ. Paris-Dauphine, France
   antonin.chambolle@ceremade.dauphine.fr
8. Pierluigi Colli, Univ. Pavia, Italy
   pier@imati.cnr.it
9. Armin Dahr, Univ. Bonn, Germany
   armin@iam.uni-bonn.de
10. Klaus Deckelnick, Otto-von-Guericke-Universität Magdeburg, Germany
    Klaus.Deckelnick@mathematik.uni-magdeburg.de
11. Camillo De Lellis, Scuola Normale Superiore di Pisa, Italy
    delellis@cibs.sns.it
12. Irina Denissova, IPME Russian Acad. Sci., St. Petersburg, Russia
    ira@hidro.ipme.ru
13. Lars Diening, Univ. Freiburg, Germany
    diening@mathematik.uni-freiburg.de
14. Fathi Dkhil, Univ. Cergy-Pontoise, France
    fathi.dkhil@math.u-cergy.fr
15. Gerhard Dziuk, Univ. Freiburg, Germany
    gerd@mathematik.uni-freiburg.de
16. Angiolo Farina, Univ. Firenze, Italy
    farina@math.unifi.it
17. Marcus Garvie, Univ. Durham, UK
    m.r.garvie@durham.ac.uk
18. Marina Ghisi, Univ. Pisa, Italy
    ghisi@gauss.dm.unipi.it

19. Gianni Gilardi, Univ. Pavia, Italy
    gilardi@dimat.unipv.it

20. Massimo Gobbino, Univ. Pisa, Italy
    m.gobbino@dma.unipi.it

21. César Gonçalves, Instituto Politécnico da Guarda, Portugal
    crg@mail.telepac.pt

22. Celine Grandmont, Univ. Paris-Dauphine, France
    grandmont@ceremade.dauphine.fr

23. Eurica Henriques, Univ. Trás-os-Montes e Alto Douro, Portugal
    eurica@utad.pt

24. Viet Ha Hoang, Univ. Cambridge, UK
    hvh21@damtp.cam.ac.uk

25. Ryo Ikota, Univ. Tokyo, Japan
    ikota@ms.u-tokyo.ac.jp

26. Helge Kristian Jenssen, S.I.S.S.A., Trieste, Italy
    jenssen@sissa.it

27. Moritz Kassmann, Univ. Bonn, Germany
    ljoscha@iam.uni-bonn.de

28. Omar Lakkis, Univ. Maryland, USA
    omar@math.umd.edu

29. Stephen Langdon, Univ. of Durham, UK
    stephen.langdon@durham.ac.uk

30. Irina Loginova, Royal Inst. Techn., Stockholm, Sweden
    irina@mech.kth.se

31. Juergen Mehnert, Univ. Freiburg, Germany
    jmehnert@mathematik.uni-freiburg.de

32. A. Meirmanov, Univ. da Beira Interior, Portugal
    meirman@ubisth.ubi.pt

33. Masayasu Mimura, Univ. Hiroshima, Japan
    mimura@math.sci.hiroshima-u.ac.jp

34. Fernando Miranda, Univ. Minho, Portugal
    fmiranda@math.uminho.pt

35. Maria Giovanna Mora, S.I.S.S.A., Trieste, Italy
    mora@sissa.it

36. Massimiliano Morini, S.I.S.S.A., Trieste, Italy
    morini@sissa.it

37. Gonoko Moussa, Univ. Nancy, France
    moussag@antares.iecn.u-nancy.fr

38. Piotr Mucha, Warsaw Univ., Poland
    mucha@hydra.mimuw.edu.pl

39. Matteo Novaga, Univ. Pisa, Italy
    novaga@mail.dm.unipi.it

40. Hermenegildo Oliveira, Univ. Algarve, Portugal
    holivei@ualg.pt

41. Emanuele Paolini, Scuola Normale Superiore di Pisa, Italy
    paolini@sns.it
42. Peter Philip, Weierstrass Inst. App. Anal. Stoch., Berlin, Germany
    philip@wias-berlin.de
43. Mahmoud Qafsaoui, LAMFA-Univ. Picardie, France
    mahmoud.qafsaoni@u-picardie.fr
44. Vincenzo Recupero, Univ. Trento, Italy
    recupero@science.unitn.it
45. Elisabetta Rocca, Univ. Pavia, Italy
    rocca@dimat.unipv.it
46. José Francisco Rodrigues, Univ. de Lisboa, Portugal
    rodrigue@ptmat.fc.ul.pt
47. Mathias Röger, Univ. Bonn, Germany
    roeger@iam.uni-bonn.de
48. Raluca Rusu, Univ. Freiburg, Germany
    raluca@mathematik.uni-freiburg.de
49. Giulio Schimperna, Univ. Pavia, Italy
    giulio@dimat.unipv.it
50. Daniel Sevcovic, Comenius Uiversity, Bratislava
    sevcovic@fmph.uniba.sk
51. Israel Michael Sigal, Institute des Hautes Études Scientifiques, France
    sigal@ihes.fr
52. Boyan Sirakov, Univ. Paris VI, France
    sirakov@ann.jussieu.fr
53. Vsvolod A. Solonnikov, Steklov Math. Inst., St. Petersburg, Russia
    solonnik@pdmi.ras.ru
54. Halil Mete Soner, Koç University, Istanbul, Turkey
    msoner@ku.edu.tr
55. Ulisse Stefanelli, I.M.A.T.I.-C.N.R., Pavia, Italy
    ulisse@imati.cnr.it
56. Nicolas van Goethem, Univ. de Pisa, Italy
    vangoeth@mail.dm.unipi.it
57. Remi Weidenfeld, Univ. Paris-Sud, France
    Remi.weidenfeld@math.u-psud.fr
58. Sandra Wieland, Univ. Bonn, Germany
    sandra@iam.uni-bonn.de

LIST OF C.I.M.E. SEMINARS

Fondazione C.I.M.E.

Centro Internazionale Matematico Estivo
International Mathematical Summer Center
http://www.math.unifi.it/∼cime
cime@math.unifi.it

2003 COURSES LIST

Stochastic Methods in Finance

July 6–13, Cusanus Akademie, Bressanone (Bolzano)
Joint course with European Mathematical Society

Course Directors:

Prof. Marco Frittelli (Univ. di Firenze), marco.frittelli@dmd.unifi.it
Prof. Wolfgang Runggaldier (Univ. di Padova), runggal@math.unipd.it

Hyperbolic Systems of Balance Laws

July 14–21, Cetraro (Cosenza)

Course Director:

Prof. Pierangelo Marcati (Univ. de L'Aquila), marcati@univaq.it

Symplectic 4-Manifolds and Algebraic Surfaces

September 2–10, Cetraro (Cosenza)

Course Directors:

Prof. Fabrizio Catanese (Bayreuth University)
Prof. Gang Tian (M.I.T. Boston)

Mathematical Foundation of Turbulent Viscous Flows

September 1–6, Martina Franca (Taranto)

Course Directors:

Prof. M. Cannone (Univ. de Marne-la-Vallée)
Prof.T. Miyakawa (Kobe University)

Printing and Binding: Strauss GmbH, Mörlenbach

4. Manuscripts should in general be submitted in English. Final manuscripts should contain at least 100 pages of mathematical text and should include
   - a general table of contents;
   - an informative introduction, with adequate motivation and perhaps some historical remarks: it should be accessible to a reader not intimately familiar with the topic treated;
   - a global subject index: as a rule this is genuinely helpful for the reader.

5. Lecture Notes are printed by photo-offset from the master-copy delivered in camera-ready form by the authors. Springer-Verlag provides technical instructions for the preparation of manuscripts. We strongly recommend that all contributions in a volume be written in the same LaTeX version, preferably LaTeX2e. Macro-packages in LaTeX2e are available from Springer's web-pages at

   http://www.springer.de/math/authors/index.html .

   Careful preparation of manuscripts will help keep production time short and ensure satisfactory appearance of the finished book. After acceptance of the manuscript authors/volume editors will be asked to prepare the final LaTeX source files (and also the corresponding dvi- or pdf-file) together with the final printout made from these files. The LaTeX source files are essential for producing a unified full-text online version of the book

   (http://www.springerlink.com/link/service/series/0304/tocs.htm).

   The actual production of a Lecture Notes volume takes approximately 12 weeks.

6. Volume editors receive a total of 50 free copies of their volume to be shared with the authors, but no royalties. They and the authors are entitled to a discount of 33.3 % on the price of Springer books purchased for their personal use, if ordering directly from Springer-Verlag.

Commitment to publish is made by letter of intent rather than by signing a formal contract. Springer-Verlag secures the copyright for each volume. Authors are free to reuse material contained in their LNM volumes in later publications: A brief written (or e-mail) request for formal permission is sufficient.

Addresses:

Professor J.-M. Morel, CMLA,
Ecole Normale Supérieure de Cachan,
61 Avenue du Président Wilson, 94235 Cachan Cedex, France
E-mail: Jean-Michel.Morel@cmla.ens-cachan.fr

Professor F. Takens, Mathematisch Instituut,
Rijksuniversiteit Groningen, Postbus 800,
9700 AV Groningen, The Netherlands
E-mail: F.Takens@math.rug.nl

Professor B. Teissier, Université Paris 7
Institut Mathématique de Jussieu, UMR 7586 du CNRS
Equipe "Géométrie et Dynamique", 175 rue du Chevaleret
75013 Paris, France
E-mail: teissier@math.jussieu.fr

Springer-Verlag, Mathematics Editorial, Tiergartenstr. 17,
69121 Heidelberg, Germany,
Tel.: +49 (6221) 487-8410
Fax: +49 (6221) 487-8355
E-mail: lnm@Springer.de